HUMAN IDENTIFICATION: THE USE OF DNA MARKERS

Contemporary Issues in Genetics and Evolution

VOLUME 4

The titles published in this series are listed at the end of this volume.

Human Identification: The Use of DNA Markers

Edited by
BRUCE S. WEIR

Contributions with an asterisk in the table of contents were first published in *Genetica*, Volume 96 no. 1–2 (1995)

Springer-Science+Business Media, B.V.

Library of Congress Cataloging-in-Publication Data

Human identification : the use of DNA markers / edited by Bruce S.
Weir.
p. cm. -- (Contemporary issues in genetic and evolution ; v.
4)
Includes bibliographical references and index.
ISBN 978-94-017-1803-5 ISBN 978-0-306-46851-3 (eBook)
DOI 10.1007/978-0-306-46851-3
1. DNA fingerprinting. I. Weir, B. S. (Bruce S.), 1943-
II. Series.
RA1057.55.H85 1995
614'.1--dc20 95-17081

Printed on acid-free paper

Contents

* Contributions indicated with an asterisk were first published in *Genetica*, Volume 96 no. 1–2 (1995).

B. S. Weir (ed.), Human Identification: The Use of DNA Markers, 1–2, 1995.

Introduction

As this volume goes to press it is not known whether O.J. Simpson will be convicted of having murdered his former wife and her friend. It is clear, however, that DNA evidence is playing a central part in the trial. The publicity given to the trial has meant that this kind of evidence has been discussed widely in the media, and so may mark a turning point—in the future there will be an expectation that DNA evidence will be sought in the same way that crime scenes are routinely examined for fingerprints.

In documents filed with the court, the Simpson defense team challenged the admissibility of DNA evidence on several grounds, including population genetics and statistics. These issues all relate to the meaning to be attached to the finding of a match between evidentiary samples and blood taken from a suspect or a victim. The more rare the matching DNA profile is in the relevant population, the more likely it is that there is a common source for the matching samples. Estimating the degree of rarity is the main theme of the papers in this volume.

Each of the contributors to the volume has been involved in the debate over the use of DNA for human identification, and although a few of the major players declined to contribute, the volume represents the diversity of current views. Although contributors were selected to provide alternative approaches, there was also an attempt to select authors who have used sound scientific reasoning and have displayed some balance in their presentations. There is no doubt that challenges to statistical and population genetic methods applied to DNA profile frequencies have led to strengthening of those methods. It is to be hoped that this volume will show past challenges to have largely been met, and to indicate what remains to be done. There are going to be situations, such as those requiring discrimination between relatives or members of small populations, where statistical issues will continue to be important but it may be that biology eventually overtakes statistics. There may well come a point where so much of the genome is examined, to determine a DNA profile, that there can be little doubt of identity. As with all forms of evidence, of course, the possibility of error or fraud may need to be considered.

A detailed synopsis of each chapter will not be given in this Introduction. I have chosen instead to add a few comments at the end of each paper. It should be mentioned that each paper went through the usual peer-review process, and that each paper was revised accordingly. All the contributors are grateful to the reviewers.

Not only is the outcome of the Simpson trial unknown at this point, but so is that of the second National Research Council (NRC) committee deliberations. The debate over the use of DNA in human identification rose to such a vocal level in early 1990 that the NRC convened a panel of distinguished scientists to examine the issues. It may not have been fully appreciated that the debate was taking place almost entirely in the courtroom, where approximately equal numbers of experts for each side were engaged. The scientific literature, a more fitting forum for scientific discourse, was relatively thin. Since that time, there has been a large literature and publication of many studies of forensic DNA frequency databases. In the normal course of events this literature would have served to shed light on the debate. Instead, attention was shifted to the NRC report that was published in 1992. Although the recommendations of the report, especially the so-called ceiling principle, were motivated by the desire to avoid courtroom battles, they often had the opposite effect. Some courts were eager to embrace a set of guidelines issued under the aegis of the National Academy of Sciences, but courts did not realize that scientists are more likely to be guided by the scientific literature than expert reports. It is expected that the second report will both avoid statistical errors and refer to published studies of data.

It would be a pity if this volume were seen merely as a response to a debate over how to calculate the frequency of a DNA profile. Instead, it should be regarded as a celebration of the power of molecular biology to identify the possible contributors to biological samples. The benefits of this technology are considerable. Determining paternity in disputed cases is probably the major use of DNA profiling. There are considerable benefits to society of ensuring that children are adequately supported. Fewer people are affected, but society also gains when violent criminals are identified and convicted—as well as when those who are wrongly suspected or wrongly convicted are exonerated by DNA profiling. Smaller numbers still are involved when identification is needed following major disasters such as war or airplane crashes, but identification of remains brings comfort and closure to friends and family. Human identification by means of DNA is playing a large role in evolutionary studies, aided by the ability to extract DNA from ancient remains.

Profiling non-human species also serves human welfare. Protection of proprietary crop species or of endangered animals is aided by DNA identification, as is quality control of cell cultures.

The new analyses presented in this volume will go a long way to ensuring that DNA identification is made on a sound scientific basis, so that the many benefits of such identification will follow.

Raleigh, N.C. *B.S. Weir*

B. S. Weir (ed.), Human Identification: The Use of DNA Markers, 3–12, 1995.

A method for quantifying differentiation between populations at multi-allelic loci and its implications for investigating identity and paternity

David J. Balding & Richard A. Nichols
School of Mathematical Sciences and School of Biological Sciences, Queen Mary & Westfield College, University of London, Mile End Road, London E1 4NS, UK

Received 16 May 1994 Accepted 26 July 1994

Key words: DNA profiles, paternity, Wright's F_{ST}, coancestry, forensic science

Abstract

A method is proposed for allowing for the effects of population differentiation, and other factors, in forensic inference based on DNA profiles. Much current forensic practice ignores, for example, the effects of coancestry and inappropriate databases and is consequently systematically biased against defendants. Problems with the 'product rule' for forensic identification have been highlighted by several authors, but important aspects of the problems are not widely appreciated. This arises in part because the match probability has often been confused with the relative frequency of the profile. Further, the analogous problems in paternity cases have received little attention. The proposed method is derived under general assumptions about the underlying population genetic processes. Probabilities relevant to forensic inference are expressed in terms of a single parameter whose values can be chosen to reflect the specific circumstances. The method is currently used in some UK courts and has important advantages over the 'Ceiling Principle' method, which has been criticized on a number of grounds.

1. Introduction

The genetic composition of human populations varies because of, among other factors, their differing evolutionary histories and patterns of dispersal and interbreeding. The magnitude of the effect of this genetic differentiation on the forensic evaluation of DNA profile evidence is controversial. It is the practice of many forensic scientists to ignore coancestry except, possibly, in cases where genetically isolated populations or close relatives are clearly involved. Some authors argue, however, that uncertainty about possible levels of differentiation may invalidate such an approach (Lewontin & Hartl, 1991; Krane *et al.*, 1992). Others take the view that typical levels of differentiation are sufficiently small that they may routinely be neglected (Chakraborty & Kidd, 1991; Roeder, 1994).

We argue for an intermediate position: even small levels of genetic differentiation can be important and the effect should not be ignored. To do so would unfairly overstate the strength of the evidence against the defendant and the error could be crucial in some cases, such as those involving partial profiles or large numbers of possible culprits, many of whom share the defendant's ethnic background. However, the forensic use of DNA profiles need not be invalidated as a consequence. One approach to allowing for population differentiation, the 'Ceiling Principle', has been proposed by the US National Research Council (NRC) (*DNA Technology in Forensic Science*, Natl. Acad. Press, Washington D.C., 1992). The principle has been widely criticized (Robertson & Vignaux, 1992; Devlin, Risch & Roeder, 1993; Morton, 1993a; Weir, 1993a). In particular, the principle is inflexible and cannot be adjusted to the circumstances of a particular case, in part because it incorporates the view that the defendant's ethnicity is irrelevant to inference. We propose a method for quantifying the effect of genetic differentiation in terms of a single parameter, which can often be interpreted in terms of coancestry. Debates about the effect of population heterogeneity in particular cases can thus be simplified to a discussion of values for the parameter appropriate to the circumstances. Our proposed method has previously been described

(Balding & Nichols, 1994) and is currently used in some UK courts. Here, we develop the justification for the method and extend its application to paternity testing. The use of DNA profile evidence when incest is alleged in paternity cases is becoming increasingly common and the proposed method is particularly appropriate in such cases.

2. Key issues in forensic inference

Although the literature on forensic identification using DNA profile evidence is now extensive, many fundamental statistical issues are still not widely appreciated. Balding and Donnelly (1995) consider the forensic identification inference problem in a general setting and their analysis clarifies several issues. In particular, they show that the weight of evidence against the defendant depends on, for each possible perpetrator other than the defendant, the ratio of the likelihood of the DNA profile data if he were the culprit, to its likelihood if the defendant were the culprit. These likelihood ratios should then be summed by the jury, weighted by their probability, based on the non-DNA evidence, that each possible culprit is the true culprit.

To facilitate the discussion, it is common to make four simplifying assumptions:

1. that the crime sample DNA is that of the culprit;
2. that matches are unequivocal;
3. that if the defendant were the culprit then the defendant and crime sample DNA profiles would be certain to match; and
4. the fact that the defendant's DNA profile was investigated is not, in itself, informative about his/her profile.

These assumptions are not valid in general, but they allow us to focus on other important issues and deviations from them can be addressed within the framework discussed here. See Balding and Donnelly (1995) for further discussion.

Under these four assumptions, each likelihood ratio is simply the conditional probability that the possible culprit has the profile given that the defendant has it, that is, the 'match probability'. Note that the match probability may also be formulated in terms of the probability that the defendant has the profile conditional on the event that the alternative culprit has it, but we find the former definition to be more convenient.

Many authors ignore the conditioning on the observed profile and take the match probability to be equivalent to the relative frequency of the defendant's profile in some population. This use of profile frequencies in place of the match probability is inappropriate for several reasons. The concept of 'match' clearly involves two profiles, not one, and there seems no logical framework for linking profile frequencies with the issue of the defendant's guilt or innocence, which is the crucial issue in court. In particular, it is unclear how to allow coherently for the possibility that the culprit is related to the defendant, or shares ancestry through common origin in a subpopulation. Perhaps most importantly, there seems no logical framework for combining the DNA evidence, quantified by a profile frequency, with the non-DNA evidence.

Correct definition of the match probability clarifies much of the current debate. A general discussion of 'reference populations' can be avoided and neither is it necessary to consider hypothetical 'random' selections of suspects. Crucially, a coherent framework becomes available for incorporating the effects of shared ancestry, on both recent and evolutionary timescales. Since match probabilities are *conditional* probabilities, they cannot be estimated directly from database relative frequencies. Correlations in profile possession must be explicitly modelled in terms of population genetic theory in addition to the available data. Consequently, the ethnicities of both defendant and possible culprits are relevant to inference. Some authorities ignore correlations in profile possession and, instead, use 'conservative' estimates of relative frequencies. The Ceiling Principle, for example, is based on this approach. However unless the correlations are specifically taken into account, it is impossible to assess what level of 'conservativeness' is appropriate.

Some of the current debate concerning population differentiation focusses on statistical tests of hypotheses of independence in forensic databases (Geisser & Johnson, 1993; Weir, 1993b). The tests are complicated by the experimental difficulties involving apparent homozygotes. It is, in any case, difficult in principle to draw conclusions relevant to forensic inference from the outcomes of such tests. Population differentiation indubitably exists, the question of interest concerns the magnitude of its effect on match probabilities. Failure to reject a null hypothesis of no differentiation reflects some combination of insufficient, or inappropriate, data, low power against the alternatives of interest and small magnitude of effect. Such tests are thus not directly helpful in forensic inference. We propose parameter estimation, both point and interval, as an alternative to hypothesis testing.

3. Likelihood ratios for identification and paternity

3.1 Identification

We consider single-locus DNA profiles, one taken from a crime sample and one from a defendant, and make the four assumptions listed in Section 2. The match probability then depends on a number of factors. In particular, it is affected by the possibility that the individuals have matching DNA profile bands through shared inheritance from a common ancestor. For some possible culprits, the amount of ancestry shared with the defendant is largely known. This can occur, for example, when the defendant's close relatives are possible culprits. (Note that 'possible culprits' is taken to include all individuals not excluded by the non-DNA evidence, not merely those on whom suspicion falls for good reason (Lempert, 1991).) More generally, the amount of shared ancestry between defendant and possible culprit will be unknown. Frequently, however, many possible culprits will have features in common with the defendant (Lempert, 1991), such as similar physical description or location of residence, and hence defendant and possible culprit may plausibly have a large level of shared ancestry compared with two 'random' individuals.

In addition to shared ancestry, the match probability is also affected by uncertainty about relative frequencies of bands. Such uncertainty occurs because forensic databases are rarely exactly appropriate for the possible culprits in a specific crime. They typically are unplanned samples from large, heterogeneous racial groups which are subject to sampling and other sources of error.

Balding and Nichols (1994) proposed the following formulae for $\Pr(AA|AA)$ and $\Pr(AB|AB)$, the single-locus match probabilities for possible culprits not known to be close relatives of the defendant in, respectively, the homozygote and heterozygote cases:

$$\Pr(AA|AA) = \frac{(2F+(1-F)p_A)(3F+(1-F)p_A)}{(1+F)(1+2F)} \tag{1}$$

$$\Pr(AB|AB) = 2\frac{(F+(1-F)p_A)(F+(1-F)p_B)}{(1+F)(1+2F)}, \tag{2}$$

in which p_A and p_B denote the relative frequencies of alleles A and B in the population from which the database is drawn, in principle that most appropriate for the possible culprit under consideration. In practice, the homozygote case is complicated by the fact that, because of experimental difficulties, some heterozygotes may be incorrectly classified as homozygotes. Balding and Nichols (1994) give match probabilities which take this difficulty into account, as well as extensions of (1) and (2) to the case that the possible culprit under consideration is known to be a close relative of the defendant.

Two distinct justifications for (1) and (2) are given in Sections 4.1 and 4.2. Equation (2) differs slightly from that originally proposed by the authors (Nichols & Balding 1991), which employed an approximation ignoring certain higher order correlations described in Section 4.

The parameter F in (1) and (2) may be interpreted as measuring the degree of uncertainty about p_A as an estimate of the match probability for a single A allele. The case $F=0$ corresponds to certainty so that the single-locus match probabilities are exactly p_A^2 and $2p_Ap_B$, which, with p_A and p_B replaced by sample relative frequencies, are the values used in the so-called 'product rule'. Absolute certainty is unrealistic in practice and thus the product rule consistently overstates the strength of the evidence against the defendant (unless $p_A+p_B \geq 2/3$, which never arises for most typing systems). For realistic values of F, the effect can be important (Balding & Nichols, 1994).

In many cases, shared ancestry between defendant and possible culprit on an evolutionary timescale may be considered the most important source of uncertainty, in which case F may be approximately the same as Wright's F_{ST}. Estimates of F_{ST} are often based on populations which are geographically closely spaced. In the forensic context, however, it is of interest to compare broad racial groups with subpopulations at varying levels of stratification. At traditional loci, collations of allele frequency estimates for disparate human populations are available. A recent survey (Cavalli-Sforza & Piazza, 1993) reports F_{ST} estimates among Europeans with a median of 0.8% while the 90th percentile is about 2.8%. The corresponding values are 2.7% and 14% among Africans and 4.3% and 12% among Asians. These results may not be directly relevant to forensic inference, since such meta-studies encompass differing methodologies, sampling may concentrate on unusual populations and the loci surveyed may be subject to geographically-varying selection. In addition, mutation rates at the these loci are typically much lower than at the VNTR loci currently used in forensic work.

More directly relevant, in view of moves to introduce short tandem repeat (STR) loci for forensic work, is the differentiation reported at two of the three loci examined in a sample of ethnic groups classified as Greek Cypriot, Gujarati, Northern European and Pakistani (Wall *et al.,* 1993). Some loci show little differentiation within the broad racial groups (European or Asian), others show dramatic differences. There are a variety of plausible explanations for the differences between loci. M. Greenhalgh (pers. comm.) has implemented accurate automatic sequencer technology to overcome technical problems with the F13A1 locus and reports substantial differentiation between Gujarati and Pakistani populations (Fig. 1). The data of Wall *et al.* (1993) from other loci show more marked differentiation within both major ethnic groups. This greater differentiation could be a consequence of a variety of processes. Geographically-varying selection on the gene containing the STR-bearing intron, or a linked locus, could cause allele frequencies to diverge. Mutation can produce either greater differentiation or, conversely, convergence in allele frequencies depending on the mechanism. It thus seems plausible that the variation within broad racial groups varies from locus to locus, but that the more variable STR loci show at least as much differentiation as traditional loci.

Appropriate surveys are not yet available at the VNTR loci which predominate in current forensic work. Direct estimation of F_{ST} from forensic databases is hampered by, among other factors, sensitivity to assumptions about apparent homozygotes and the ill-defined sampling frame of the databases. In addition, the ethnic origin of individuals in databases is often not known in sufficient detail to permit the investigation of population differentiation at the finer levels of stratification which may be appropriate for forensic inference. Reliable estimates will require substantial surveys of individuals of known ethnicity at varying levels of stratification. The preliminary evidence which is available suggests that F_{ST} at VNTR loci may typically be smaller than at traditional loci, which is plausible in view of the higher mutation rates, and that values may differ substantially from locus to locus. Morton (1993b) gave average point estimates of around 0.1% for US Caucasians and 1% for US Blacks. Because of the difficulties discussed above, and below, the values appropriate for forensic inference are likely to be substantially larger.

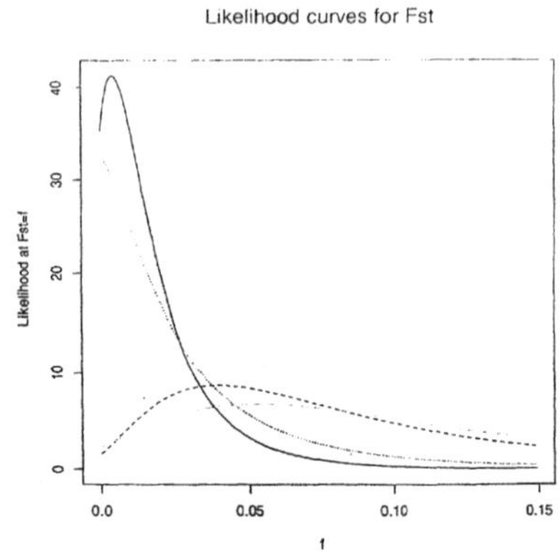

Fig. 1. Likelihood curves for F_{ST} measuring the differentiation at the STR loci CD4 and F13A1. For each locus the differentiation of a Cypriot sample from a North European database (C-N) and a Gujarati sample from a Pakistani (G-P) database is shown. At the origin ($f = 0$) the highest pair of curves relate to the F13A1 locus, and the lowest to CD4. Within each pair the G-P curve is higher (at $f = 0$). The CD4 data are from Wall *et al.*, (1993) and the F13A1 data from M. Greenhalgh (unpublished). Sample sizes for the Pakistani, North European, Gujarati and Greek Cypriot populations were 186, 58, 66, 50 (F13A1) and 50, 88, 44, 80 (CD4). The y-axis is scaled so that the curves can be directly interpreted as posterior densities with respect to a uniform prior for F_{ST}.

Morton (1992) proposed the use of formulae due to Yasuda (1968) which agree with (1) and (2) up to terms in F. Ignoring terms of order F^2 is reasonable when F is small compared with p_A and p_B, such as occurs with most traditional loci, but is often not appropriate for VNTR data.

Equations (1) and (2) deal with the single-locus case. In principle, it is not reasonable to assume independence across loci because of a sequential effect similar to that described by Donnelly (1995). If, as is usually the case, there is some uncertainty about the amount of shared ancestry between defendant and possible culprit, each successive single-locus match makes a higher level of shared ancestry more plausible, and hence a subsequent match is somewhat less surprising than the first. This feature of forensic inference differs from the usual use of F_{ST} in population genetics. The effect can in principle be accounted for by regarding F in equations (1) and (2) as having a distribution of possible values. Final single-locus match probabilities should then be obtained by integration with respect to this distribution. Four-locus match probabilities based on (1) and (2) involve powers of F up to order eight and the upper tail of the distribution will thus have a

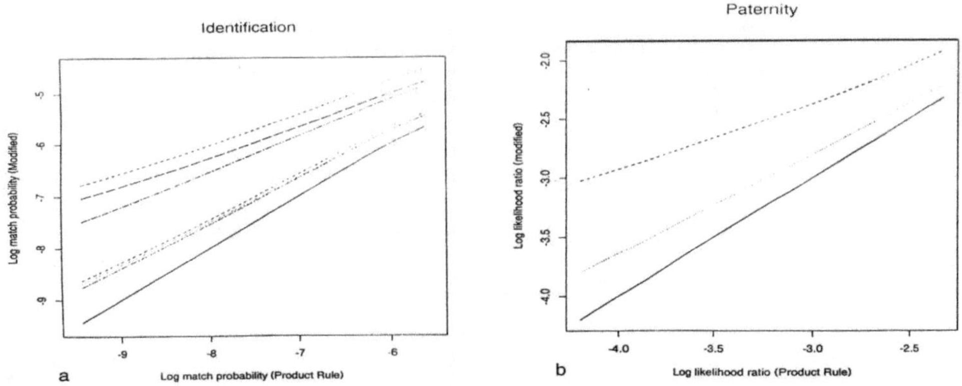

Fig. 2. (a) Match probabilities ($\log_{10}$) for four-locus, all-heterozygote profiles when the possible culprit under consideration is not known to be a relative of the defendant. The straight line $y = x$ gives the result of the unmodified product rule, which corresponds to $F = 0$. The upper set of three curves is for $F = 5\%$, the lower set is for $F = 1\%$. Within each set of three curves, the middle curve represents the modified match probabilities based on (2), the higher and lower curves represent, respectively, the approximations of Nichols and Balding (1991) and Morton (1992). The four pairs of allele frequencies are $(\alpha, 2\alpha)$, $(9\alpha, 5\alpha)$, $(6\alpha, 8\alpha)$ and $(7\alpha, 7\alpha)$ for $0.01 \leq \alpha \leq 0.03$. (b) Likelihood ratios ($\log_{10}$) for four-locus profiles when the alternative father under consideration is not known to be a relative of the alleged father and the genotypes of mother, child and alleged father at each locus are of the form AB, BC and CD. From lower to higher, the three lines represent the likelihood ratio with, respectively, $F = 0$, $F = 1\%$ and $F = 5\%$. The frequencies of the child's paternal alleles at the four loci are α, 9α, 6α, and 7α for $0.01 \leq \alpha \leq 0.03$.

large influence on the resulting match probability. This effect may be approximately accounted for by using an 'effective' value of F in the upper tail of the distribution.

As an illustration, suppose that F has probability density proportional to $(1-f)^{100}$ at $F = f$, so that the mean and standard deviation of F are both near 1% while the median is close to 0.7%. The shape of this distribution is chosen to reflect a high probability that the amount of shared ancestry between defendant and possible culprit, and hence F, is small and a small probability for F to be large. Ignoring this distribution and wrongly assuming that $F = 1\%$ leads to a three-fold error in the match probability for a four-locus, all-heterozygote profile with $p = 5\%$ for each band. When $p = 1\%$, the error is 34-fold. The appropriate 'effective' values of F are, respectively, 1.9% and 2.7%. Point estimates of the mean or median of F_{ST} are thus not directly relevant to forensic inference.

Figure 2(a) illustrates the effect of allowing for realistic levels of population differentiation. For a range of four-locus, all-heterozygote profiles, it compares match probabilities calculated using the product rule with those obtained from (1) and (2), and two approximations (Nichols & Balding, 1991; Morton, 1992), when F takes values 1% and 5%. With $F = 5\%$, the product rule can understate the appropriate match probability by two orders of magnitude which can be important, especially in cases involving little or no evidence other than the DNA profiles. Balding and Donnelly (1995) show that a small match probability does not necessarily provide convincing proof of guilt and thus a change of one or two orders of magnitude, even in a very small match probability, can be crucial in some cases.

3.2 Paternity

The principles which lead to (1) and (2) can be extended to allow for uncertainty, including that due to possible shared ancestry, in paternity testing. Suppose that we have single-locus DNA profiles for each of mother, child and alleged father. Evaluating the probability that the alleged father is the true father requires, for every other possible father, the ratio of the likelihood of the observed DNA profiles if he were the true father to their likelihood if the alleged father were the true father. These likelihood ratios each depend on the amount of shared ancestry among the mother, alleged father and alternative possible father. If the alternative father under consideration is not directly related to either the mother or the alleged father, but has a similar level of shared ancestry with both of them, then, under assumptions analogous to those which led to (1) and (2), the single-locus likelihood ratios for the pos-

Table 1. Single-locus likelihood ratios for paternity when the mother's genotype is AB. Blank entries indicate that the alleged father is excluded.

Alleged Father	Child AA	Child AB	Child AC
AA	$\frac{3F+(1-F)p_A}{1+3F}$	$\frac{4F+(1-F)(p_A+p_B)}{1+3F}$	
AB	$2\left(\frac{2F+(1-F)p_A}{1+3F}\right)$	$\frac{4F+(1-F)(p_A+p_B)}{1+3F}$	
AC	$2\left(\frac{2F+(1-F)p_A}{1+3F}\right)$	$2\left(\frac{3F+(1-F)(p_A+p_B)}{1+3F}\right)$	$2\left(\frac{F+(1-F)p_C}{1+3F}\right)$
CC			$\frac{2F+(1-F)p_C}{1+3F}$
CD			$2\left(\frac{F+(1-F)p_C}{1+3F}\right)$

Table 2. Single-locus likelihood ratios for paternity when the mother's genotype is AA. Blank entries indicate that the alleged father is excluded.

Alleged Father	Child AA	Child AB
AA	$\frac{4F+(1-F)p_A}{1+3F}$	
AB	$2\left(\frac{3F+(1-F)p_A}{1+3F}\right)$	$2\left(\frac{2F+(1-F)p_B}{1+3F}\right)$
BB		$\frac{2F+(1-F)p_B}{1+3F}$
BC		$2\left(\frac{F+(1-F)p_B}{1+3F}\right)$

sible observed genotypes are given in Tables 1 and 2. If the alternative father has substantial shared ancestry with either mother or alleged father but not both, then $1+F$ may be more appropriate in the denominator of the likelihood ratio, in place of $1+3F$. The difference will, however, usually be unimportant.

Most current practice employs the values in Tables 1 and 2 but with $F = 0$, corresponding to no shared ancestry and complete certainty about band relative frequencies. An inappropriate assumption of certainty thus leads, for realistic values of the parameters, to an overstatement of the probability that the alleged father is the true father. The magnitude of the overstatement for a four-locus profile is illustrated in Figure 2(b). With $F = 5\%$, ignoring uncertainty can lead to an order of magnitude overstatement of the likelihood ratio.

Finally, we consider the case that the alternative father under consideration is known to be a close relative of the alleged father (but the DNA profile of the former is not available). Let r denote the probability that an allele drawn from the alternative father matches one of the alleged father's alleles at that locus through inheritance from the known ancestors, so that $r = 1/2$ when they are brothers and $r = 1/4$ for either uncle-nephew or half-brothers. The selection of an allele from the alternative father is exactly equivalent to selecting with probability r an allele from the alleged father and with probability $1-r$ an allele from an apparently unrelated person, typically in the same subpopulation. The single-locus likelihood ratio is thus

$$r + (1-r)LR, \qquad (3)$$

where LR denotes the appropriate value from Tables 1 or 2.

4. Derivation of likelihood ratios

If two individuals are drawn from a randomly-mating subpopulation then, when the subpopulation frequencies are known, the probability of observing any four specified alleles can be expressed in terms of the product of the corresponding frequencies. For example, two AA homozygotes are observed with probability $\tilde{p}_A^4$, where $\tilde{p}_A$ is the subpopulation frequency of A alleles, while two AB heterozygotes are observed with probability $4\tilde{p}_A^2\tilde{p}_B^2$ (the constant 4 occurs because of the two possible orderings of each of the two AB pairs). When the subpopulation frequencies are unknown, the probability is given by the expectation of the product. The probabilities given at (1) and (2) can thus be expressed in the form

$$\Pr(AA|AA) = \frac{\mathrm{E}(\tilde{p}_A^4)}{\mathrm{E}(\tilde{p}_A^2)} \quad (4)$$

$$\Pr(AB|AB) = 2\frac{\mathrm{E}(\tilde{p}_A^2\tilde{p}_B^2)}{\mathrm{E}(\tilde{p}_A\tilde{p}_B)}. \quad (5)$$

The expectations in (4) and (5) must be based on a model for the evolution of the population at each locus. The evolution of VNTR loci is complicated and traditional population genetic models may not accurately describe their behaviour (Harding, 1992; Jeffreys *et al.*, 1994). Here, we formulate expressions for moments such as (4) and (5) which are valid under a range of evolutionary models. We develop justifications for these expressions using two approaches, the first based directly on a specific evolutionary model and the second using more general statistical arguments.

4.1 Genetical derivation

To specify fully the expected frequencies of pairs of diploid genotypes would require nine parameters (Cockerham, 1971). It is clearly not feasible to estimate all these parameters for each of the possible culprits relevant to a particular case. Here we specify a genetic model under which each of the nine parameters, and hence the expected frequencies, can be expressed in terms of a single parameter. The model is reasonably general and the simplification to a single, readily interpreted parameter is very helpful in a court environment, in which the use of more complicated, multi-parameter models may be inappropriate.

We consider a randomly-mating subpopulation, partly isolated from a large population, in which migration and mutation events occur independently and at constant rates. We write $\theta/(2N)$ for the sum of the two rates, where N denotes the subpopulation size (number of alleles). The probability F that two alleles are identical by descent (ibd) through an ancestor in the same subpopulation is simply the probability that, in tracing back the two lineages, an immigration or mutation event does not occur prior to the lineages coalescing in a common ancestor. Coalescences occur independently of migrations and mutations at rate $1/N$ while the total rate at which mutations or migrations occur on the two lineages is θ/N and thus we have

$$F = \frac{1/N}{1/N + \theta/N} = \frac{1}{1+\theta}.$$

The probability that two alleles drawn from the subpopulation are both type A is given by the familiar formula (Crow & Kimura, 1970)

$$\Pr(AA) = \mathrm{E}(\tilde{p}_A^2) = F\pi_A + (1-F)\pi_A^2, \quad (6)$$

in which we introduce π_A for the probability that a migration or mutation event produces an allele of type A. The first term in (6) is the probability that the two alleles are ibd and the most recent common ancestor was of type A, while the second term gives the probability that the two alleles are not ibd and are, in effect, the results of independent draws from a mechanism which generates A alleles with probability π_A. If the subpopulations are in equilibrium, the value of π_A is naturally estimated by p_A, the population relative frequency of A alleles, and henceforth we replace π_A with p_A. The probability that the two alleles are distinct, of types A and B say, is

$$\Pr(AB) = 2\mathrm{E}(\tilde{p}_A\tilde{p}_B) = 2(1-F)p_Ap_B.$$

Consider next $\mathrm{E}(\tilde{p}_A^3)$, the probability that three alleles chosen randomly from the subpopulation are all of type A. Tracing the three lineages backward in time, the rate at which any two coalesce is $3/N$, while the total rate at which mutations or migrations occur on the three lineages is $3\theta/(2N)$. The probability that the first event is a coalescence is then

$$\frac{3/N}{3/N + 3\theta/(2N)} = \frac{2}{2+\theta} = \frac{2F}{1+F}.$$

Continuing backwards in time, the two remaining lineages coalesce prior to a mutation or migration event with probability F. The probability that all three alleles are ibd is thus $2F^2/(1+F)$. Similarly, it can be seen that the probabilities that precisely one and zero pairs of alleles are ibd are, respectively, $3F(1-F)/(1+F)$ and $(1-F)^2/(1+F)$. The probability that all three alleles are of type A is given by the above terms multiplied by the probability that a type A allele is generated at each mutation or migration event, and hence

$$\begin{aligned} E(\tilde{p}_A^3) &= p_A\frac{2F^2}{1+F} + p_A^2\frac{3F(1-F)}{1+F} + p_A^3\frac{(1-F)^2}{1+F} \\ &= \frac{p_A}{1+F}(F + p_A(1-F))(2F + p_A(1-F)). \end{aligned} \quad (7)$$

Extending this argument to arbitrary numbers of alleles leads to recursive formulae of the following form:

$$E(\tilde{p}_A^{r+1}\tilde{p}_B^s\tilde{p}_C^t\tilde{p}_D^u) = \mathrm{E}(\tilde{p}_A^r\tilde{p}_B^s\tilde{p}_C^t\tilde{p}_D^u) \times \left(\frac{rF + p_A(1-F)}{1 + (r+s+t+u-1)F}\right), \quad (8)$$

for all integers $r, s, t, u \geq 0$. Similar formulae apply for more than four distinct alleles. Note that equation (6) is valid for F corresponding to shared ancestry either through known ancestors, such as parents and grandparents, or on an evolutionary timescale. In general, however, the derivation of (8) is only valid when F has the latter interpretation. Shared ancestry through known relatives is discussed by Balding and Nichols (1994).

Equations (1) and (2) follow from (4), (5) and (8). The formulae of Tables 1 and 2 also follow from (8). For example, consider the case that mother, child, and alleged father's genotypes are, respectively, AB, AC, and CD. If the alleged father were the true father then the conditional probability of the child's genotype, given the parent's genotypes, would be simply 1/4. Further, the parent's genotypes represent outcomes A, B, C and D in four draws from the subpopulation and thus have likelihood $\mathrm{E}(\tilde{p}_A\tilde{p}_B\tilde{p}_C\tilde{p}_D)$. The the joint likelihood is therefore $\mathrm{E}(\tilde{p}_A\tilde{p}_B\tilde{p}_C\tilde{p}_D)/4$. If the alternative father were the true father then the likelihood of the child's maternal allele, given the mother's genotype, is 1/2 and we have, under this hypothesis, observed two C alleles, one from the alleged father and one from the true father. The joint likelihood in this case is thus $\mathrm{E}(\tilde{p}_A\tilde{p}_B\tilde{p}_C^2\tilde{p}_D)/2$. Substituting from (8), the ratio of these joint likelihoods gives

$$2\left(\frac{F+(1-F)p_C}{1+3F}\right),$$

as given in Table 1.

Note that the moments (8) are exactly those which follow from assuming that $(\tilde{p}_A, \tilde{p}_B, \tilde{p}_C, \tilde{p}_D, 1-\tilde{p}_A-\tilde{p}_B-\tilde{p}_C-\tilde{p}_D)$ is jointly Dirichlet distributed with parameter vector

$$(\theta p_A-1, \theta p_B-1, \theta p_C-1, \theta p_D-1, \theta(1-p_A-p_B-p_C-p_D)-1), \tag{9}$$

and $\theta = (1-F)/F$. Although our derivation does not start from the Dirichlet assumption, it would in any case be a natural family of distributions to consider for modelling uncertainty about relative frequencies.

4.2 Statistical derivation

Equation (6) was interpreted in terms of a specific evolutionary model which may not be accurate for VNTR loci. However, it follows from the results of Lindley (1990) that equation (6) is more general than the above derivation suggests. Sufficient conditions for (6) are that $\tilde{p}_A = p_A$ whenever p_A is either zero or one and that both the expectation and variance of $\tilde{p}_A$ given $p_A, p_B, p_C, \ldots$, are twice differentiable functions of p_A and do not depend on $p_B, p_C, \ldots$. In addition to (6), it follows immediately from these assumptions that

$$\mathrm{E}(\tilde{p}_A|p_A, p_B, p_C, \ldots) = \mathrm{E}(\tilde{p}_A|p_A) = p_A.$$

We henceforth suppress the explicit conditioning in the moments and write, for example, $\mathrm{E}(\tilde{p}_A)$ in place of $\mathrm{E}(\tilde{p}_A|p_A)$.

We now extend Lindley's argument to derive (8) for $r+s+t+u \leq 4$. In addition to the assumptions in the previous paragraph, we require that the third, fourth and fifth moments of $\tilde{p}_A$ are each sufficiently differentiable functions of p_A. Further, we assume that joint moments of $\tilde{p}_A$ and $\tilde{p}_B$ are differentiable functions of p_A and p_B, and similarly for more complicated joint moments up to order five. Finally, we assume that the even and odd central moments of $\tilde{p}_A$ are functions of p_A which are, respectively, symmetric and anti-symmetric about $p_A = 1/2$. This assumption is natural because of the arbitrary labelling of the alleles.

Let $h(p_A) = \mathrm{E}((\tilde{p}_A-p_A)^3)$. Then

$$\begin{aligned} &h(p_A+p_B+p_C+p_D) = \\ &h(p_A+p_B+p_C) + h(p_A+p_B+p_D) \\ &+h(p_A+p_C+p_D) + h(p_B+p_C+p_D) \\ &-h(p_A+p_B) - h(p_A+p_C) - h(p_A+p_D) \\ &-h(p_B+p_C) - h(p_B+p_D) - h(p_C+p_D) \\ &+h(p_A) + h(p_B) + h(p_C) + h(p_D). \end{aligned} \tag{10}$$

Since, by the assumptions above, no term on the RHS of (10) is a function of each of p_A, p_B, p_C and p_D, we have

$$\frac{\partial^4}{\partial p_A \partial p_B \partial p_C \partial p_D} h(p_A+p_B+p_C+p_D) = \frac{\partial^4}{\partial p_A^4} h(p_A) = 0,$$

so that $h(p_A)$ is a polynomial of degree three in p_A. The boundary conditions imply that this polynomial has roots at 0 and 1 and hence p_A and $(1-p_A)$ are both factors. By symmetry, the third factor must be $(1-2p_A)$ and hence

$$h(p_A) = \kappa p_A(1-p_A)(1-2p_A), \tag{11}$$

for some constant κ. Expanding (11) and substituting from (6) we obtain

$$\mathrm{E}(\tilde{p}_A^3) = \kappa p_A + 3(F-\kappa)p_A^2 + (1-3F+2\kappa)p_A^3.$$

In order to assign the value of κ, we note that

$$\frac{\partial}{\partial p_A}\mathrm{E}(\tilde{p}_A^3)\,|_{p_A=0} = \kappa. \tag{12}$$

The LHS can be interpreted as the probability that three alleles are of the same type in the limit as the number of distinct alleles increases and the relative frequency of mutation to each allele vanishes in an isolated randomly-mating population. Thus κ should agree with the value for the probability that three random alleles are of the same type given by the Ewens Sampling Formula (Ewens, 1979, equation (3.76)) for the infinite alleles model (in which every mutation is to a distinct type). This gives $\kappa = 2F^2/(1+F)$ and (7) follows.

The other third-order moments follow from (7). For example,

$$\frac{\partial^3}{\partial p_A \partial p_B \partial p_C} \mathrm{E}((\tilde{p}_A + \tilde{p}_B + \tilde{p}_C)^3) = \frac{\partial^3}{\partial p_A^3} \mathrm{E}(\tilde{p}_A^3)$$
$$= 6\frac{(1-F)^2}{1+F}. \quad (13)$$

Expanding the LHS of (13), the only term which is a function of each of p_A, p_B and p_C, and hence does not vanish in the differentiation, is $6\mathrm{E}(\tilde{p}_A\tilde{p}_B\tilde{p}_C)$ and thus

$$\frac{\partial^3}{\partial p_A \partial p_B \partial p_C} E(\tilde{p}_A\tilde{p}_B\tilde{p}_C) = \frac{(1-F)^2}{1+F}. \quad (14)$$

Using (14) and the boundary conditions, it follows that

$$\mathrm{E}(\tilde{p}_A\tilde{p}_B\tilde{p}_C) = \frac{(1-F)^2}{1+F} p_A p_B p_C.$$

Turning now to fourth order moments, $\mathrm{E}((\tilde{p}_A - p_A)^4)$ can similarly be shown to be a polynomial of degree four in p_A which, from the boundary conditions and symmetry, is of the form

$$\mathrm{E}((\tilde{p}_A - p_A)^4) = p_A(1-p_A)(\kappa + \lambda p_A(1-p_A)),$$

for some constants κ and λ. Invoking again the Ewens Sampling Formula we have

$$\frac{\partial}{\partial p_A} \mathrm{E}(\tilde{p}_A^4)\,|_{p_A=0} = \kappa = \frac{6F^3}{(1+F)(1+2F)}.$$

To obtain a value for λ, we argue as at (13) and (14) that

$$\frac{1}{24}\frac{\partial^4}{\partial p_A^4} \mathrm{E}(\tilde{p}_A^4) = \frac{\partial^4}{\partial p_A \partial p_B \partial p_C \partial p_D} \mathrm{E}(\tilde{p}_A\tilde{p}_B\tilde{p}_C\tilde{p}_D)$$
$$= \frac{(1-F)^3}{(1+F)(1+2F)}, \quad (15)$$

the final expression being the probability that four randomly-drawn alleles are distinct in an infinite-alleles model.

Similarly for fifth-order moments, we proceed from the observation that $\mathrm{E}((\tilde{p}_A - p_A)^5)$ is a polynomial of degree five which, from the boundary conditions and symmetry, is of the form

$$\mathrm{E}((\tilde{p}_A - p_A)^5) = p_A(1-p_A)(1-2p_A)(\kappa + \lambda p_A(1-p_A)).$$

We omit the further details.

4.3 Discussion

The substantive assumption in the derivation of Section 4.2 is that, given p_A, moments of $\tilde{p}_A$ up to order five are conditionally independent of $p_B, p_C, \ldots$, and similarly for joint moments. If nothing were known about mutation, the value of p_B, for example, might be informative about it and hence about $\tilde{p}_A$. This dependence may, however, be unimportant given partial knowledge about mutation. Similarly, p_B may be informative about the genealogy of the whole population, but this may also be unimportant for the very large population sizes of the major racial groups into which forensic databases are usually classified.

Equations (1) and (2) are appropriate for a range of genetic typing systems. The most common such system, based on VNTR loci, is problematic because of their complicated evolution. For example, mutation events at VNTR loci frequently generate new alleles of a similar length to the progenitor allele (Jeffreys *et al.*, 1988). This constraint on mutation seems to produce patterns that persist over evolutionary time scales. Waye and Eng (1994) investigated a thalassaemia deletion linked to a VNTR locus. There was an atypical set of VNTR allele lengths on haplotypes that bore the deletion. These alleles were rare on other chromosomes, and had a narrow distribution of lengths. Presumably the newly arisen deletion (or one of its early descendants) bore a rare VNTR allele, and its present day descendants have a narrow range of lengths produced by the subsequent mutations.

Even the more general assumptions of Section 4.2 may fail to encompass exactly the complicated behaviour of VNTR loci. In particular, as a consequence of the mutation process, the frequencies of alleles of similar lengths are positively correlated (Nichols & Balding, 1991) and this effect is not accounted for in (2). When more detailed knowledge of VNTR evolution becomes available, it may prove possible to improve the proposed method to include such length dependent correlations, possibly by using only one additional parameter. Given present knowledge, however, we believe that the proposed method captures the primary effects of coancestry and other sources of uncertainty. In court, the single parameter has proved a common currency in which experts can attempt to quantify their disagreement. The calculations presented here have been used to assess the consequences for the evaluation of the DNA evidence.

Acknowledgements

We are grateful to Wilson Wall and Matthew Greenhalgh for providing access to the data for Figure 1 and Peter Donnelly for valuable discussions. Work supported in part by the UK Science and Engineering Research Council under grants GR/F 98727 (DJB) and GR/G 11101 (RAN).

References

Balding, D.J. & P. Donnelly, 1995. Inference in forensic identification. To appear J. Roy. Statist. Soc. 158.

Balding, D.J. & R.A. Nichols, 1994. DNA profile match probability calculation: how to allow for population stratification, relatedness, database selection and single bands. Forensic Sci. Inter. 64: 125-140.

Cavalli-Sforza, L.L. & A. Piazza, 1993. Human genomic diversity in Europe: a summary of recent research and prospects for the future. Eur. J. Hum. Genet. 1: 3-18.

Chakraborty, R. & K.K. Kidd, 1991. The utility of DNA typing in forensic work. Science 254: 1735-1739.

Cockerham, C.C., 1971. Higher order probability functions of identity of alleles by descent. Genetics 69: 235-246.

Crow, J.F. & M. Kimura, 1970. An Introduction to Population Genetics Theory. New York: Harper and Row.

Devlin, B., N. Risch & K. Roeder, 1993. Statistical evaluation of DNA fingerprinting: a critique of the NRC's report. Science 259: 748, 749, 837.

Donnelly, P., 1995. The non-independence of matches at different loci in DNA profiles: quantifying the effect of close relatives on the match probability. (to appear) Heredity.

Ewens, W.J., 1979. Mathematical Population Genetics. Berlin: Springer-Verlag.

Geisser, S. & W. Johnson, 1993. Testing independence of fragment lengths within VNTR loci. Am. J. Hum. Genet. 53: 1103-1106.

Harding, R.M., 1992. VNTRs in review. Evol. Anthrop. 1: 62-71.

Jeffreys, A.J., N.J. Royale, V. Wilson & Z. Wong, 1988. Spontaneous mutation rates to new alleles at tandem repetitive hypervariable loci in human DNA. Nature 332: 278-281.

Jeffreys, A.J., K. Tamaki, A. MacLeod, D.G. Monckton, D.L. Neil & J.A.L. Armour, 1994. Complex gene conversion events in germline mutation at human minisatellites. Nature Genetics 6: 136-145.

Krane, D.E., R.W. Allen, S.A. Sawyer, D.A. Petrov & D.L. Hartl, 1992. Genetic differences at four DNA typing loci in Finnish, Italian and mixed Caucasian populations. Proc. Nat. Acad. Sci. USA 89: 10583-10587.

Lempert, R., 1991. Some caveats concerning DNA as criminal identification evidence: with thanks to the Reverend Bayes. Cardozo Law Rev. 13: 303-341.

Lewontin, R.C. & D.L. Hartl, 1991. Population genetics in forensic DNA typing. Science 254: 1745-1750.

Lindley, D.V., 1990. The present position in Bayesian statistics. Statist. Sci. 5: 44-89.

Morton, N.E., 1992. Genetic structure of forensic populations. Proc. Nat. Acad. Sci. USA 89: 2556-2560.

Morton, N.E., 1993a. DNA in court. Eur. J. Hum. Genet. 1: 172-178.

Morton, N.E., 1993b. Kinship bioassay on hypervariable loci in blacks and caucasians. Proc. Nat. Acad. Sci. USA 90: 1892-1896.

Nichols, R.A. & D.J. Balding, 1991. Effects of population structure on DNA fingerprint analysis in forensic science. Heredity 66: 297-302.

Robertson, B. & T. Vignaux, 1992. Why the NRC report on DNA is wrong. New Law J.: 1619-1621.

Roeder, K., 1994. DNA fingerprinting: a review of the controversy. Statist. Sci. 9: 222-278.

Wall, W.J., R. Williamson, M. Petrou, D. Papaioannou & B.H. Parkin, 1993. Variation of short tandem repeats within and between populations. Hum. Molec. Genet. 2: 1023-1029.

Waye, J.S. & B. Eng, 1994. Allelic stability of a VNTR locus 3'αHVR: Linkage disequilibrium with the common α-thalassaemia-1 deletion of South-East Asia (–SEA/). Hum. Heredity 44: 61-67.

Weir, B.S., 1993a. Forensic Population Genetics and the National Research Council (NRC). Am. J. Hum. Genet. 52: 437-440.

Weir, B.S., 1993b. Independence tests for VNTR alleles defined as quantile bins. Am. J. Hum. Genet. 53: 1107-1113.

Yasuda, N., 1968. An extension of Wahlund's principle to evaluate mating type frequency. Am. J. Hum. Genet. 20: 1-23.

Note added in proof

Subsequent work (M. Greenhalgh, pers. comm.) suggests that the data from the CD4 locus (Fig. 1) may be affected by laboratory error. Recent data continue to indicate different values of F_{ST} at different loci, in some cases showing as much variation as at traditional loci.

Editor's comments

The authors' work offers a sound approach to accommodating the effects of population structure, based on use of Wright's F_{ST}. Their equations 1 and 2 are very convenient, and are good approximations to the exact results given by Weir (1994). As they point out, good estimates of F_{ST} are needed. The comments about the 'generally mixed' results of independence tests may be met, in part, by the paper of Maiste and Weir in this volume. The authors cite Krane *et al.* (1992) but had not seen the subsequent rebuttal by Budowle *et al.* (1994). The work of Wall *et al.* (1993) contained errors, as noted in Greenhalgh *et al.* (1994).

B. S. Weir (ed.), Human Identification: The Use of DNA Markers, 13–19, 1995.

The effect of relatedness on likelihood ratios and the use of conservative estimates

John F.Y. Brookfield
Department of Genetics, University of Nottingham, Queens Medical Centre, Nottingham NG7 2UH, UK

Received 20 April 1994 Accepted 10 June 1994

Key words: conservativeness, DNA profiling, forensic science, relatedness, population substructure

Abstract

DNA profiling can be used to identify criminals through their DNA matching that left at the scene of a crime. The strength of the evidence supplied by a match in DNA profiles is given by the likelihood ratio. This, in turn, depends upon the probability that a match would be produced if the suspect is innocent. This probability could be strongly affected by the possibility of relatedness between the suspect and the true source of the scene-of-crime DNA profile. Methods are shown that allow for the possibility of such relatedness, arising either through population substructure or through a family relationship. Uncertainties about the likelihood ratio have been taken as grounds for the use of very conservative estimates of this quantity. The use of such conservative estimates can be shown to be neither necessary nor harmless.

Introduction

Considerable use is made of DNA profile information in human identification. In 1985 Alec Jeffreys and his co-workers discovered the presence of 'minisatellite' loci scattered around the human chromosomes, which show hypervariability in the number of tandem repeats that they contain (Jeffreys, Wilson & Thein, 1985a, 1985b). These minisatellite loci can be revealed using a 'core' probe, which simultaneously detects 50–100 loci. Alternatively, individual unlinked loci can be probed at high stringency, to reveal, typically, a heterozygous pattern of two bands, but sometimes a single homozygous band (Wong *et al.*, 1987). Successive reprobings can reveal information from four to six loci. In a criminal case in which DNA evidence is used, a match will typically have been found between a defendant and a blood or a semen stain from the scene of the crime. This match could be either a match in one or two multi-locus profiles (sometimes referred to as 'DNA fingerprints' (Jeffreys, Wilson & Thein, 1985b)), or with a set of single-locus profiles. In each case, the match between the two sets of profiles constitutes immensely powerful evidence that they are derived from the same individual. (DNA profiling has also proved very useful in the establishment of paternity (Jeffreys, Brookfield & Semeonoff, 1985; Jeffreys, Turner & Debenham, 1991), but I will not consider this further here).

The likelihood ratio and its calculation

Likelihood ratios

Likelihood ratios assess the strength of evidence

The strength of any evidence can, in principle, be assessed using a likelihood ratio. This is the ratio of the probabilities that the evidence would have been obtained given that the suspect is guilty and innocent respectively. In other words,

$$\text{Likelihood Ratio} = \\ = \frac{\text{Probability of the evidence}|\text{Guilt}}{\text{Probability of the evidence}|\text{Innocence}}$$

These probabilities may both be very low, in that it could be that the particular evidence obtained was unlikely under both hypotheses. This is, in itself, no cause for concern, since it is the ratio between them which is of relevance. While it is true that for many types of evidence it is impossible in practice to arrive at numerical values of the two probabilities, the formula illustrates, for example, the reason why evidence in the form of a confession is less powerful than is popularly supposed. One view of confession evidence is that it must be powerful since it seems highly unlikely that an innocent man should confess to a crime that he did not commit. The likelihood ratio expression reminds us that the probability that a guilty man will confess may also be low, and, since it is only the ratio of the probabilities of confession under the two hypotheses which gives us evidence to decide between these hypotheses, the likelihood ratio arising from confession evidence is unlikely to be large. In DNA profiling, the most typical situation is where there is a DNA sample (from blood or semen) left at the scene of the crime, and the circumstances of the case make it highly likely that this DNA came from the perpetrator. A defendant has a matching DNA profile. The probability that such a match would have been found, given that the defendant is guilty will, in most cases, be little less than one. The major issue therefore becomes the determination of the probability that the defendant will match the scene-of-crime DNA, given that he is innocent, and therefore not the source of the scene-of-crime sample. Clearly, in some improbable cases, it may be that the defendant is the source of the scene of crime DNA, and yet is innocent. Thus, the guilty man may not be the source of the scene-of-crime DNA sample. However, for simplicity, I will here regard the defendant being the source of the scene-of-crime DNA and the defendant being guilty as equivalent.

Likelihood ratios and odds of guilt

Bayes' theorem shows that the likelihood ratio can be related to the prior odds of guilt and of innocence in the following way, in order to arrive at the posterior odds of guilt:

$$\text{Prior odds of Guilt} \times \text{Likelihood ratio} = \\ = \text{Posterior odds of guilt} \qquad (1)$$

In this expression the odds of guilt are the probability that the suspect is guilty of the crime divided by the probability that he is innocent. The prior odds of guilt are the probability of guilt based on all the evidence and information relevant to the case, but excepting the evidence based on the DNA. The posterior odds, which is what is directly of interest to the court, includes the DNA evidence.

Formula (1) reminds us that almost any likelihood ratio is consistent with almost any posterior odds, and that DNA evidence cannot be used directly to provide the posterior odds. In almost all cases, it is impossible for the jury, or any others concerned with a case, to assess the prior odds in a quantitative form. Thus in a trial in the UK, for example, a jury confronted by evidence of a quantitative kind has to perform the complicated mental gymnastics involved in assessing, in some way, the strength of the non-quantitative evidence that contributes to the prior odds, incorporating, in some way, the likelihood ratio into this assessment, and then deciding whether the posterior odds suggesting guilt are sufficiently high to constitute 'beyond reasonable doubt'.

This raises difficulties. Consider a jury asked to decide on the issue of the guilt of a defendant who has a likelihood ratio from the DNA of one million against him. What prior odds of guilt should the jury assume prior to the DNA evidence? Indeed, what prior odds should they assume prior to hearing any evidence at all? The assumption of innocence would seem to require that, prior to any evidence being presented, the defendant's probability of guilt should be no more than that of anyone else. (It is absurd to interpret the presumption of innocence as implying a prior probability of guilt as zero, since, from (1) above, this would necessary mean that the posterior odds would correspondingly remain zero, however strong the evidence.) One way of looking at cases in which DNA forms the most powerful evidence against a defendant is to consider the prior odds in terms of the size of some potential suspect population. Balding and Donnelly (1994), for example, consider a hypothetical case in which a population of 500,000 possible perpetrators, which equal prior odds, leads to a posterior probability of innocence of about one third even though the DNA evidence yields a likelihood ratio of one million.

This raises an interesting question. Suppose that there are 500,000 individuals in a town each of whom could, in a 'hypothetical' scenario raised by a defense expert witness, be guilty. If we take one in 500,000 as the prior odds of guilt, then the posterior probability of guilt, of two thirds, could not constitute certainty beyond reasonable doubt. However, if the defendant's DNA profile was obtained because the police regarded him to be a suspect in this particular case, and not through a screen of a pre-existing database, then we know that the investigators would not have sampled this individual's DNA if they believed that he had only a one in 500,000 chance of being guilty. Usually, in addition to the DNA evidence, there is other evidence, some of which may not be revealed to the court, which also points to the guilt of a suspect. This could be evidence of prior convictions, for example, which is not admissible in UK courts. The police may have used this other evidence in deciding who to ask for a DNA sample. Now suppose that very little evidence other than DNA is presented to the court. The jury will probably believe that police and forensic scientists would not be so foolish as to screen individuals with a very low probability of guilt. They may indeed be aware that the success rate in finding matches with DNA is consistently (across cases) high. Certainly, it is very much higher than would be expected if the individuals screened had a one in 500,000 probability of guilt. The jury may also know that, in this case, the match was found with, for example, only the fourth suspect examined. Thus, the court may be in a position to assess empirically the justifiability of the grounds for police suspicions against a suspect without knowing what those grounds are. In these circumstances, it would be foolish for the jury to regard suggested prior odds of one in 500,000 as likely to be accurate. Nevertheless, the jury's view does include an element of their trusting the police to be able to identify criminals, which might be seen to be inconsistent with the view that the jury should make up its mind entirely on the basis of the evidence presented in court.

More commonly, furthermore, there will be other admissible evidence against the suspect, such as the absence of a convincing alibi, which will considerably increase the prior odds relative to any odds based on the concept of a 'population of possible perpetrators'.

The effects of population substructure

We have seen that the probability of an innocent suspect matching the DNA sample from the scene of the crime is the major determinant of the likelihood ratio. The assessment of this probability has caused some controversy (reviewed by, among others, Weir, 1992; Kaye, 1993). The most logical way to assess this probability would be to estimate the frequency in the population of a multi-locus profile or a collection of single-locus profiles on the basis of its sample frequency. Thus the number of times the profile has been seen in some database could be divided by the total number of individuals in the database. This has an obvious problem that, almost always, the profile whose frequency is to be assessed will only ever have been seen twice, once in the scene-of-crime sample and once in the suspect. Furthermore, since the existence of a trial including DNA evidence is conditional upon the profile having been found twice in this way, the sample frequency will be biased upwards to an unquantifiable degree. This has led to the use of a calculation of the probability of a multi-locus profile or a collection of single locus profiles on the basis of what has been called the 'product rule'. Here the frequencies of the individual alleles in the database are coupled with an assumption that the presence or absence of any given band in a profile is held to be independent of that of any other. Thus, in the case of a suspect's DNA sample that matches a scene-of-crime DNA sample at k heterozygous loci, the probability of this match is calculated as

$$2^k \prod_{i=1}^{2k} p_i$$

Here the p_i values represent the frequencies of the $2k$ bands seen in the relevant population databases. Corresponding formulae exist for cases in which some of the loci are homozygous. Databases are available for Caucasians and Afro-Caribbeans in the USA and in the UK, and also for 'Hispanics' in the USA. The relevant population database, as pointed out by Weir and Evett (1992), is that of the truly guilty person, assuming the defendant is not the source of the DNA. Thus, the racial background of the defendant is irrelevant unless there is convincing evidence that, if the defendant is innocent, this background will be shared by the truly guilty man.

Human populations are, at least to some extent, divided into subgroups in addition to broad racial categories. Two alleles sampled from the same population subgroup will share more recent common ancestry than two alleles sampled from different subpopulations, and thus have a higher chance of identity by descent. This will lead to at least a low level of linkage disequilibrium between alleles, and thus some inaccuracy in the product rule. The fear of major inaccuracies arising in the use of the product rule was the basis of the critique of Lewontin and Hartl (1991). Since this time there have been numerous rebuttals of their argument (e.g. Chakraborty & Kidd, 1991; Risch & Devlin, 1992; Morton, 1992) and data sets have been generated showing that the effect of population substructure is likely to be very small (Devlin & Risch, 1992; Weir, 1992; Morton, Collins & Balazs, 1993; Budowle & Monson, 1993; Balazs, 1993).

Nevertheless, the United States National Research Council report on DNA profiling (National Research Council, 1992) recommended that, to counter fears concerning population substructure, a 'ceiling principle' be adopted. To apply this, a large number of homogeneous ethnic subgroups would be randomly sampled, and, for each allele, the highest frequency seen in any of the subgroups would be used as the allele frequency, assuming that this frequency was above 5%. If the allele had a lower frequency than 5% in all ethnic subgroups, then 5% would be used. However, until the data are available, the committee recommended, in what was called the 'interim ceiling principle', that the frequency used for each allele should be either 10%, or the upper 95% confidence limit of the frequency in the major race with the largest frequency, whichever is higher. One advantage of using such extremely high values is that they will remain conservative (i.e. biased in favor of the defendant) whatever the racial mix of the suspect population, (but see Cohen, 1992, for arguments why the NRC recommendations may yet fail to be conservative).

The effect of population substructure can, slightly more realistically, be quantified using Wright's F_{ST} values (Wright, 1943), and the use of this approach in forensic work has been suggested by Nichols and Balding (1991). They suggested that, for any locus, the probability of a match between an innocent suspect and a heterozygous DNA profile from the scene of a crime is not $2p_ip_j$, but

$$2(p_i + F_{ST} - p_iF_{ST})(p_j + F_{ST} - p_jF_{ST}),$$

where p_i and p_j are the frequencies of the two alleles seen in the database, and F_{ST} quantifies the degree of population subdivision. These authors illustrate the strength of the effect by choosing hypothetical values for p_i, p_j and F_{ST} of 0.01, 0.01 and 0.05 respectively. The formula is only approximate and these numerical values are unrealistically extreme ones which maximise the impact of population substructure on the match probability at the locus. The analysis also makes the assumption that, if the suspect is innocent, the true source of the scene-of-crime DNA came from the same subpopulation as the suspect. The use of this formula with extremely conservative values (i.e. ones which favor the defendant) was designed to produce an upper bound for the effect of population substructure. There is, however, little point in producing an unrealistic upper bound for a quantity and then concluding that its interesting quality is that it is very large. Trying to use more reasonable values of F_{ST} and the allele frequencies, and relaxing the assumption that defendant and truly guilty person were from the same subpopulation, Brookfield (1992) found that, in hypothetical data sets, allowing for population substructure made very little difference to the likelihood ratio.

The effects of relatives

A much larger potential impact on the likelihood ratio stems from the possibility that, if the suspect is innocent, the true source of the scene-of-crime DNA is a close relative of the suspect. This situation has been considered by Evett (1992), and also more recently (Brookfield, in press). The argument in the latter manuscript is that, if a suspect is innocent, the probability that he will match the scene-of-crime DNA, which is effectively the probability of the evidence given innocence, will be

$$\sum_{k=1}^{j} P(E|S,k) \,.\, P(k) \tag{2}$$

Here there are *j* possible relationships between the suspect and the source of the scene-of-crime DNA sample, and the probability of the *k*th such relationship, given that the suspect is not the source of the scene-of-crime sample, is *P*(*k*). *S* represents the suspect's DNA profile and *E* the scene-of-crime DNA profile. If the relationship is *k*, then the probability of *E*, i.e. the DNA evidence, given *k* and *S*, is *P*(*E*|*S*, *k*). The main conclusion of the analysis is that *P*(*E*|*S*, *k*) is very much greater when k = siblings than for any other possible relationship (except identical twins). This means that assessment of the likelihood ratio depends greatly on the empirical determination of *P*(*k*) for k = siblings, and thus it is very useful to test the siblings of the suspect, and to demonstrate that they do not match the scene-of-crime DNA profile. Doing so shows *P*(*k*) to be zero. If this is true then, typically, the effect of all possible other types of relationship will probably be small unless there is evidence that such relatives are more likely, prior to the DNA evidence, to have committed the crime than are other members of the population.

Individual specificity

When DNA fingerprints were discovered, they were claimed to individual-specific with the exception of identical twins (e.g. Jeffreys, Wilson & Thein, 1985b). It has been said that these initial claims for the individual-specificity of DNA fingerprinting have subsequently been retracted. In reality, what has happened has been a methodological change in which the multilocus profiling which was initially used, and for which individual-specificity was postulated, has, in UK and

US courts, been replaced by collections of single-locus probes. These generally give likelihood ratios of a few million, and thus are obviously not individual-specific. This lack of individual specificity has been thought to be of great relevance in some courtroom situations, and one might wonder why the potential individual specificity of multi-locus probes has not been seen as an overwhelming reason for their continued use. In fact, the quest for absolute individual-specificity is irrational. The probability that two multi-locus profiles are identical can be calculated in a number of ways, with varying degrees of conservativeness built into their assumptions. The match probability under one set of assumptions might be 10^{-30}. A very much more conservative set of assumptions might give a match probability of 10^{-15}. Each would constitute exceptionally powerful evidence against a suspect. These two different match probabilities have, however, very different implications in terms of the individual-specificity of the technique. With 5×10^9 people on earth, and thus 1.25×10^{19} pairwise comparisons, a probability of a chance match of 10^{-15} in each comparison would be expected to give 12,500 matching pairs in the world if all individuals were compared. In a formal sense, therefore, DNA profiles would not be individual-specific. However, this is completely irrelevant to any forensic investigation. The expected number of pairs in which both members of the pair are British, for example, is about one hundredth of one pair. Also, we must recognise that the probability of a match being found through a laboratory error or deception, in which the suspect's DNA is substituted for that from the scene-of-crime, will be 10^{-9} at the very least, which makes irrelevant all questions of whether the match probability is 10^{-15} or 10^{-30}.

Conservative estimates of the likelihood ratio

Why use conservative estimates?

The principle of the prior assumption of innocence has implications in terms of the ways in which evidence in court is evaluated. Sometimes it is felt that a numerical assessment of the likelihood ratio cannot be produced reliably. In these circumstances a value of the likelihood ratio may be given which is deliberately conservative. In other words, the interpretation of the evidence against the defendant embodied in the likelihood ratio is that the evidence is less powerful than it truly is. There are three possible reasons for such a procedure. The first is a rather non-specific concept of 'the benefit of the doubt' arising from a prior assumption of innocence. However, the likelihood ratios are ratios of probabilities and there is no obvious reason why one should seek to revise upwards a probability on no better grounds than its being a probability. Seemingly more rational is the concept that, since the probability that is calculated is itself based upon empirical evidence that is incomplete, the underlying true probability could have any of a range of values, and we should, if we are to avoid incriminating the suspect by the use of an inaccurately low probability, give the upper bound of what the 'true' probability might turn out to be. This is, at least in part, the grounds for the 'ceiling principle' advocated by the National Research Council (1992). It is important to remember, however, that a probability statement is, by its nature, an encapsulation of our partial knowledge about a situation. Given this, it is paradoxical to believe that, in principle, we cannot produce a probability statement due to a lack of empirical knowledge. If a range of probability values are possible, each of which can itself be given a probability of being correct, then a weighted average can be used to replace the range. However, while in principle we should be able to arrive at such an overall probability based on our knowledge of a case, we may, in practice, not know how to do this. There may, in these circumstances, be utility in evaluating the standard error and confidence interval of a probability, as suggested by Chakraborty, Srinivasan and Daiger (1993), notwithstanding the paradox that this engenders. It is also true that additional information gathered after a trial has resulted in conviction, such as the details of allele frequencies within ethnic groups, could be legally viewed as significant new evidence in the case if the probability calculated including this extra information were more favorable to the defendant than that used at the original trial. Use of conservative estimates may avoid this outcome.

The third argument for the use of conservative likelihood ratios is the very pragmatic one that, if the prosecution uses a highly conservative value of the likelihood ratio, then this cannot honestly be revised downwards by the defense. In many cases prosecution and defense experts have produced large but different likelihood ratios, usually because the defense have made conservative assumptions that the prosecution have not. It is probable that the benefit of this situation to the defendant will not be confined to the acceptance by the jury of the defense expert's likelihood ratio. In addition, the divergence in estimates may create the impression that the allocation of numbers to this evidence is entirely arbitrary, and that the true strength of the evidence cannot be assessed. This may cause the DNA evidence to be grossly undervalued in comparison to its true weight.

What are the effects of the use of conservative likelihood ratios?

The use of conservative likelihood ratios for DNA evidence has been justified on the grounds that they are safe estimates which cannot be prejudicial to defendants, and yet remain sufficiently impressively large that they will not lead to false acquittals through juries underestimating the power of the DNA evidence. However, the logic which makes it possible for the use of conservative likelihood ratios to prevent false convictions without risking false acquittals has never been clearly explained and remains mysterious. It is unquestionably laudable that court procedures should seek to minimise the rate of false conviction, but it is also important to remember two general truths about criminal trials. The first is that, in any case, the guilt of the defendant can never be certain, in the sense that the posterior probability of guilt is exactly one. This follows from the necessarily finite collection of information available about the unique event that was the crime. It is also true that for the criminal justice system to serve any function to society, some defendants must be convicted. If so, some risk of false convictions is inevitable. If, however, we were to consider the optimisation of a system of justice in terms of the minimisation of the number of individuals who are falsely convicted, such an optimisation can, in principle, be carried out in two ways. (What we must first assume is that it is possible, in principle, to assess the strength of the evidence available concerning a suspect and to arrive at a probability of guilt of the suspect based on this evidence). Firstly, there could be an increase in the probability of guilt required before the probability could be defined as 'shown beyond reasonable doubt'. How high the probability should be is not a scientific question, but rather a question of social policy. The higher that it is defined as being, the lower will be the rate of false conviction, and the higher will be the rate of false acquittal. A second way of minimising the rate of false convictions is by calculating as accurately as possible the probability of guilt, and then comparing this probability to the threshold by which 'reasonable doubt' is defined. This produces the outcome that all those convicted are, indeed, more likely to be guilty than are all of those who are acquitted. The use of conservative estimates of the strength of certain types of evidence is a third method, but one that has no logic in its favor. In this method, for an effectively random subset of defendants (those with DNA evidence against them), the stated probability of guilt is arbitrarily reduced relative to its true value. This will prevent some false convictions, but, in general, the defendants whose convictions it will prevent will not be those who are most likely to be innocent.

Discussion

DNA evidence provides courts with evidence in criminal identification that is of unprecedented power. Some have found this power disconcerting, and have been particularly concerned that extremely large likelihood ratios, of perhaps tens of millions, arising from DNA evidence, result in a very high probability of conviction, whatever the rest of the evidence. If the likelihood ratios have been correctly arrived at, however, and if the prior odds of guilt are not exceptionally small, it is quite appropriate for convictions to generally follow from the existence of DNA evidence against the defendant. We have seen that the likelihood ratios calculated from DNA may be greatly affected if, given that the suspect is innocent, the truly guilty man is likely to be a very close relative, particularly a brother. It is also important to remember that circumstances could exist, such as if the DNA match was identified in a screen of a database of very large numbers of individuals, where the prior odds of guilt are very low.

It is also essential to realise that the probabilities of laboratory error, or indeed fabrication of evidence, are likely to be greater than the probability of a chance match with, for example, five or six single locus probes (which might have a combined probability of 10^{-8}). This will probably be true even when there is no evidence in a case that suggests that error or deception has occurred. Undetected and unsuspected human error or mischief is potentially a problem in almost all criminal cases, yet no convictions could follow if this possibility was consistently deemed to lead to reasonable doubt. The lower probability of a match arising through the 10^{-8} chance that suspect and criminal, while different individuals, have matching profiles, is comparatively unlikely to lead to false conviction.

References

Balazs, I., 1993. Population genetics of 14 ethnic groups using phenotypic data from VNTR loci, pp. 193–210 in DNA Fingerprinting: State of the Science, Edited by S.D.J. Pena, R. Chakraborty, J.T. Epplen and A.J. Jeffreys. Birkhäuser Verlag Basel Boston Berlin.

Balding, D.J. & P. Donnelly, 1994. How convincing is DNA evidence? Nature 368: 285–286.

Brookfield, J.F.Y., 1992. The effect of population subdivision on estimates of the likelihood ratio in criminal cases using single-locus DNA probes. Heredity 69: 97–100.

Brookfield, J.F.Y. The effect of relatives on the likelihood ratio associated with DNA profile evidence in criminal cases. J. Forensic Science Soc. (In press).

Budowle, B. & K.L. Monson, 1993. The forensic significance of various reference population databases for estimating the variable number of tandem repeat (VNTR) loci profiles, pp. 177–192 in DNA Fingerprinting: State of the Science, Edited by S.D.J. Pena, R. Chakraborty, J.T. Epplen and A.J. Jeffreys. Birkhäuser Verlag Basel Boston Berlin.

Chakraborty, R. & K.K. Kidd, 1991. The utility of DNA typing in forensic work. Science 254: 1735–1739.

Chakraborty, R., M.R. Srinivasan & S.P. Daiger, 1993. Evaluation of standard error and confidence interval of estimated multilocus genotype probabilities, and their implications in DNA forensics. Am. J. Hum. Genet. 52: 60–70.

Cohen, J.E., 1992. The ceiling principle is not always conservative in assigning genotype frequencies for forensic DNA testing. Am. J. Hum. Genet. 51: 1165–1168.

Devlin, B. & N. Risch, 1992. Ethnic differentiation of VNTR loci, with special reference to forensic applications. Am. J. Hum. Genet. 51: 534–548.

Evett, I.W., 1992. Evaluating DNA profiles in a case where the defence is 'It was my brother'. J. Forensic Science Soc. 32: 5–14.

Jeffreys, A.J., J.F.Y. Brookfield & R. Semeonoff, 1985. Positive identification of an immigration test-case using human DNA fingerprints. Nature 317: 818–819.

Jeffreys, A.J., M. Turner & P. Debenham, 1991. The efficiency of multilocus DNA fingerprinting probes for individualization and establishment of family relationship, determined from extensive casework. Am. J. Hum. Genet. 48: 824–840.

Jeffreys, A.J., V. Wilson & S.L. Thein, 1985a. Hypervariable 'minisatellite' regions in human DNA. Nature 314: 67–73.

Jeffreys, A.J., V. Wilson & S.L. Thein, 1985b. Individual-specific 'fingerprints' of human DNA. Nature 316: 76–79.

Kaye, D.H., 1993. DNA evidence: probability, population genetics, and the courts. Harvard J. Law Tech. 7: 101–172.

Lewontin, R.C. & D.L. Hartl, 1991. Population genetics in forensic DNA typing. Science 254: 1745–1751.

Morton, N.E., 1992. Genetic structure of forensic populations. Proc. Natl. Acad. Sci. USA 89: 2556–2560.

Morton, N.E., A. Collins & I. Balazs, 1993. Kinship bioassay on hypervariable loci in Blacks and Caucasians. Proc. Natl. Acad. Sci. USA 90: 1892–1896.

National Research Council, 1992. DNA technology and forensic science. National Academy Press. Washington DC.

Nichols, R.A. & D.J. Balding, 1991. Effects of population structure on DNA fingerprinting analysis in forensic science. Heredity 66: 297–302.

Risch, N. & B. Devlin, 1992. On the probability of matching DNA fingerprints. Science 255: 717–720.

Wong, Z., V. Wilson, I. Patel, S. Povey & A.J. Jeffreys, 1987. Characterisation of a panel of highly variable minisatellites cloned from human DNA. Ann. Hum. Genet. 51: 269–288.

Weir, B.S., 1992. Independence of VNTR alleles defined by fixed bins. Genetics 130: 873–887.

Weir, B.S., 1993. Population genetics in the forensic DNA debate. Proc. Natl. Acad. Sci. USA 90: 11654–11659.

Weir, B.S. & I.W. Evett, 1992. Whose DNA? Am. J. Hum. Genet. 50: 869.

Wright, S., 1943. Isolation by distance. Genetics 28: 114–138.

Editor's comments

The author makes a good case for the use of likelihood ratios in presenting evidence. The author cites Nichols and Balding (1991) for the match probability involving F_{ST}, and readers will also need to consult Balding and Nichols (1994 and this volume) and Weir (1994). In light of current concern over laboratory errors or fraud, the author's comment that 'Undetected and unsuspected human error or mischief is potentially a problem in almost all criminal cases, yet no convictions could follow if this possibility was consistently deemed to lead to reasonable doubt' is well taken.

B. S. Weir (ed.), Human Identification: The Use of DNA Markers, 21–25, 1995.

The effects of inbreeding on DNA profile frequency estimates using PCR-based loci

Bruce Budowle
Forensic Science Research and Training Center, FBI Academy, Quantico, VA 22135, USA

Received 12 May 1994 Accepted 15 July 1994

Key words: African American, Caucasian, Oriental, Hispanic, inbreeding, population substructure, PCR, Hardy-Weinberg Expectations, DNA profile

Abstract

Estimates of inbreeding were determined using Wright's F_{ST} for loci used for PCR-based forensic analyses. The populations analyzed were African Americans, Caucasians, Hispanics, and Orientals. In most cases the F_{ST} values at each locus were less than 0.01. The F_{ST} values over all loci for African Americans, Caucasians, and Orientals ranged from 0.0015 to 0.0048. No substantial differences were observed for DNA profile frequency estimates when calculated under the assumption of independence or with the incorporation of F_{ST}.

Introduction

DNA typing has become an important tool in the analysis of forensic biological evidence. When a match occurs between the DNA profiles of the evidentiary sample and an exemplar from the suspect, or depending on the case, the victim, an estimate of the likelihood of occurrence of the DNA profile in a general reference population is provided (Budowle *et al.*, 1991; Chakraborty & Kidd, 1991; Weir, 1992). Some have proffered that estimates of DNA profile frequencies derived from general population groups, such as United States Caucasians and African Americans, may overstate the rarity of the DNA profile compared with the DNA profile frequency estimated in a particular subgroup (Lander, 1989; Lewontin & Hartl, 1991; Nichols & Balding, 1991; Balding & Nichols, 1994). In contrast, population data to date demonstrate that subgroups tend to have very few differences among them, either genetically or for forensic applications (Budowle & Stafford, 1991; Budowle *et al.* 1994a; 1994b; Budowle, Monson & Giusti, 1994; Devlin & Risch 1992a; 1992b; Devlin, Risch & Roeder, 1993; 1994; Chakraborty & Kidd, 1991; Hartl & Lewontin, 1993; Morton, Collins & Balazs, 1993; Weir, 1992). Furthermore, Li and Chakravarti (1994) showed theoretically that for realistic models of population heterogeneity, the use of the product rule for calculating DNA profile frequencies is conservative when population substructure is present but ignored. Thus, extant data demonstrate that there is little evidence of wrongful bias in forensic applications when estimates are provided using the product rule and general population databases.

Recently, Crow (1993), Morton (1992), and Nichols and Balding (1991) have suggested that potential wrongful bias effects due to population substructure can be addressed by considering the degree of inbreeding in the population. The degree of relatedness in a population can be estimated using Wright's F_{ST} statistic (Wright, 1922, 1965). Although not considered in the current analysis, Nei's coefficient of gene diversity (G_{ST}) also can be used (Nei, 1973, 1977; Chakraborty & Jin, 1992). Two questions arise for the application of F_{ST}: these are 1) when is it necessary or appropriate to modify a DNA profile frequency estimate with F_{ST}: and 2) what is a realistic value for F_{ST}?

The appropriate application of F_{ST} is when there is evidence that the potential contributors of the DNA sample are from one subpopulation (Crow, 1993; Weir & Hill, 1993; Weir, 1994). Ideally, in this situation, population data from the particular subgroup should be used, and the use of F_{ST} would not have to be invoked. If no population data were available for the particular subgroup, then general population data frequencies and F_{ST} could be used. Realistically, in most forensic situations there rarely is evidence to suggest that one subpopulation can be the only contributor of the evidence sample. The determination of the appropriate subgroup is confounded because: 1) the ethnic background of the suspect, who is presumed innocent, should not be used to determine the appropriate population subgroup; 2) even if the suspect's ethnicity were considered meaningful for determining the appropriate subgroup, the suspect may not know his true ethnicity, he may fabricate his ethnic make-up, or he may choose not to tell the court his ethnicity; and 3) many people in the United States, particularly Caucasians, African Americans, Hispanics, and even Native Americans, are not ethnically pure. When the population of contributors cannot be assigned to a particular subgroup, then the current forensic approach of using general population reference databases and the product rule under the assumption of independence is appropriate (Crow, 1993; Weir, 1994).

Balding and Nichols (1994) have suggested that an F_{ST} of 0.05 should be used to correct for population subdivision. Nichols and Balding (1991), however, acknowledge that this estimate corresponds to severe inbreeding associated with a tradition of uncle-niece marriages, and smaller values are more appropriate for situations where extreme inbreeding does not occur. Morton (1992), Chakraborty and Jin (1992), Brookfield (1992) and Li and Chakravarti (1994) advocate using more realistic F_{ST} values, generally less than 0.01. Recently, Weir (1994) estimated the F_{ST} for the variable number of tandem repeat loci in United States Caucasians and African Americans and found it ranged from −0.002 to 0.002. Even for southeastern and southwestern Hispanics, which would not be pooled into a single database, all estimates were less than 0.01. With realistic F_{ST} values, such as those reported by Weir (1994), an estimate of a DNA profile frequency would differ little from that derived by the current approaches used in forensic analyses (Brookfield, 1992).

The use of genetic markers whose analysis is based on the polymerase chain reaction (PCR) is becoming more prevalent in human identity testing. There are no estimates for F_{ST} values for these genetic markers used routinely in forensic analyses. This paper presents F_{ST} values for the following loci: HLA-DQα (Gyllensten & Erlich, 1988), low density lipoprotein receptor (LDLR) (Yamamoto *et al.*, 1984), glycophorin A (GYPA) (Siebert & Fakuda, 1987), hemoglobin G gammaglobin (HBGG) (Slightom, Blechl & Smithies, 1980), D7S8 (Horn *et al.*, 1990), and group-specific component (Gc) (Yang *et al.*, 1985). Therefore, a realistic F_{ST} value can be considered with general population data when the appropriate situation arises.

Materials and methods

Population data on the six PCR-based loci HLA-DQα, LDLR, GYPA, HBGG, D7S8, and Gc were derived from Budowle *et al.* (1994c and manuscript in preparation), Roche Molecular Systems, Alameda, CA, Perkin-Elmer, 1994), Alabama Department of Forensic Sciences (manuscript in preparation), Hochmeister *et al.* (1994), and Huang and Budowle (1994). The Japanese data were also provided by Rebecca Reynolds at Roche Molecular Systems (Alameda, CA). The values for F_{ST} were determined as described by Weir and Cockerham (1984; Weir, 1994) using the DIPLOID.FOR program kindly provided by Weir (1990).

Results and discussion

Table 1 displays the F_{ST} values for Caucasians (three United States Caucasians and Swiss and in the adjacent column two United States, Swiss, Israeli, and Basques), African Americans (three sample populations), Hispanics (southeastern and southwestern), and Orientals (Chinese and Japanese) for six PCR-based loci. The estimates displayed in Table 1 are consistent with those described by Morton (1992). The F_{ST} values at each locus for African Americans, Caucasians, and Orientals are much less than 0.05 and in most cases less than 0.01. The F_{ST} values over all loci for the three major population groups ranged from 0.0015 to 0.0048. The Basque and Israeli databases should not be considered as relevant databases for the population of potential perpetrators in the United States. However, under the highly unlikely assumption of no gene flow among subgroups in the United States, these population groups can be used to gain insight on the degree of inbreeding within subgroups and the effects on forensic DNA statistical estimates. When the Basque and Israeli databases were added to the Caucasian sample populations the overall increase in F_{ST} was minimal. As expected, the estimates for the Hispanics are slightly higher than those for African Americans, Caucasians, and Orientals. Southeastern and southwestern Hispan-

Table 1. F_{ST} values for six PCR-based loci in different population groups[a].

Locus	African American (n = 3)[b]	Caucasian (n = 4)[c]	Caucasian (n = 5)[d]	Hispanic (n = 2)	Oriental (n = 2)
LDLR	−0.0005	−0.0035	−0.0014	0.0374	−0.0020
GYPA	0.0003	0.0036	−0.0007	0.0257	0.0106
HBGG	0.0049	0.0027	0.0108	0.0053	0.0136
D7S8	−0.0011	−0.0012	−0.0002	0.0150	−0.0047
Gc	0.0025	0.0015	−0.0006	−0.0049	−0.0032
HLA-DQα	0.0048	0.0040	0.0089	0.0124	NA[c]
F_{ST} Over all loci	0.0023	0.0015	0.0034	0.0142	0.0024

[a]F_{ST} values were determined according to Weir and Cockerham (1984) and Weir (1994).
[b]n = the number of sample populations
[c]The four Caucasian sample populations were from the FBI, Alabama, Roche, and Switzerland.
[d] The five Caucasian sample populations were from the FBI, Alabama, Switzerland, Israel, and French Basques.
[e] NA = data not available.

ics have different genetic admixtures, and in forensic applications would not be pooled into one database. Since both of the Hispanic databases are available, use of the F_{ST} values in Table 1 would not be necessary.

In forensics, the assumption of Hardy-Weinberg Equilibrium (HWE) is used to estimate a DNA profile frequency at a single locus. When estimating the likelihood of occurrence of a homozygote or a heterozygote, respectively, under the assumptions of HWE, the following formulae are used

$$P_{ii} = p_i^2 \qquad P_{ij} = 2p_i p_j, \quad i \neq j$$

Departure from HWE can be caused by the occurrence of population substructure. The subsequent formulae incorporate F_{ST} (Weir, 1994; Li & Chakravarti, 1994).

$$\begin{aligned} P_{ii} &= p_i^2 + p_i(1 - p_i) F_{ST} \\ P_{ij} &= 2p_i p_j(1 - F_{ST}), \quad i \neq j \end{aligned}$$

DNA profile frequencies of selected single locus genotypes of the PCR-based loci were calculated under the assumption of independence and by incorporating F_{ST} to demonstrate the effects of population subdivision on forensic estimates (Table 2). These are genotype frequencies and not the conditional frequencies discussed by Weir (1994). No substantial differences in DNA profile frequencies were observed when calculated with or without F_{ST}. Nichols and Balding (1991) have claimed that if the population is subdivided, the probability calculated will be underestimated and wrongful bias will occur. The data in Table 2 demonstrate that for realistic F_{ST} values there will be negligible effects on DNA profile frequency estimates. For the PCR-based loci, some F_{ST} values are positive and some are negative, with an F_{ST} over all loci resulting in a value close to zero. Moreover, when estimating DNA profile frequencies using F_{ST}, the frequency of heterozygotes is decreasing while the frequency of homozygotes in increased. Thus, across multiple PCR-based loci the overall effect of inbreeding would be close to zero.

In conclusion, if one were to argue incorrectly that it is appropriate to consider that the innocent suspect and true perpetrator were drawn from the same subpopulation (Nichols & Balding, 1991; Balding & Nichols, 1994), there would still be little concern for wrongful bias if HWE was assumed and no F_{ST} values were included. Additionally, Balding and Nichols (1994) have proposed that even under the hypothesis that the defendant is not the source of the evidence sample, the genotypes of the defendant and the true contributor may be positively correlated; thus, the degree of correlation will be large if the actual contributor has a similar ethnic background to the defendant. Extant data on forensically-employed DNA markers demonstrate that at racial levels there are fewer correlations than at ethnic levels within a race, especially for forensic estimates of the likelihood of occurrence of a DNA profile, However, the estimates of DNA profile frequencies vary little among different subgroup databases within a major population group, either ethnically or geo-

Table 2. DNA profile frequency estimates for selected single locus profiles estimated under the assumptions of HWE and by incorporation of F_{ST}.

African American	Genotype Frequency[a]	Frequency with locus specific F_{ST}	Frequency with over all loci F_{ST}
LDLR AA	0.0502	0.0501	0.0506
LDLR AB	0.3476	0.3478	0.3468
LDLR BB	0.6022	0.6021	0.6026
HBGG AA	0.2570	0.2582	0.2576
HBGG AB	0.1998	0.1988	0.1993
HBGG BB	0.0388	0.0396	0.0392
HBGG AC	0.3012	0.2997	0.3005
HBGG BC	0.1170	0.1164	0.1167
HBGG CC	0.0882	0.0892	0.0887
Gc AA	0.0106	0.0108	0.0108
Gc AB	0.1456	0.1453	0.1453
Gc BB	0.4998	0.5003	0.5003
Gc AC	0.0391	0.0390	0.0390
Gc BC	0.2687	0.2680	0.2680
Gc CC	0.0361	0.0365	0.0356
HLA-DQα 4,1.3	0.0372	0.0370	0.0371
HLA-DQα 1.3,1.3	0.0030	0.0032	0.0031
HLA-DQα 2,1.3	0.0107	0.0106	0.0106
HLA-DQα 3,2	0.0153	0.0153	0.0153
HLA-DQα 4,1.2	0.2123	0.2112	0.2118
HLA-DQα 2,2	0.0094	0.0098	0.0096
HLA-DQα 4,4	0.1142	0.1152	0.1147
	Chinese		
GYPA AA	0.3660	0.3685	0.3666
GYPA AB	0.4780	0.4729	0.4768
GYPA BB	0.1560	0.1585	0.1566
HBGG AA	0.0590	0.0615	0.0594
HBGG AB	0.3679	0.3629	0.3670
HBGG BB	0.5730	0.5755	0.5734

[a]Frequencies estimated under the assumption of HWE

graphically (Brookfield, 1992; Budowle *et al.* 1994a, 1994b; Chakraborty & Jin, 1992; Li & Chakravarti, 1994). The data to date demonstrate that there is little evidence for wrongful bias to a defendant when general population databases are employed.

This is publication number 94–14 of the Laboratory Division of the Federal Bureau of Investigation. Names of commercial manufacturers are provided for identification only, and inclusion does not imply endorsement by the Federal Bureau of Investigation.

References

Balding, D.J. & R.A. Nichols, 1994. DNA profile match probability calculation: how to allow for population stratification, relatedness, database selection and single bands. Forensic Science International 64: 125–140.

Brookfield, J., 1992. The effect of population subdivision on estimates of the likelihood ratio in criminal cases using single-locus DNA probes. Heredity 69: 97–100.

Budowle, B. & J. Stafford, 1991. Response to 'population genetic problems in the forensic use of DNA profiles' by R.C. Lewontin submitted in the case of United States versus Yee. Crime Laboratory Digest 18: 109–112.

Budowle, B., A.M. Giusti, J.S. Waye, F.S. Baechtel, R.M. Fourney, D.E. Adams, L.A. Presley, H.A. Deadman & K.L. Monson,

1991. Fixed bin analysis for statistical evaluation of continuous distributions of allelic data from VNTR loci for use in forensic comparisons. Amer. J. Hum. Genet. 48: 841–855.

Budowle, B., K.L. Monson, A.M. Giusti & B. Brown, 1994a. The assessment of frequency estimates of Hae III-generated VNTR profiles in various reference databases. J. Forens. Sci. 39: 319–352.

Budowle, B., K.L. Monson, A.M. Giusti & B. Brown, 1994b. Evaluation of Hinf I-generated VNTR profile frequencies determined using various ethnic databases. J. Forens. Sci. 39: 94–112.

Budowle, B., J.A. Lindsey, J.A. DeCou, B.W. Koons, A.M. Giusti & C.T. Comey, 1994c. Validation and population studies of the loci LDLR, GYPA, HBGG, D7S8, and Gc (PM loci), and HLA-DQα using a multiplex amplification and typing procedure. J. Forens. Sci. (in press).

Budowle, B., K.L. Monson & A.M. Giusti, 1994. A reassessment of frequency estimates of Pvu II-generated VNTR profiles in a Finnish, and Italian, and a general United States Caucasian database: No evidence for ethnic subgroups affecting forensic estimates. Amer. J. Hum. Genet. 55: 533–539.

Chakraborty, R. & L. Jin, 1992. Heterozygote deficiency, population structure and their implications in DNA fingerprinting. Human Genetics 88: 267–272.

Chakraborty, R. & K.K. Kidd, 1991. The utility of DNA typing in forensic work. Science 254: 1735–1739.

Crow, J.F., 1993. Population genetics as it relates to human identification. In: The Fourth International Symposium on Human Identification, Promega Corporation, Madison, WI (in press).

Devlin, B. & N. Risch, 1992a. A note on Hardy-Weinberg equilibrium of VNTR data using the FBI's fixed bin method. Amer. J. Hum. Genet. 51: 549–553.

Devlin, B. & N. Risch, 1992b. Ethnic differentiation at VNTR loci, with special reference to forensic applications. Amer. J. Hum. Genet. 51: 534–548.

Devlin, B., N. Risch & K. Roeder, 1993. Statistical evaluation of DNA fingerprinting: a critique of the NRC's report. Science 259: 748–750.

Devlin, B., N. Risch & K. Roeder, 1994. Comments on the statistical aspects of the NRC's report on DNA typing. J. Forens. Sci. 39: 28–40.

Gyllensten, U.B. & H.A. Erlich, 1988. Generation of single-stranded DNA by the polymerase chain reaction and its application to direct sequencing of the HLA-DQ alpha locus. Proc. Natl. Acad. Sci. USA 85: 7652–7656.

Hartl, D.L. & R.C. Lewontin, 1993. Response to Devlin et al. Science 260: 473–474.

Hochmeister, M.N., B. Budowle, U.V. Borer & R. Dirnhofer, 1994. Swiss population data on the loci HLA-DQα, LDLR, GYPA, HBGG, D7S8, Gc and D1S80. Forens. Sci. Int. 67: 175–184.

Horn, G.T., B. Richards, J.J. Merrill & K.W. Klinger, 1990. Characterization and rapid diagnostic analysis of DNA polymorphisms closely linked to the cystic fibrosis locus. Clin. Chem. 36: 1614–1619.

Huang, N.E. & B. Budowle, 1994. Chinese population data on the PCR-based loci HLA-DQα, LDLR, GYPA, HBGG, D7S8, and Gc. Human Heredity (in press).

Lander, E.S., 1989. Population genetic considerations in the forensic use of DNA typing, pp. 143–156 in Banbury Report 32: DNA Technology and Forensic Science. Cold Spring Harbor Laboratory Press, Cold Spring Harbor, N.Y.

Lewontin, R.C. & D.L. Hartl, 1991. Population genetics in forensic DNA typing. Science 254: 1745–1750.

Li, C.C. & A. Chakravarti, 1994. DNA profile similarity in a subdivided population. Human Heredity 44: 100–109.

Morton, N.E., 1992. Genetic structure of forensic populations. Proc. Natl. Acad. Sci. USA 89: 2556–2560.

Morton, N.E., A. Collins & I. Balazs, 1993. Kinship bioassay on hypervariable loci in Blacks and Caucasians. Proc. Natl. Acad. Sci. USA 90: 1892–1896.

Nei, M., 1973. Analysis of gene diversity in subdivided populations. Proc. Natl. Acad. Sci. USA 70: 3321–3323.

Nei, M., 1977. F-statistics and analysis of gene diversity in subdivided populations. Ann. Hum. Genet. 41: 225–233.

Nichols, R.A. & D.J. Balding, 1991. Effects of population substructure on DNA fingerprint analysis in forensic science. Heredity 66: 297–302.

Perkin Elmer: AmpliType PM PCR Amplification and Typing Kit Manual, Part No. N808-0057, 1994, pp 3.

Siebert, P.M. & M. Fukuda, 1987. Molecular cloning of human glycophorin B cDNA: nucleotide sequence and genomic relationship to glycophorin A. Proc. Natl. Acad. Sci. USA 84: 6735–6739.

Slightom, J.L., A.E. Blechl & O. Smithies, 1980. Human fetal $^{G}\gamma$- and $^{A}\gamma$-globin genes: complete nucleotide sequences suggest that DNA can be exchanged between these duplicated genes. Cell 21: 627–638.

Weir, B.S. & C.C. Cockerham, 1984. Estimating F-statistics for the analysis of population structure. Evolution 38: 1358–1370.

Weir, B.S., 1990, In Genetic Data Analysis, Sinauer Associates, Inc., Sunderland, Massachusetts, pp. 145–162.

Weir, B.S., 1992. Independence of VNTR alleles defined as fixed bins. Genetics 130: 873–887.

Weir, B.S. & W.G. Hill, 1993. Population genetics of DNA profiles. J. Forens. Sci. Soc. 33: 218–225.

Weir, B.S., 1994. The effects of inbreeding on forensic calculations. Ann. Rev. Genet. 28: 597–621.

Wright, S., 1922. Coefficients of inbreeding and relationship. Amer. Nat. 56: 330–338.

Wright, S., 1965. The interpretation of population strucuture by F-statistics with special regard to systems of mating. Evolution 19: 395–420.

Yamamoto, T., C.G. Davis, M.S. Brown, W.J. Schneider, M.J. Casey, J.L. Goldstein & D.W. Russell, 1984. The human LDL receptor: A cysteine-rich protein with multiple Alu sequences in its mRNA. Cell 39: 27–38.

Yang, F., J.L. Brune, S.L. Naylor, R.L. Apples & K.H. Naberhaus, 1985. Human group-specific component (Gc) is a member of the albumin family. Proc. Natl. Acad. Sci. USA 82: 7994–7998.

Editor's comments

The author provides an empirical study to accompany the papers by Balding and Nichols and Brookfield in this volume. As may have been expected, F_{ST} values for large heterogeneous populations such as 'Caucasian' or 'African American' are very small.

B. S. Weir (ed.), Human Identification: The Use of DNA Markers, 27–36, 1995.

Correlation of DNA fragment sizes within loci in the presence of non-detectable alleles

Ranajit Chakraborty & Zhaojue Li
Human Genetics Center, The University of Texas Houston Health Science Center, P.O. Box 20334, Houston, TX 77225, USA

Received 1 July 1994 Accepted 29 July 1994

Abstract

At present most forensic databases of DNA profiling of individuals consist of DNA fragment sizes measured from Southern blot restriction fragment length polymorphism (RFLP) analysis. Statistical studies of these databases have revealed that, when fragment sizes are measured from RFLP analysis, some of the single-band patterns of individuals may actually be due to heterozygosity of alleles in which fragment size resulting from one allele remains undetected. In this work, we evaluate the effect of such allelic non-detectability on correlation of fragment sizes within individuals at a locus, and its impact on the inference of independence of fragment sizes within loci. We show that when non-detectable alleles are present in a population at a locus, positive correlations of fragment sizes are expected, which increase with the proportion of non-detectable alleles at the locus. Therefore, a non-zero positive correlation is not a proof of allelic dependence within individuals. Applications of this theory to the current forensic RFLP databases within the US show that there is virtually no evidence of significant allelic dependence within any of the loci. Therefore, the assumption that DNA fragment sizes within loci are independent is valid, and hence, the population genetic principles of computing DNA profile frequencies by multiplying binned frequencies of fragment sizes are most likely to be appropriate for forensic applications of DNA typing data.

Introduction

DNA profiling of individuals by Southern blot restriction fragment length polymorphism (RFLP) analysis of several Variable Number of Tandem Repeat (VNTR) loci has been demonstrated to be an efficient method of human identification and for determining familial relationship of individuals (Budowle *et al.*, 1991a; Jeffreys, Turner & Debenham, 1991; Chakraborty & Jin, 1993). For forensic identification of individuals, when two DNA samples provide evidence of a match of DNA profiles at several such loci, the significance of the match is judged from the frequency with which such DNA profiles occur in a population. Since the VNTR loci conform to the Mendelian mode of segregation, the frequency of any specific multilocus DNA profile is generally computed by assuming that the alleles at each locus combine independently to form genotypes at all loci, and the genotype frequencies at each locus are in turn independent of each other (Chakraborty & Kidd, 1991). DNA fragment sizes scored by Southern blot RFLP analysis also involve measurement errors and incomplete resolution of similar size alleles (Devlin, Risch & Roeder, 1990; Budowle *et al.*, 1991a). Discrete alleles are not recognized, and hence the allele frequency computations from such data involve categorizing the fragment sizes in groups, called bins, that are generally wider than the measurement errors. While several studies demonstrated that grouped into such bins, the assumptions of independence of fragment sizes within as well as across loci are valid for the current forensic databases for the VNTR loci scored by RFLP analysis (Devlin, Risch & Roeder, 1990; Chakraborty & Kidd, 1991; Weir, 1992a, b; Risch & Devlin, 1992; Chakraborty, Srinivasan & de Andrade, 1993), there were some discussions suggesting that these assumptions may not be entirely correct (Lander, 1989, 1991; Cohen, 1990; Lewontin & Hartl, 1991; Geisser & Johnson, 1992, 1993; NRC, 1992).

Biological as well as statistical evidence, given in Budowle *et al.* (1991a), Jeffreys, Turner and Debenham (1991), Chakraborty (1991), Chakraborty *et al.*

(1992, 1993, 1994), Chakraborty and Jin (1992), Weir (1992c, 1993), and Devlin and Risch (1993), indicate that such concerns against using the population genetic principles in computing estimates of DNA profile frequencies are based on faulty and incomplete understanding of DNA databases generated by Southern blot RFLP analysis. In particular, when correlation of DNA fragment sizes within loci is shown by grouping the fragment sizes in quantile bins (Geisser & Johnson, 1992, 1993), an important feature that the RFLP protocol occasionally fails to detect alleles of unusually small or large sizes was neglected. In contrast, statistical (Devlin & Risch, 1992; Chakraborty *et al.*, 1992; Chakraborty & Jin, 1992; Weir, 1992a; Chakraborty *et al.*, 1994) as well as experimental data (Jeffreys, Turner & Dabenham, 1991; Budowle *et al.*, 1991a; Steinberger, Thompson & Hartmann, 1993; Chakraborty *et al.*, 1994) indicate that non-detectable alleles do in fact occur at many VNTR loci, and even small frequencies of such alleles may cause pseudo-dependence of DNA fragment sizes within as well as across loci. In addition, there is also evidence that nondetectability of alleles is not restricted to Southern blot RFLP analysis of DNA samples. Even when discretized alleles are recognized by the polymerase chain reaction method, differential amplification as well as sequence polymorphism at the flanking sites surrounding the repeat regions can cause allelic non-detectability (Tully, Sullivan & Gill, 1993; Callen *et al.*, 1993; Koorey, Bishop & McCaughan, 1993). Therefore, the effect of allelic non-detectability should constitute a critical aspect of testing the assumption of independence of allele sizes in any analysis of DNA typing data.

Weir (1992a) and Chakraborty, Srinivasan and de Andrade (1993) have shown that tests of independence of DNA fragment sizes scored from Southern blot RFLP analysis can be made by evaluating intraclass and interclass correlation of allele sizes within databases. In their analyses, they showed that the current forensic databases of DNA fragment sizes scored by Southern blot RFLP analysis provide little evidence of significant correlations within as well as across loci. Chakraborty, Srinivasan and de Andrade (1993) further showed that the correlations of fragment sizes within and across loci are only trivially affected by measurement errors. However, since no explicit relationship between correlations and the frequency of non-detectable alleles has been given in these works, it could not be established whether the occasional departure from the independence assumptions can be explained by the presence of non-detectable alleles at one or more of the respective loci.

The purpose of this research is to determine to what extent non-detectable alleles would affect the correlation of allelic sizes within a database. We obtain this relationship analytically and show that positive intraclass correlation of DNA fragment sizes is expected in the presence of allelic non-detectability, and this correlation is an increasing function of the frequency of non-detectable alleles. In contrast, the inter-class correlation of fragment sizes between two loci is unaffected by the presence of non-detectable alleles at one or both loci. Applications of this theory to the current forensic RFLP databases of DNA fragment sizes at six loci show that with a single exception (at the 5% level), no intraclass correlation at six loci in four broadly defined samples (US Caucasians, US Blacks, Southeastern Hispanics, and Southwestern Hispanics) collected by the Federal Bureau of Investigation (Budowle *et al.*, 1991a, b) is statistically significant, once the presence of non-detectable alleles is taken into account. Furthermore, the observed positive intraclass correlations within these samples provide alternative estimates of the frequencies of non-detectable alleles within these samples. These correspond fairly well with estimates obtained from other methods (e.g., Weir, 1992a, Chakraborty *et al.*, 1994). Finally, implications of these results in forensic computations of DNA profile frequencies are discussed, suggesting the appropriateness of product and multiplication rules in estimating multi-loci DNA profile frequencies.

Intraclass fragment length correlation and the effect of non-detectable alleles

For single locus VNTR probes, the *i*th individual's DNA profile may be represented by a pair (x_{i1}, x_{i2}) of fragment sizes where x_{ij} may have a quasi-continuous distribution of complex shape (multimodal, skewed, etc.) due to unknown etiology. When the DNA profile of an individual exhibits a double-banded pattern, two different fragment sizes $(x_{i1} \neq x_{i2})$ are recorded in a database, but when the individual exhibits a single-banded profile, the single fragment size is duplicated $(x_{i1} = x_{i2})$ in the database to record such observations. Following Weir (1992a) and Chakraborty, Srinivasan and de Andrade (1993), three expectations of functions of the x_{ij} values define the intraclass correlation (ρ_x) between fragment sizes; $E(x_{ij}^2) = \mu_x^2 + \sigma_x^2$,

$E(x_{ij}x_{ij'}) = \mu_x^2 + \rho_x \sigma_x^2$; for $j \neq j'$ and $E(x_{ij}x_{i'j'}) = \mu_x^2$; for all $i \neq i'$, where x_{ij}, $j = 1, 2$ are assumed to have been randomly drawn from a distribution with mean μ_x and variance σ_x^2.

These authors also showed that the intraclass correlation (ρ_x) can be estimated either by the analysis of variance (ANOVA) approach, or by using a generalized nonparametric correlation (Karlin, Cameron & Williams, 1981), by defining the three sample statistics

$$\bar{x} = \sum_{i=1}^{n} \sum_{j=1}^{2} x_{ij}/2n, \tag{1a}$$

$$s_x^2 = \sum_{i=1}^{n} \sum_{j=1}^{2} x_{ij}^2 - 2n\bar{x}^2, \tag{1b}$$

and

$$s_{xx} = \sum_{i=1}^{n} x_{i1}x_{i2} - n\bar{x}^2; \tag{1c}$$

where n is the number of individuals in the sample. Chakraborty, Srinivasan and de Andrade (1993) also showed that in large samples (i.e., when $n \to \infty$) both estimators of the intraclass correlation reduce to

$$\hat{\rho}_x = 2s_{xx}/s_x^2, \tag{2}$$

so that the asymptotic value of the expected intraclass correlation under any model of DNA fragment size distributions in the population can be evaluated by studying the sampling properties of the two statistics s_x^2 and s_{xx}. Equations (1a) ~ (1c) indicate that these, in turn, are determined by the sampling properties of x_{ij}^2, $x_{i1}x_{i2}$, and $\bar{x}^2$.

Let us now assume that the measured fragment sizes are determined by an unknown number of discrete alleles $A_0, A_1, A_2, \ldots, A_r$, where the true size of the kth allele is a_k $(k = 0, 1, 2, \ldots, r)$, but the allele A_0 is the class of all non-detected alleles, so that whenever A_0 occurs in an individual's DNA profile, no fragment size is reported for this specific allele. Let us represent the unknown frequencies of these alleles in the population by $p_0, p_1, p_2, \ldots, p_r$, where

$$0 \leqslant p_0, p_k \leqslant 1; \text{ and } p_0 + \sum_{k=1}^{r} p_k = 1. \tag{3}$$

Furthermore, let us assume that the database does not consist of any individual for which both alleles are A_0. Note that when no band is detected for an individual's profile, an experimentalist cannot be sure whether this is truly due to homozygosity of two non-detectable alleles, or if it is an experimented failure due to lack of enough DNA or incomplete hybridization with the locus-specific probe used in DNA profiling. When such individuals are excluded from the database, the sample may be represented as being from the truncated distribution of genotypes, in which the genotype A_0A_0 is not represented in the sample, and all genotypes of the form A_0A_k (for $k = 1, 2, \ldots, r$) are classified as A_kA_k. Under the assumption of allelic independence within such loci, the paired allele sizes for individuals in a sample, (x_{i1}, x_{i2}), $i = 1, 2, \ldots, n$, will have the independent and identical distribution given by

$$(x_{i1}, x_{i2}) = \begin{cases} (a_k, a_k) \text{ with probability} \\ \quad (p_k^2 + 2p_0p_k)/(1 - p_0^2), \\ \text{for } k = 1, 2, \ldots, r; \\ (a_k, a_l) \text{ with probability} \\ \quad 2p_kp_l/(1 - p_0^2), \\ \text{for } k > l = 1, 2, \ldots, r. \end{cases} \tag{4}$$

Following the algebra of Chakraborty, Srinivasan and de Andrade (1993), under the assumption of random association of alleles within a locus,

$$E(x_{ij}) = M_1/(1 - p_0), \tag{5a}$$

$$E(x_{ij}^2) = M_2/(1 - p_0), \tag{5b}$$

and

$$E(x_{i1}x_{i2}) = [M_1^2 + 2p_0M_2]/(1 - p_0^2), \tag{5c}$$

where $M_1 = \sum_{k=1}^{r} a_kp_k$ and $M_2 = \sum_{k=1}^{r} a_k^2p_k$. Furthermore, since

$$\begin{aligned} \bar{x}^2 &= \left[\sum_{i=1}^{n}(x_{i1} + x_{i2})/2n\right]^2 \\ &= \left[\sum_{i=1}^{n}(x_{i1} + x_{i2})^2 \right. \\ &\quad \left. + \sum_{i \neq i', i=1}^{n} \sum_{i'=1}^{n}(x_{i1} + x_{i2})(x_{i'1} + x_{i'2})\right] /4n^2, \end{aligned}$$

the same algebra shows that

$$\begin{aligned} E(\bar{x}^2) &= \left(\frac{2M_1^2 + (2 + 6p_0)M_2}{(1 - p_0^2)} \right. \\ &\quad \left. + \frac{4(n-1)M_1^2}{(1 - p_0)^2}\right) /(4n). \end{aligned} \tag{5d}$$

These in turn may be substituted in taking expectations of the expressions (1b) and (1c) to get

$$E(s_x^2) = \frac{2nM_2}{(1-p_0)} - \frac{M_1^2 + (1+3p_0)M_2}{(1-p_0^2)} - \frac{2(n-1)M_1^2}{(1-p_0)^2}, \quad (6a)$$

and

$$E(s_{xx}) = \frac{n[M_1^2 + 2p_0M_2]}{(1-p_0^2)} - \frac{M_1^2 + (1+3p_0)M_2}{2(1-p_0^2)} - \frac{(n-1)M_1^2}{(1-p_0)^2}. \quad (6b)$$

Writing $E(\hat{\rho}_x) \simeq 2E(s_{xx})/E(s_x^2)$, we therefore obtain the asymptotic expectation of the sample intraclass correlation

$$E(\hat{\rho}_x) \cong \frac{\left[\frac{(M_1^2 + 2p_0M_2)}{(1-p_0^2)} - \frac{M_1^2}{(1-p_0)^2}\right]}{\left[\frac{M_2}{(1-p_0)} - \frac{M_1^2}{(1-p_0)^2}\right]} = 2p_0/(1+p_0). \quad (7)$$

In other words, in the presence of non-detectable alleles, even when the DNA fragment sizes within a locus are independently distributed in individuals, positive intraclass correlation of fragment lengths estimated from databases with no A_0A_0 homozygotes is expected, and the correlation is an increasing function of the frequency of the non-detectable alleles at a locus. Chakraborty, Srinivasan and de Andrade (1993) showed that, when nonrandom association exists among fragment sizes, and the nonrandomness is represented by a single parameter, *f*, Wright's fixation index (Robertson & Hill, 1984; Weir & Cockerham, 1984), the asymptotic value of expected intraclass correlation becomes $E(\hat{\rho}_x) \cong f$. Therefore, the model of random association of fragment sizes in the presence of non-detectable alleles in truncated data (i.e., no homozygote for non-detectable alleles is recorded) is mathematically equivalent to the model of nonrandom association with the identity $f = 2p_0/(1+p_0)$. In this sense, the mathematical equivalence of these two models established for discrete allelic typing (Gart & Nam, 1988) also holds for quasi-continuous data on DNA fragment lengths.

Equation (7) further shows that a positive non-zero intraclass correlation provides a moment estimator of the frequency of non-detectable alleles,

$$\hat{p}_0 = \hat{\rho}_x/(2-\hat{\rho}_x). \quad (8)$$

When the frequencies of non-detectable alleles are available, either from experimental data such those reported by Steinberger, Thompson and Hartmann (1993) and Chakraborty *et al.* (1994), or from statistical analysis of binned genotype data (Weir, 1992a; Chakraborty *et al.*, 1994), one may employ the large sample Fisher's *z*-transformation statistic (Rao, 1973) to test whether the positive non-zero observed intraclass correlation may be explained by only the presence of nondetectable alleles.

Logic parallel to the above formulation also shows that, when two VNTR loci are scored simultaneously for each individual, the interclass correlation (ρ_{xy}) of fragment sizes (x_{ij}, y_{ij}), estimated by

$$\hat{\rho}_{xy} = s_{xy}/[2(s_x^2 s_y^2)^{1/2}],$$

where

$$s_{xy} = \sum_{i=1}^{n}(x_{i1}+x_{i2})(y_{i1}+y_{i2}) - 4n\bar{x}\bar{y},$$

and $\bar{x}$, $\bar{y}$, s_x^2 and s_y^2 are defined as in (1a) and (1b), has an asymptotic expectation of zero, even when at one or both of these loci nondetectable alleles exist. This is so because $E(s_{xy}) = 0$ even when non-detectable alleles exist at one or both of these loci. In other words, the presence of non-detectable alleles at any locus affects only the intraclass correlation of fragment sizes within a locus, but the interclass correlation of fragment lengths between loci is not affected.

Applications to forensic DNA typing databases

Budowle *et al.* (1991a, b) described the details of data collection, laboratory typing methods, and statistical summary of current forensic DNA typing databases generated by the Federal Bureau of Investigation Forensic Science Academy, where DNA samples from 2,046 individuals grouped into four groups (Caucasians, Blacks, Southeastern and Southwestern Hispanics) were typed for six VNTR loci (D1S7, D2S44, D4S139, D10S28, D14S13, and D17S79) after digestion with the *Hae*III restriction enzyme. These data have been already analyzed for testing independence of fragment lengths by various methods (see Weir, 1992a; Chakraborty, Srinivasan & de Andrade, 1993; Chakraborty *et al.*, 1994). In this work, we reexamined the intraclass and interclass fragment size correlations in these four samples to examine whether any of the departures from zero of these correlations can be explained by non-detectability of alleles in this database. Note that Chakraborty *et al.* (1994) have already demonstrated that *Hae*III digested fragment lengths occasionally involve non-detectable alleles. Our purpose here is to examine: (i) whether the deviant intraclass correlations can be explained by the presence of non-detectable alleles; and (ii) to examine the correspondence of statistical estimates of frequencies of non-detectable alleles from intraclass correlations and binned genotype data of fragment lengths (as obtained by Weir, 1992a and Chakraborty *et al.*, 1994).

Table 1 shows the summary results of intraclass correlations, where the $\hat{\rho}_x$ values are computed by using equation (2) from which the predicted frequencies of non-detectable alleles are computed by using equation (8). Alternative maximum likelihood estimates of the frequencies of non-detectable alleles from fixed bin fragment sizes, as obtained in Chakraborty *et al.* (1994), are also shown. The last two columns of this table show the levels of significance for tests of independence with and without consideration of non-detectable alleles. When non-detectable alleles are ignored, the test is based on the permutation procedure as described in Chakraborty, Srinivasan and de Andrade (1993). Briefly, the fragment sizes in each sample are randomly shuffled across individuals to form new DNA profiles. For each replication of shuffling, intraclass correlations were computed, and the proportion of 2000 replications in which correlations exceeding (in absolute values) those observed in the data gave the empirical levels of significance. In contrast, the deviation of the observed intraclass correlation from that predicted in the presence of non-detectable alleles $[\rho_x = 2p_0/(1+p_0)]$ was determined by the large sample test based on Fisher's z-transformed statistic (Rao, 1973), in which the expected intraclass correlation in the presence of non-detectable alleles was computed by using the estimated frequencies of non-detectable alleles from binned data (shown in the fifth column of the table). Two-sided levels of significance were determined by assuming that this test statistic follows a standard normal distribution.

As seen from these computations, of the 24 locus-population combinations, three intraclass correlations were significantly different (at the 5% level) from zero when non-detectable alleles were ignored. These three discrepancies are: Blacks and Southwestern Hispanics for the D2S44 locus, and Blacks for the D14S13 locus. When non-detectability of alleles is taken into account, only one (Southwestern Hispanics at the D2S44) locus intraclass correlation remains significant at the 5% level ($P = 0.012$). Recall that the data presented in Table 1 represents testing for independence of fragment sizes within loci for 24 independent population-locus combinations. At the 5% level of significance we expect to find at least one significant departure in 24 tests with a probability of 71%, and hence the single discrepancy from the independence assumption (once the non-detectable alleles are taken into account) is not unexpected. For a 5% level of significance the actual multiple-testing adjusted significance level (Bonferroni correction) is 0.0021, and since none of the test results reaches this low value, we contend that the present forensic databases provide no evidence of any significant intraclass fragment size correlation.

As mentioned before, there are other methods of estimating the frequency of non-detectable alleles, apart from the intraclass correlation approach developed here. Chakraborty *et al.* (1992) showed that if nondetectability of *Hae*III-digested fragment sizes is the only cause of an observed departure from the independence assumption, the frequency of non-detectable alleles may be computed from the proportional heterozygote deficiency, where heterozygosity is estimated from binned fragment sizes. They estimated such frequencies for the same database, using fixed bin (Bodowle *et al.*, 1991a) definition of alleles. Weir (1992a) used an alternative method of estimation where he assumed that a constant fraction of all single-banded DNA profiles at a locus are truly due to non-detectable alleles. His reported estimating equation (see p. 882 of Weir, 1992) has an error (B.S. Weir, personal communication), because the frequency of double-banded profiles in which both fragment sizes are within the same fixed bin must be subtracted from the expected number of homozygotes to determine the constant fraction (γ) of all single-fragment patterns that are due to non-detectable alleles. With this correction, Weir's estimator of non-detectable alleles reduces to

$$\tilde{p}_0 = (\text{number of apparent homozygotes} \\ -\text{expected homozygosity})/2n,$$

which is parallel to the theory of Chakraborty *et al.* (1992). The other statistical method of estimating the frequency of non-detectable alleles from binned fragment size data is a maximum likelihood method, discussed in Chakraborty *et al.* (1994), where a score-statistic, originally proposed by Gart and Nam (1984), is used for deriving the estimating equation. While a sampling error cannot be readily assigned to Weir's

Table 1. Intraclass fragment size correlations at six VNTR loci in the current forensic DNA typing databases and the estimated frequencies of non-detectable alleles.

Locus	n	$\hat{\rho}_x$	Estimate of p_0 (in %) from		Levels of significance of the test of independence based on	
			$\hat{\rho}_x$	binned data*	$\rho_x = 0$	$\hat{\rho}_x = 2\hat{p}_0/(1+\hat{p}_0)$**
Caucasians						
D1S7	595	0.067	3.46	0.34 ± 0.54	0.096	0.129
D2S44	792	0.009	0.44	0.51 ± 0.67	0.789	0.984
D4S139	594	0.031	1.55	0.85 ± 0.75	0.454	0.726
D10S28	429	0.032	1.60	1.90 ± 0.96	0.518	0.928
D14S13	751	−0.022	–	0.43 ± 0.65	0.520	0.412
D17S79	776	0.045	2.31	1.05 ± 1.30	0.198	0.484
Blacks						
D1S7	359	0.038	1.93	2.68 ± 0.93	0.454	0.810
D2S44	475	0.157	8.52	3.73 ± 1.08	<0.001	0.051
D4S139	448	−0.077	–	–	0.087	0.107
D10S28	288	0.070	3.61	1.84 ± 1.06	0.241	0.555
D14S13	524	0.104	5.51	2.86 ± 1.34	0.013	0.254
D17S79	550	0.037	1.89	4.24 ± 2.06	0.381	0.312
Southeastern Hispanics						
D1S7	305	−0.052	–	–	0.370	0.389
D2S44	300	−0.003	–	0.56 ± 0.87	0.949	0.825
D4S139	311	−0.088	–	–	0.115	0.128
D10S28	230	0.092	4.84	0.27 ± 0.80	0.168	0.180
D14S13	307	−0.011	–	-	0.847	0.872
D17S79	314	0.043	2.18	1.73 ± 1.96	0.429	0.976
Southwestern Hispanics						
D1S7	216	0.028	1.42	0.22 ± 0.76	0.666	0.712
D2S44	215	0.172	9.44	0.30 ± 0.91	0.011	0.012
D4S139	211	−0.004	–	0.52 ± 1.20	0.950	0.865
D10S28	210	0.011	0.57	2.21 ± 1.46	0.879	0.674
D14S13	187	0.037	1.89	1.48 ± 1.18	0.613	0.888
D17S79	207	0.020	1.02	1.08 ± 1.08	0.775	0.996

*Estimates of non-detectable allele frequencies from binned fragment size data are from Chakraborty *et al.* (1994);
**This adjusts for the presence of non-detectable alleles in which the hypothesized expected intraclass correlations are obtained by substituting the frequencies of non-detectable alleles as obtained from binned data (shown in the fifth column).

estimator, the standard error of the maximum likelihood estimator (MLE) can be computed for large samples (see Chakraborty *et al.*, 1994). In addition to these statistical methods, direct estimate of frequencies of *Hae*III non-detectable fragments can also be estimated by digesting the same DNA samples with an alternative restriction enzyme. This has been done for the same database for five locus-population combinations, details of which are given in Chakraborty *et al.* (1994). With *Pvu*II-digestion of DNA, the Southeastern Hispanic and Black samples were tested for D2S44 and D17S79 loci, and the Southwestern Hispanic sample was tested for the D17S79 locus. Thus, for these five locus-population combinations, we have four different estimators of non-detectable allele frequencies (Weir, MLE, intraclass correlation, and direct experiment).

Table 2. Estimates of frequencies of non-detectable alleles by four different approaches.

Locus	Estimates of non-detectable alleles in *Hae*III data (in %)				
	Populations	from binned data		Intraclass	Direct
		Weir	MLE	correlation	experiment*
D2S44	Blacks	3.61	3.73 ± 1.08	8.52	2.70 ± 1.57
	Hispanics (SE)	1.16	0.56 ± 0.87	0.00	0.33 ± 0.60
D17S79	Blacks	4.45	4.24 ± 2.06	1.89	6.49 ± 1.77
	Hispanics (SE)	1.66	1.73 ± 1.96	2.18	2.98 ± 1.25
	Hispanics (SW)	2.73	1.08 ± 1.08	1.02	3.70 ± 2.11

*The standard errors of these estimates are derived from the binomial proportions of the single-banded *Hae*III fragment patterns that resolved into 2-banded patterns when an alternative restriction enzyme was used.

Table 2 provides a comparison of these estimates. As seen from these comparisons, the estimates are consistent with each other, especially when their sampling errors are taken into account. There are situations where the intraclass correlation estimate of non-detectable allele frequencies can be higher than that obtained by other methods (e.g., for D2S44 in Blacks). We shall discuss a possible reason for this later.

Finally, to demonstrate that the presence of non-detectable alleles does not affect the interclass fragment size correlations between pairs of loci, Table 3 shows the estimated interclass correlations ($\hat{\rho}_{xy}$) for all pairs of loci in the four samples along with the sample sizes (*n*, the number of individuals for which fragment size data exists for the specific pairs of loci). Only three of the total 60 correlation estimates are significant at the 5% level. None has a level of significance below 1%. This is exactly what is expected, given the level of test procedure (5%), and we conducted 60 independent tests. Furthermore, of the three significance interclass correlations, two are negative, which cannot be ascribed to population heterogeneity within the samples (Chakraborty, Srinivasan & de Andrade, 1993). Therefore, we conclude that the analysis of interclass fragment size correlations between these pairs of loci offers little, if any, evidence of departure from the assumption of independence of DNA fragment sizes across loci.

Discussions and conclusion

The theory developed here provides a quantitative assessment of the effect of non-detectability of fragment sizes correlations within and between loci. The theoretical predictions are: when certain fragment sizes remain undetected in RFLP analysis of DNA, positive intraclass correlations are expected, while interclass correlations will not be affected in the presence of non-detectable fragment sizes at any of the loci. Analysis of the current forensic RFLP database within the US shows that when the presence of non-detectability is taken into account, there is little, if any, departure from the independence assumption. One might argue that an absence of correlation is not a formal proof of independence. While hypothetical examples of dependence in the absence of significant correlations can be constructed, the most common cause of dependence (i.e., population heterogeneity within the forensic databases) is expected to produce positive non-zero correlations (Chakraborty, Srinivasan & de Andrade, 1993). Therefore, the present results indicate that there is no significant effect of any population substructure on the DNA profiles as revealed in these RFLP databases.

Even though the theory developed here is specifically formulated in terms of fragment sizes scored by Southern blot RFLP analysis, it is directly applicable to discrete allele data at Short Tandem Repeat (STR) or microsatellite loci detected by the polymerase chain reaction (PCR) method, or for dot-blot genotype data. For example, sequence variation at the flanking PCR-primer regions may occasionally fail to detect certain alleles. Koorey, Bishop and McCaughan (1993) demonstrated that a single base substitution polymorphism in the priming sequence can cause allele non-amplification at the L5.62CA locus on human chromosome 5*q*. Callen *et al.* (1993) found more extensive

Table 3. Interclass correlation of *Hae*III fragment sizes in four databases.

Locus-Pair	Populations							
	Caucasians		Blacks		Hispanics (SE)		Hispanics (SW)	
	n	$\hat{\rho}_{xy}$	n	$\hat{\rho}_{xy}$	n	$\hat{\rho}_{xy}$	n	$\hat{\rho}_{xy}$
D1S7–D2S44	583	0.020	343	−0.010	279	0.021	208	0.029
D1S7–D4S139	575	−0.025	351	0.044	288	0.003	206	−0.047
D1S7–D10S28	417	−0.013	277	−0.010	218	−0.007	202	0.040
D1S7–D14S13	570	−0.003	351	0.013	288	−0.037	183	0.002
D1S7–D17S79	556	−0.012	336	−0.060*	292	0.002	199	−0.010
D2S44–D4S139	579	0.018	412	−0.034	285	0.012	203	−0.001
D2S44–D10S28	420	−0.003	277	−0.015	208	−0.027	202	0.012
D2S44–D14S13	722	−0.004	457	−0.002	281	−0.029	185	0.006
D2S44–D17S79	746	0.002	451	0.002	295	0.024	198	0.021
D4S139–D10S28	413	0.017	281	−0.006	216	−0.002	197	0.024
D4S139–D14S13	569	0.045*	433	−0.047*	289	−0.008	177	−0.019
D4S139–D17S79	559	0.023	424	−0.003	300	0.002	195	0.043
D10S28–D14S13	408	−0.002	283	−0.051	214	0.024	177	0.018
D10S28–D17S79	407	−0.012	278	−0.010	220	−0.048	193	−0.020
D14S13–D17S79	705	0.026	494	0.008	293	−0.045	170	−0.049

*$0.01 < P < 0.05$, based on the shuffling test.

occurrences of such phenomena at seven of the 16 CA-repeat loci they examined on human chromosome 16. They ascribed these to an 8-bp deletion in the priming sequence. Tully, Sullivan and Gill (1993) also noted allelic non-detectability caused by preferential amplification of small size alleles at several minisatellite loci (e.g., D19S20 and D16S83) in their PCR-based studies. For dot-blot genotyping at the HLA-DQα locus, the phenomenon of allele drop-out has been noted in several populations (Imanishi *et al.*, 1992). For such discrete allele systems, the test of independence through intraclass correlations of allele sizes is statistically equivalent to the global test of Hardy-Weinberg genotypic proportions, although when the number of segregating alleles is small (say less than six, as in the case of HLA-DQα), a likelihood-ratio test procedure for testing independence from genotypic proportions is more direct and is simpler to implement.

In this work we assumed that non-detectability of fragment sizes is the only cause of departure from independence. For the RFLP data, when the fragment sizes are used directly in computing the intraclass and interclass correlations, one other factor may influence the estimated correlations. Devlin, Risch and Roeder (1990) showed that RFLP-based fragment sizes are not always distinguishable when both alleles have nearly similar sizes. This process, called coalescence, will artificially inflate correlations, since nearly equal dissimilar allele sizes would be regarded as being equal. Coalescence is more plausible for loci where fragment size distribution has a narrower range of variation, as in the case of the D2S44 locus. Note that the two highest estimates (17.2% in Southwestern Hispanics, and 15.7% in Blacks) of intraclass correlations (Table 1) both occur at the D2S44 locus. Therefore, we conclude that higher estimates of nondetectable allele frequencies, obtained from such intraclass correlations, can be also due to the presence of coalescent alleles in these specific samples.

In summary, we conclude that an understanding of the technical features of DNA typing protocols is essential for appropriate interpretation of such data. Failure to account for the technical aspects may lead to wrong conclusions. For example, in partial analysis of the databases analyzed here, Geisser and Johnson (1993) claimed evidence of fragment size dependence within as well as between loci, but they failed to account for coalescence or nondetectability of alleles. In contrast, we show that consideration of these features reveal no significant presence of allele size dependence. This, along with other analyses of binned fragment size data (e.g., Weir, 1992a; Chakraborty *et al.*, 1994), shows that the population genetic principles of computing DNA profile frequencies by using the assumptions of independence (within and across) loci is justifiable.

Acknowledgements

This work was supported by US Public Service research grants GM 41399 from the National Institutes of Health and NIJ-92-CX-K024 from the National Institute of Justice. We thank Dr. B. Budowle for making available the current forensic databases generated by the Federal Bureau of Investigations for this analysis. The opinions expressed here are those of authors, and these are not necessarily endorsed by the granting agencies. We thank Dr. Yixi Zhong for his help during data analysis. Dr. B. Weir's comments were particularly helpful in improving the presentation.

References

Budowle, B., A.M. Giusti, J.S. Waye, F.S. Baechtel, R.M. Fourney, D.E. Adams, L.A. Presley, H.A. Deadman & K.L. Monson, 1991a. Fixed-bin analysis for statistical evaluation of continuous distributions of allelic data from VNTR loci, for use in forensic computations. Am. J. Hum. Genet. 48: 841–855.

Budowle, B., K.L. Monson, K.S. Anoe, F.S. Baechtel, D.L. Bergman, E. Buel, P.A. Campbell, M.E. Clement, H.W. Coey, L.A. Davis, A. Dixon, P. Fish, A.M. Giusti, T.L. Grant, T.M. Gronert, D.M. Hoover, L. Jankowski, A.J. Kilgore, W. Kimoto, W.H. Landrum, H. Leone, A.J. Longwell, D.C. MacLaren, L.E. Medlin, S.D. Narveson, M.L. Piarson, J.M. Polloack, R.J. Raquel, J.M. Reznicek, G.S. Rogers, J.E. Smerick & R.M. Thompson, 1991b. A preliminary report on binned general population data on six VNTR loci in Caucasians, Blacks and Hispanics from the United States. Crime Lab. Digest 18: 9–26.

Callen, D.F., A.D. Thompson, Y. Shen, H.A. Phillips, R.I. Richards, J.C. Mulley & Sutherland, G.R., 1993. Incidence and origin of 'null' alleles in the (AC)n microsatellite markers. Am. J. Hum. Genet. 52: 922–927.

Chakraborty, R., 1991. Statistical interpretation of DNA typing data. Am. J. Hum. Genet. 49: 895–897.

Chakraborty, R., M. de Andrade, S.P. Daiger & B. Budowle, 1992. Apparent heterozygote deficiencies observed in DNA typing data and their implications in forensic applications. Ann. Hum. Genet. 56: 45–57.

Chakraborty, R. & L. Jin, 1992. Heterozygote deficiency, population substructure and their implications in DNA fingerprinting. Hum. Genet. 88: 267–272.

Chakraborty, R. & L. Jin, 1993. A unified approach to study hypervariable polymorphisms: statistical considerations of determining relatedness and population distances. In DNA Fingerprinting: State of the Science. pp. 153–175. ed by S.D.J. Pena, R. Chakraborty, J.T. Epplen and A.J. Jeffreys. Birkhäuser, Basel.

Chakraborty, R. & K.K. Kidd, 1991. The utility of DNA typing in forensic work. Science 254: 1735–1739.

Chakraborty, R., M.R. Srinivasan & M. de Andrade, 1993. Intraclass and interclass correlations of allele sizes within and between loci in DNA typing data. Genetics 133: 411–419.

Chakraborty, R., Y. Zhong, L. Jin & B. Budowle, 1994. Nondetectability of restriction fragments and independence of DNA-fragment sizes within and between loci in RFLP typing of DNA. Am. J. Hum. Genet. 55: 391–401.

Cohen, J.E., 1990. DNA fingerprinting for forensic identification: potential effects on data interpretation of subpopulation of heterogeneity and band number variability. Am. J. Hum. Genet. 46: 358–368.

Devlin, B. & N. Risch, 1992. A note on Hardy-Weinberg equilibrium of VNTR data by using the Federal Bureau of Investigation's fixed-bin method. Am. J. Hum. Genet. 51: 549–553.

Devlin, B. & N. Risch, 1993. Physical properties of VNTR data, and their impact on a test of allelic independence. Am. J. Hum. Genet. 53: 324–329.

Devlin, B., N. Risch & K. Roeder, 1990. No excess of homozygosity at loci used for DNA fingerprinting. Science 249: 1416–1420.

Gart, J.J. & J. Nam, 1984. A score test for the possible presence of recessive alleles in generalized ABO-like genetic systems. Biometrics 40: 887–894.

Gart, J.J. & J. Nam, 1988. The equivalence of two tests and models for HLA data with no observed double blanks. Biometrics 44: 869–873.

Geisser, S. & W. Johnson, 1992. Testing Hardy-Weinberg equilibrium on allelic data from VNTR loci. Am. J. Hum. Genet. 51: 1084–1088.

Geisser, S. & W. Johnson, 1993. Testing independence of fragment lengths within VNTR loci. Am. J. Hum. Genet. 53: 1103–1106.

Imanishi, T., T. Akaza, A. Kimura, K. Tokunaga & T. Gojobori, 1992. Allele and haplotype frequencies for HLA and complement loci in various ethnic groups. In: HLA 1991, Vol. 1. pp. 1065–1220. ed. by K. Tsuji, M. Aizawa and T. Sasazuki. Oxford Univ. Press, Oxford.

Jeffreys, A.J., M. Turner & P. Dabenham, 1991. The efficiency of multilocus DNA fingerprinting probes for individualization and establishment of family relationships, determined from extensive casework. Am. J. Hum. Genet. 48: 824–840.

Karlin, S., E.C. Cameron & P.T. Williams, 1981. Sibling and parent-offspring correlation estimation with variable family size. Proc. Natl. Acad. Sci. USA 78: 2664–2668.

Koorey, D.J., G.A. Bishop & G.W. McCaughan, 1993. Allele non-amplification: a source of confusion in linkage studies employing microsatellite polymorphisms. Hum. Mol. Genet. 2: 289–291.

Lander, E.S., 1989. DNA fingerprinting on trial. Nature 339: 501–505.

Lander, E.S., 1991. Research on DNA typing catching up with courtroom applications. Am. J. Hum. Genet. 48: 819–823.

Lewontin, R.C. & D.L. Hartl, 1991. Population genetics in forensic DNA typing. Science 254: 1745–1750.

National Research Council (NRC) 1992. DNA Technology in Forensic Science. National Academy Press, Washington DC.

Rao, C.R., 1973. Linear Statistical Inference and Its Applications. Wiley, New York.

Risch, N. & B. Devlin, 1992. On the probability of matching DNA fingerprints. Science 255: 717–720.

Robertson, A. & W.G. Hill, 1984. Deviation from Hardy-Weinberg proportions: sampling variances and use in estimation of inbreeding coefficients. Genetics 107: 703–718.

Steinberger, E.M., L.D. Thompson & J.M. Hartmann, 1993. On the use of excess homozygosity for subpopulation detection. Am. J. Hum. Genet. 52: 1275–1277.

G. Tully, K.M. Sullivan & P. Gill, 1993. Analysis of 6 VNTR loci by 'multiplex' PCR and automated fluorescent detection. Hum. Genet. 92: 554–562.
Weir, B.S., 1992a. Independence of VNTR alleles defined as fixed bins. Genetics 130: 873–887.
Weir, B.S., 1992b. Independence of VNTR alleles defined as floating bins. Am. J. Hum. Genet. 51: 992–997.
Weir, B.S., 1992c. Population genetics in the forensic DNA database. Proc. Natl. Acad. Sci. USA 89: 11654–11659.
Weir, B.S., 1993. Independence tests for VNTR alleles defined by quantile bins. Am. J. Hum. Genet. 53: 1107–1113.
Weir, B.S. & C.C. Cockerham, 1984. Estimating F-statistics for the analysis of population structure. Evolution 38: 1358–1370.

Editor's comments

The presence of non-detectable alleles for VNTR loci has plagued the use of these highly-discriminating systems in human identification. The authors take explicit account of these alleles, and are able to show independence of the frequencies of detectable alleles. They raise the troubling issue of how to account for occasional significant results when multiple tests are performed. By invoking Bonferroni corrections, they regard all tests, even those performed on different loci, as addressing the same hypothesis—the absence of dependence at any VNTR locus.

B. S. Weir (ed.), Human Identification: The Use of DNA Markers, 37–49, 1995.

Inference of population subdivision from the VNTR distributions of New Zealanders

Andrew G. Clark[1], J. Frances Hamilton[2] & Geoffrey K. Chambers[3]
[1]*Department of Biology, Pennsylvania State University, University Park, PA 16802, USA*
[2]*Institute of Environmental Science and Research Limited, Wellington Science Centre, P.O. Box 30-547, Lower Hutt, New Zealand*
[3]*Biochemistry and Genetics Research Unit, School of Biological Sciences, Victoria University of Wellington, P.O. Box 600 Wellington, New Zealand*

Received 1 July 1994 Accepted 29 July 1994

Key words: population subdivision, VNTR, Maori, Polynesian

Abstract

A population sample from people of diverse ethnic origins living in New Zealand serves as a database to test methods for inference of population subdivision. The initial null hypothesis, that the population sample is homogeneous across ethnic groups, is easily rejected by likelihood ratio tests. Beyond this, methods for quantifying subdivision can be based on the probability of drawing alleles identical by descent (F_{ST}), probabilities of matching multiple locus genotypes, and occurrence of unique alleles. Population genetic theory makes quantitative predictions about the relation between F_{ST}, population sizes, and rates of migration and mutation. Some VNTR loci have mutation rates of 10^{-2} per generation, but, contrary to theory, we find no consistent association between the degree of population subdivision and mutation rate. Quantification of population substructure also allows us to relate the magnitudes of genetic distances between ethnic groups in New Zealand to the colonization history of the country. The data suggests that the closest relatives to the Maori are Polynesians, and that no severe genetic bottleneck occurred when the Maori colonized New Zealand. One of the central points of contention regarding the application of VNTR loci in forensics is the appropriate means for estimating match probabilities. Simulations were performed to test the merits of the product rule in the face of subpopulation heterogeneity. Population heterogeneity results in large differences in estimates of multilocus genotype frequencies depending on which subpopulation is used for reference allele frequencies, but, of greater importance for forensic purposes, no five locus genotype had an expected frequency greater than 10^{-6}. Although this implies that a match with an innocent individual is unlikely, in a large urban area such chance matches are going to occur.

Introduction

A persistent problem in the analysis of minisatellite loci (VNTRs) is quantification of the degree of population subdivision. The reason standard population genetic techniques are not easily applied is that these loci exhibit a very large number of alleles, and for technical reasons, allelic designation is not exact. Allelic differences are identified by differences in band size, but alleles can be so similar in size that they approach the limit of resolution of standard methods of molecular separation by gel electrophoresis. In practice, investigators have lumped fragments of discrete size ranges together and considered them as alleles. This process of 'binning' may introduce errors by pooling alleles that are truly different, and by splitting some alleles that are right on a bin boundary into either of two adjacent bins. In practice, binning has been shown to be reasonably effective, and many population genetic principles still hold true under several different binning schemes (Weir 1992a, b, 1993). It is nevertheless important to consider the consequences of binning when quantifying population subdivision of VNTR alleles.

Despite the claim by some investigators that individual genotypes are so rare that population subdivision does not have a big effect on estimates of match probability (Budowle & Monson, 1993), the magnitude of population subdivision remains a controversy in forensic application of VNTR loci. Subdivision first became an issue when it was recognized that VNTR loci often exhibit an excess of homozygotes, an observation that was interpreted to be caused by the Wahlund effect (Lander, 1989; Cohen, 1990). The prospect of population heterogeneity in VNTR allele frequencies led Lewontin and Hartl (1991) to point out that estimates of match probability (the probability that a random individual chosen from the population matches a forensic sample) would be profoundly affected if the wrong reference population is used to estimate allele frequencies. Devlin and Risch (1992) claim that alleles that are common in one population are generally common in all populations, and similarly that rare alleles are either rare in all populations or are not even found in a sample, and the generality of this claim needs to be tested. In their analysis of VNTR data from Caucasians, Blacks and Hispanics in the United States, Devlin and Risch (1992) found that 6.9% of the diversity is attributable to variation among ethnic groups, very close to the figure of 6.3% obtained from blood group data (Lewontin, 1982).

Population subdivision is just one of several causes of an excess of homozygotes, and Chakraborty and Jin (1992) found that the Wahlund effect was not sufficient to explain all of the excess. Other causes of excess homozygosity are that some single-banded individuals may be heterozygous, either for two very similar sized alleles which 'coalesce' into one apparent band, or the other alleles may result in a fragment that is so small it runs off the gel (i.e., effectively a null allele). Although the method of Devlin, Risch and Roeder (1990) reveals no significant tendency towards coalescence in data that we examine here (Hamilton, 1994), the smaller population size of a well defined population subgroup (New Zealand Maori) leaves open the possibility for excess homozygosity by inbreeding. In all these tests we need to be aware that different statistical approaches have different power, so that failure to detect coalescence leaves open the possibility that the phenomenon occurs but there are insufficient data to detect it as significant.

Forensic applications provide just one of the motivations to study population subdivision. The high degree of polymorphism at VNTR loci makes them ideal for examining genetic distance between populations and for examining historical relationships among groups of people. Here we examine several different aspects of population subdivision among four major ethnic divisions in a sample of people living in New Zealand. Note that there is always a degree of arbitrariness in specifying such divisions, and that one expects to find some heterogeneity within whatever division is specified. In our sample, for instance, our classification of 'Asians' includes both Orientals and Indians.

Before the arrival of people from Europe and Asia, the Maori had made New Zealand their home. The time that the Maori have been residing in New Zealand is currently the subject of some debate. Conventional ^{14}C dating dates their arrival at 950 AD (see Davidson, 1987). However, technical problems with these estimates were recently identified, and new work puts the initial settlement of the Maori nearer 1300 AD (Anderson & McFadgen, 1990). Maori culture is enriched by many oral traditions regarding their settlement in New Zealand, but anthropologists know very little for certain about the number of waves of immigration or the number of individuals that arrived.

Flint *et al.* (1989) and Martinson *et al.* (1993) found that Polynesians show reduced VNTR diversity compared with that revealed in European databases, and they conclude that the loss of variability could be caused by repeated bottlenecks during the colonization of the Pacific. A population known from historical records to have undergone a population bottleneck of 120 individuals is that on Rapa Island (Martinson *et al.*, 1993), yet they still show no significant decrease in the level of VNTR diversity relative to other Polynesian populations. This suggests that bottlenecks that resulted in the reduced VNTR diversity of Polynesians as a whole must have been quite severe (Martinson *et al.*, 1993). VNTR diversity is either rapidly regained or more difficult to lose by bottlenecks, because the African Cheetah, a species known to be virtually lacking in any blood group, protein, or RFLP variation, does exhibit VNTR variation (Menotti-Raymond & O'Brien, 1993).

The above studies strongly suggest that VNTR loci can rapidly accumulate variability after a population bottleneck through mutation. By examining parent-offspring sets, several authors have directly estimated mutation rates of VNTRs, and several loci exhibit mutation rates on the order of 10^{-2} per generation (Smith *et al.*, 1990; Jeffreys *et al.*, 1988; P. Stapleton, pers. com. and A. Eisenberg, pers. com.). This has some important theoretical consequences for the magnitude of population subdivision. High mutation rates serve to erase distant history (Martinson *et al.*, 1993), depending on the magnitude of mutations, migration, and random genetic drift, and whether there is a stationary distribution of fragment lengths that is approached with increasing mutation rate, a higher mutation rate can reduce the between-population portion of variance.

By fitting population genetic models to patterns of VNTR variation, it is possible to test the significance of forces that modify the patterns of variation. Roe (1993) shows that empirical distributions of allele fragment sizes are smoothest in the case of loci with the highest mutation rate. Theoretically, one would expect that, in the absence of factors that constrain the variance in copy number, this variance would grow without bound. The loci with the highest mutation rates do not have the highest variance in fragment size, so it is likely that the smoother distributions are caused by convergence to a steady state distribution. VNTR loci mutate by changing numbers of copies of repeats, suggesting the application of the classical stepwise mutation model. When applied to microsatellite variation, the conclusion is that these models fit adequately with a mutation rate that is independent of allele size (Valdes, Slatkin & Freimer, 1993), but for minisatellites, additional factors such as copy-number dependent mutation rates are constraining the distribution of alleles (Shriver *et al.*, 1993). Finally, we have to recognize that allelic variants can also be generated by inter-allelic recombination, and that most of what are scored as mutations (non-parental alleles) are actually caused by unequal exchange events. Harding, Boyce and Clegg (1992) simulated the process of unequal sister chromatid exchange and obtained relationships between the array size and persistence time, the rate of recombination, and the rules for misalignment and exchange. With additional empirical data, particularly parent-offspring data, it will be possible to test the fit of models of this sort and to estimate rate parameters and infer rules of recombinational exchange.

Our goal here is to quantify and test the significance of population substructure among New Zealanders. The New Zealand data serve as a model system for the problem of population substructure of VNTR loci.

Materials and methods

Sample collection and ethnic origin. Blood samples for the population survey were obtained from unrelated volunteer donors through the Blood Transfusion Services of Wellington and Auckland Hospitals, DNA Diagnostics, Auckland, and The Hepatitis Foundation, Whakatane. Volunteers were interviewed to ascertain their ethnic identity with either Caucasian, Maori and Polynesian ethnic groups, and self-declared ethnic identity was accepted in each case.

Blood collection and DNA extraction. Whole blood was collected into Vacutainer tubes with EDTA anticoagulant, and stored at 4°C until processed (usually within 14 days). High molecular weight DNA was extracted from leucocytes and ethanol precipitated by a standard phenol-chloroform procedure (Sambrook, Fritsch & Maniatis, 1989) with details presented elsewhere (Hamilton *et al.*, in prep.). The DNA samples were stored frozen in TE buffer (pH 8.0) at −20°C.

Single locus analyses. Detailed protocols will appear elsewhere (Hamilton *et al.*, in prep.), but briefly, DNA samples were digested to completion with restriction endonuclease *Hinf* I for 5 h at 37°C according to the supplier's (BRL) instructions. The digested DNA was ethanol precipitated, resuspended in TE (pH 8.0), and quantitated spectrofluorometrically. Aliquots (1–2 μg) of the *Hinf* I digested DNA samples were electrophoresed in 0.7% agarose with TBE buffer until the 2.0 kb fragment of a *Hind*III digested lambda DNA visual marker had run 18 cm from the origin. The gels were depurinated, alkali washed and neutralized prior to capillary transfer (Southern, 1975) to uncharged nylon membranes (Amersham, HYBOND N). The digested DNA was fixed to the nylon membranes by baking for 2 h at 80°C. The nylon membranes were then hybridized with radioactive probes as described.

DNA probes and hybridization. Probes and their corresponding loci are listed in Table 1. Probe DNA

Table 1. Minisatellite VNTR loci examined and their corresponding probes and mutation rate estimates.

Probe	Locus	Mutation rate
MS1	D1S7	0.052 (A.J. Jeffreys *et al.*, 1988)
YNH24	D2S44	<1/600 (P. Stapleton, pers. com.)
pH30	D4S139	0.060 (P. Stapleton, pers. com.)
pLH1	D5S110	0.009 (A. Eisenberg, pers. com.)
MS43a	D12S11	0.007 (J.C. Smith *et al.*, 1990)

Table 2. Number of DNA samples from each ethnic group examined with the minisatellite probes.

	Locus				
Population	D1S7	D2S44	D4S139	D5S110	D12S11
Asian	64	62	61	63	65
Caucasian	172	182	179	86	183
Maori	174	177	181	183	182
Polynesian	151	142	148	156	148
Total	561	563	569	588	578

Table 3. Number of unique alleles, in each ethnic group binned according to the 2.5% rule (fragments considered to be the same allele if estimates of size are within 2.5%).

Locus	Asian	Caucasian	Maori	Polynesian
D1S7	0	4	1	0
D2S44	1	5	0	0
D4S139	0	5	1	1
D5S110	0	0	1	0
D12S11	2	0	3	4

(20 ng) was radioactively labeled with ^{32}P by the random primer method of Feinberg and Vogelstein (1983). Nylon membranes were hybridized to individual probes at 65°C overnight as described in Church and Gilbert (1984). The membranes were washed twice for 15 min in 3X SSC, 1% SDS at 65°C; twice for 15 min in 1X SSC, 1% SDS at 65°C, followed by two stringent washes for 45 min at 0.1X SSC, 0.1% SDS at 65°C. Damp membranes were wrapped in plastic wrap and exposed to autoradiography film (KODAK AR) with two intensifying screens at −70°C. Exposures were made up to two weeks, or until sufficient band density was obtained. Membranes were stripped of radioactive probes and stored for reprobing.

Scoring autoradiographs. The final data consisted of estimates of the fragment sizes of the one or two bands seen in each sample lane probed with each of the five probes. Fragment lengths were estimated by running each gel with four lanes of a ^{32}P labeled DNA molecular weight ladder (Amersham), and interpolating fragment sizes using a MacGel System (S. Kessell *et al.*, pers. com.) and digitizing tablet. Details of data analysis are given in the text.

Results

There were 608 individuals included in the population survey, but not all DNA samples were probed with all five probes (Table 2). This lack of balance required care in performing some tests, and tests involving multiple loci were performed by only using data on individuals for which all five loci were scored. The data included 508 individuals for which all five loci were scored, 58 with four loci, 15 with three loci, 15 with two loci and 12 with one locus scored. Each of the four ethnic groups had alleles which were found uniquely in that ethnic group, and these unique alleles were evenly spread among the group (Table 3). Asians appear to have fewer unique alleles. However, the sample size of Asians was smaller, and a chi-square test shows that the counts of unique alleles are homogeneous across the four ethnic groups. Unique alleles contribute little to the measures of subdivision because of their rarity.

Genetic distance measures produce consistent genealogies. In the case of single locus VNTR probes, the standard calculation of Nei's genetic distance statistic D (Nei, 1972) can be applied to binned allele frequency data. One concern is that estimates of genetic distance may be influenced by the binning. However it is a simple matter to estimate distances for a variety of bin sizes to test the robustness of any conclusions. Not surprisingly, as the bin sizes increase, the estimates of genetic distance between ethnic groups decrease (Table 4). The program NEIGHBOR in Felsenstein's *Phylip* package (Felsenstein, 1989) was used to infer the genealogical relations among the four ethnic groups based on genetic distance measures. We found that data from five loci yielded the same genealogy (Fig. 1), whose topology can be expressed as [(A,C),(M,P)].

Table 4. Estimates of Nei's genetic distance statistics D for each locus.

bp/bin	i	j	D1S7	D2S44	D4S139	D5S110	D12S11
200	A	C	0.238	0.174	0.257	0.254	0.333
	A	M	0.220	0.350	0.419	0.394	0.458
	A	P	0.227	0.335	0.436	0.454	0.588
	C	M	0.419	0.320	0.392	0.411	0.350
	C	P	0.436	0.284	0.333	0.353	0.335
	M	P	0.333	0.104	0.189	0.204	0.284
400	A	C	0.180	0.165	0.179	0.180	0.187
	A	M	0.342	0.302	0.345	0.343	0.353
	A	P	0.320	0.288	0.325	0.327	0.337
	C	M	0.345	0.275	0.304	0.306	0.303
	C	P	0.326	0.251	0.271	0.275	0.288
	M	P	0.271	0.098	0.109	0.109	0.251
600	A	C	0.139	0.131	0.138	0.139	0.141
	A	M	0.262	0.254	0.266	0.264	0.269
	A	P	0.243	0.235	0.247	0.248	0.251
	C	M	0.266	0.237	0.247	0.248	0.255
	C	P	0.247	0.214	0.222	0.224	0.235
	M	P	0.222	0.076	0.081	0.081	0.214
800	A	C	0.114	0.110	0.113	0.113	0.115
	A	M	0.213	0.202	0.215	0.214	0.217
	A	P	0.199	0.192	0.202	0.203	0.204
	C	M	0.215	0.210	0.218	0.218	0.203
	C	P	0.202	0.191	0.195	0.197	0.192
	M	P	0.195	0.064	0.067	0.067	0.191

The data were binned with constant bin sizes of 200 to 800 bp. The populations are A: Asian, C: Caucasian, M: Maori, P: Polynesian. Table entries are Nei's D (1972).

The only peculiarities in these trees are that the length of the branch to Asians is remarkably short in the case of *D1S7*, and long in the case of *D12S11*. The branching order of these trees is unaffected by binning, as is illustrated in Figure 2. Note that the Asian-Caucasian and Maori-Polynesian branches remain the shortest for all bin sizes, resulting in the same topology as the trees in Figure 1. The best way to estimate standard errors of genetic distance is to calculate jackknife statistics over all the loci, but we have insufficient loci to do that.

Likelihood ratio test for population heterogeneity. Likelihood ratio tests have proven usefulness for testing independence of VNTR alleles (Hernandez & Weir, 1989; Weir 1992a, b). The need for applying likelihood methods, as opposed to doing chi-square contingency table tests of homogeneity, is that VNTR loci have so many alleles, and so many of them are rare, that contingency tables are very sparse. Sparse tables require continuity correction, and so the tests lose power. Let p_{ik} be the count of allele i in the kth subpopulation, n_k be the number of individuals scored in subpopulation k where n is the total number of individuals, and let p_i be the count of allele i in all subpopulations. Under the assumption of independence of alleles within and between loci the log likelihood of the parameters p_{ik} is defined as

$$\log(L_1) = \sum_i \sum_k p_{ik} \log(p_{ik}/2n_k)$$

and the log likelihood under the null hypothesis is

$$\log(L_0) = \sum_i \sum_k p_{ik} \log(p_i/2n)$$

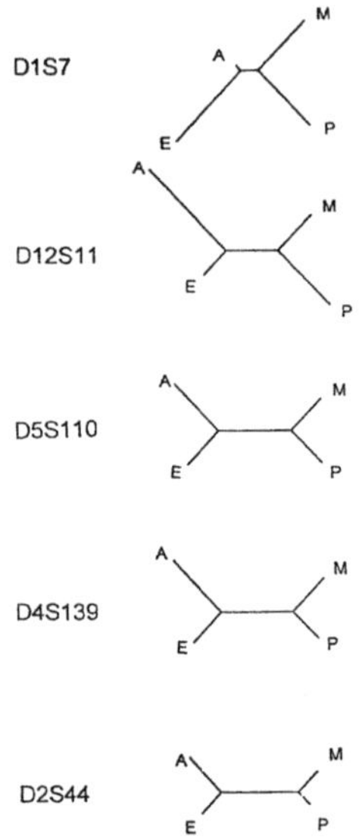

Fig. 1. Neighbor-joining trees of the four major ethnic groups. A:Asian, C:Caucasian, M:Maori, and P:Polynesian inferred from estimates of genetic distance using allele frequency data of each locus. The observation that all trees produce the same topology clearly supports the significance of differences between the groups.

We expect that

$$G^2 = -2[\log(L_0) - \log(L_1)]$$

should be approximately χ^2 distributed, but rather than depending on this as a parametric test, we can simulate the null distribution of G^2 by shuffling the alleles among the ethnic groups, and for each shuffling, calculate G^2 on the pseudodata. This procedure is known as a permutation test, and its statistical properties are well known (Manly, 1991). Table 5 reports the results of this procedure, and shows that the observed G^2 is greater than the simulated G^2 in an overwhelming majority of cases for all five loci. By this measure, there is significant heterogeneity among the ethnic groups, and we reject the null hypothesis that the alleles represent samples drawn from the same population. The same test can also be applied to genotype frequency data, but note that these two tests are not independent of one another. G^2 is calculated in much the same way:

Table 5. Likelihood ratio tests of population heterogeneity.

Locus	Allele Tail Prob*	Genotype Tail Prob
D1S7	.04	<.01
D2S44	<.01	<.01
D4S139	<.01	<.01
D5S110	<.01	<.01
D12S11	<.01	<.01

Tail prob refers to the tail probability, determined from the fraction of reshuffled likelihood ratio test statistics that were larger than the observed test statistics. In all cases the reshuffling test indicates significant population substructure. One hundred reshuffled samples were analyzed.

$$\log(L_1) = \sum_i \sum_j \sum_k g_{ijk} \log(g_{ijk}/n_k)$$

and

$$\log(L_0) = \sum_i \sum_j \sum_k g_{ijk} \log(g_{ij}/n)$$

where g_{ijk} is the count of genotype *i,j* in the kth subpopulation and g_{ij} is the count of genotype *i,j* in the entire population. The G^2 statistic is calculated as before, and significance is tested by simulation as before. Again, all five loci exhibit significant heterogeneity among ethnic groups (Table 5). There is need for further analysis of the properties of this test in order to determine its statistical power and the relationship between power and sample size.

Excess of multiple locus homozygotes. Examination of the genotypic counts reveals that the Maori have more two and three-locus homozygotes than do the other ethnic groups (Table 6). This leads us to inquire whether the Maori have more multiple-locus homozygotes than would be predicted from the single locus homozygosity. We can define a statistic to measure this as

$$G = F_{11} - F^2$$

where F_{11} is the observed frequency of two locus homozygotes, and F is the observed frequency of one

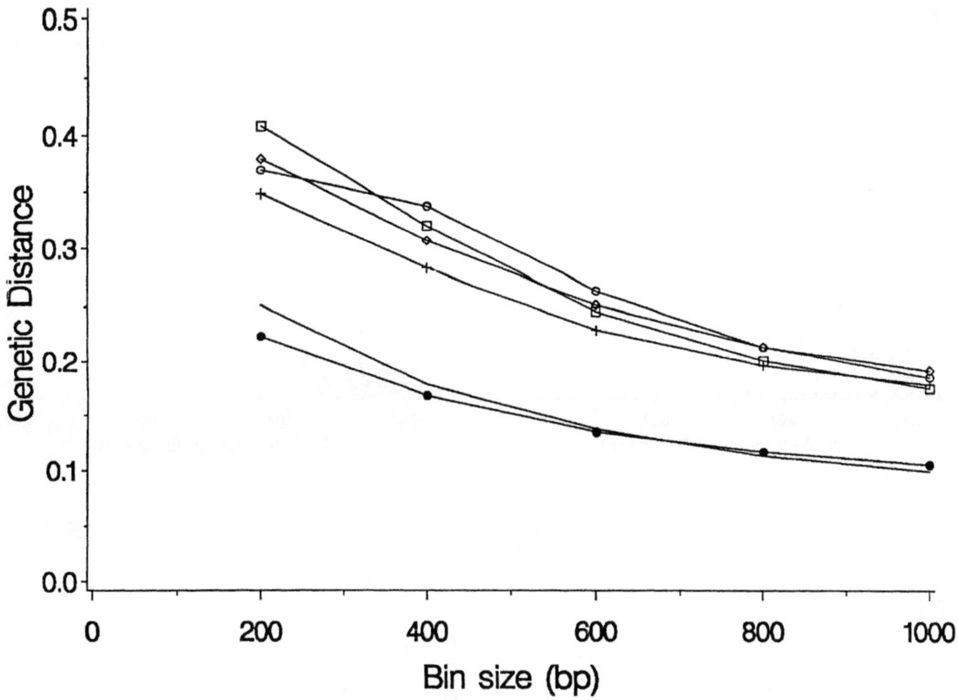

Fig. 2. Genetic distance as a function of bin size. In all cases, estimates of D decline with larger bin sizes, but the predicted tree topology remains the same because the Asian-Caucasian distance (dots) and Maori-Polynesian distance (no symbols) remain less than all four other pairwise distances.

Table 6. Counts and frequencies of one-, two-, and three-locus homozygotes, and the coefficient of excess two-locus homozygosity (*G*) calculated over all pairs of loci.

	One-locus homozygotes							
Population	D1S7	D2S44	D4S139	D5S11	D12S11	2-locus	3-locus	G
Asia	7	0	10	2	6	4	0	0.00114
Caucasian	10	7	12	8	20	4	0	−0.00157
Maori	33	20	11	6	34	13	3	0.00060
Polynesian	27	15	9	9	23	15	1	0.00089
Total	77	42	42	25	42	35	4	0.00029

'2-locus' and '3-locus' refer to counts of 2-locus and 3-locus homozygotes. *G* measures the departure of genotype frequencies from that predicted by random mating.

locus homozygotes. (Note that *G* is analogous to η, a population parameter known as identity disequilibrium). *G* is positive in Asian, Maori and Polynesian populations, but negative in Caucasians. To test whether *G* is significantly different from zero, we performed a resampling analysis drawing single locus genotypes at random, shuffling them to form two-locus genotypes, and calculating *G* from each shuffled sample.

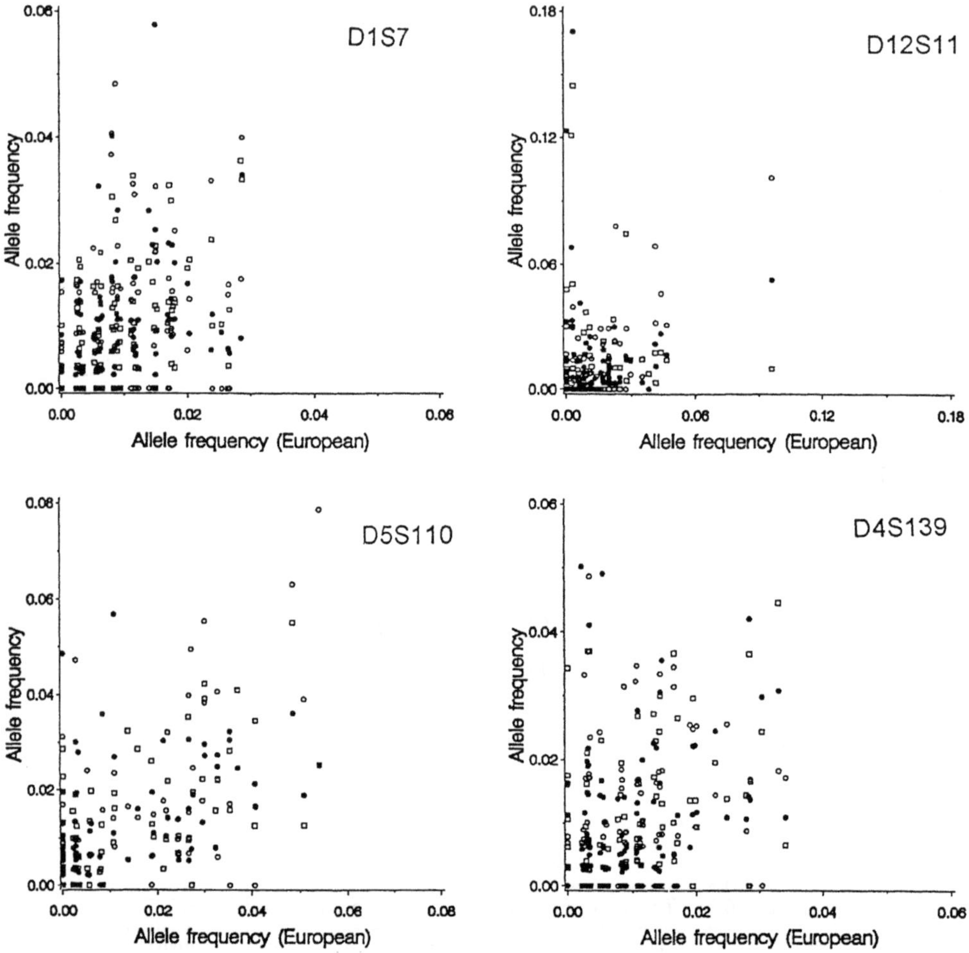

Fig. 3. Scattergrams of allele frequencies in Europe (x-axis) against allele frequency in the other three populations (circle:Asian, dot:Maori, square:Polynesian). Note that the most common Caucasian allele is not necessarily the most common allele in all populations, and that Caucasian allele frequencies appear to be more even than those of the other groups.

The observed G values fell well within the range of G values obtained by these simulations, so we conclude that the excess two-locus homozygosity is not significant.

Population genetic theory: relation between population subdivision and mutation rate. Under the infinite alleles island model, Cockerham and Weir (1993) show that a neutral gene will have equilibrium correlation of genes in subpopulations, β, as:

$$\beta = \frac{1}{1 + 4N\mu + 4Nm\alpha}$$

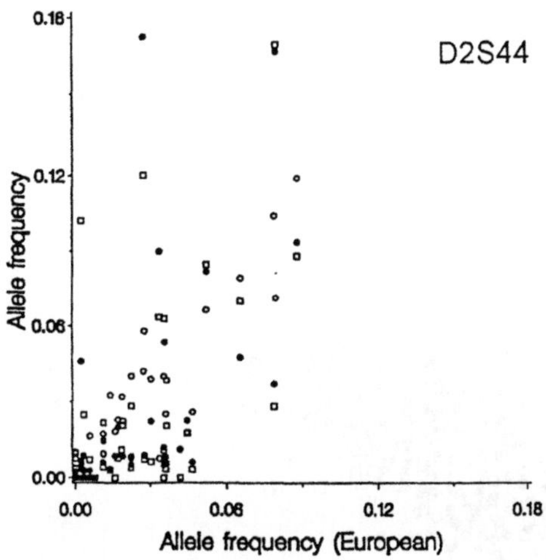

Fig. 3. continued

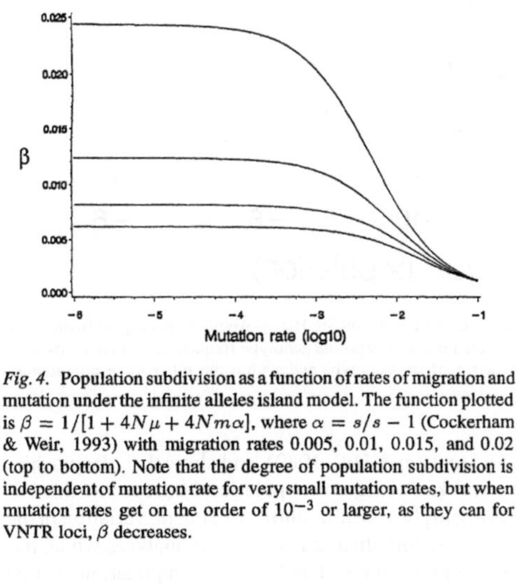

Fig. 4. Population subdivision as a function of rates of migration and mutation under the infinite alleles island model. The function plotted is $\beta = 1/[1 + 4N\mu + 4Nm\alpha]$, where $\alpha = s/s - 1$ (Cockerham & Weir, 1993) with migration rates 0.005, 0.01, 0.015, and 0.02 (top to bottom). Note that the degree of population subdivision is independent of mutation rate for very small mutation rates, but when mutation rates get on the order of 10^{-3} or larger, as they can for VNTR loci, β decreases.

where m is migration, μ is mutation rate, $\alpha = s/(s-1)$ and s is the number of subpopulations. If the number of islands is very large, this converges to Crow and Aoki's (1984) G_{st}. This expression shows that a higher mutation rate reduces G_{st} if it is of the same order of magnitude as the migration rate. In the case of VNTR loci this is probably true, and in some cases, $\mu \gg m$. Hence, for VNTR loci, there may be cases in which the degree of population subdivision is dominated by mutation. The time to equilibrium is on order of $1/(m + \mu)$, which may also be dominated by mutation, and with mutation rates as high as 0.06 (Table 1), equilibrium may be attained as fast as 14 generations. Assuming constant migration rates, if the Maori arrived in New Zealand even as late as 1300 A.D. (Anderson & McFadgen, 1990), then VNTR loci like *D1S7* ($\mu = 0.052$) might be expected to be in equilibrium already.

Consequences of subdivision on match probability. The issue associated with subpopulation heterogeneity that is most relevant to forensic applications is match probability. Here we take each genotype and ask: what is its expected frequency in the population in which it was found, and in each of the other three populations? To the extent that allele frequencies differ among subpopulations, these estimates will also vary, and the best estimate for forensic purposes would make use of the true allele frequencies. On the other hand, if all alleles are rare, even in the face of significant population heterogeneity, genotype frequency estimates may be so low that heterogeneity is not important. Figure 5 shows that the heterogeneity in genotype frequency estimates is such that genotypes are generally more rare in the 'wrong' population than they are in the true population, and that the discrepancy between the estimates can be as large as 10^{-4}. On the other hand, no five-locus genotype was predicted to be more common than 10^{-6}, so an observed match is a very unlikely event. In fact, there were no five-locus matches in the dataset.

Discussion

Genetic distances and genealogies. Classical genetic distances calculated from single locus VNTR data can provide useful measures of establishing relationships between human populations. Our finding that all five loci produced the same topology for the genealogy connecting Asians, Caucasians, Maori and Polynesians lends some confidence to the conclusion that the Maori and Polynesians had common ancestry. However, there is a large stochasticity associated with gene genealogies, and one does not expect that every gene tree will have the same topology as the true pattern of population splitting. The stochasticity is reflected

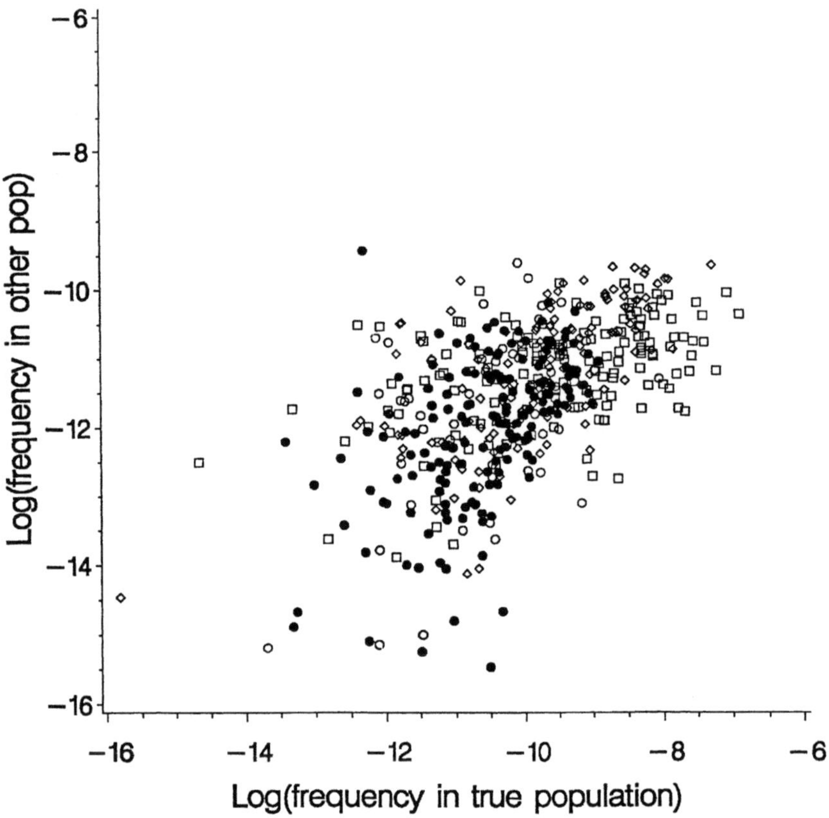

Fig. 5. Scatter plot of 5-genotypic match probabilities estimated under the product rule. Assuming Hardy-Weinberg genotypic frequencies at each locus, and linkage equilibrium (Hamilton *et al.* in prep.), the product rule produced expected genotype frequencies for each observed genotype. Expected frequencies were estimated within the population of origin of the genotype and within the other three populations. (circle:Asian, dot:Caucasian, square:Maori, diamond:Polynesian).

in the variation in branch lengths of the trees (Fig. 1). The consistency of the tree topologies is all the more surprising given that the loci differ by orders of magnitude in mutation rate. A common origin for Polynesians and Maori is inferred from other genetic loci as well. Several RFLP and VNTR genotypes have been found to be restricted to Melanesian and Polynesian populations (Hill, O'Shaughnessy & Clegg, 1989; Flint *et al.*, 1989). HLA studies have placed the origin of the New Zealand Maori in an East Polynesian cluster with Easter Island and Hawaii (Serjeantson, 1989). These genetic data also provide information about past bottlenecks that occurred in the colonization of the Pacific (Serjeantson & Hill, 1989), and show that, despite historical knowledge of population bottlenecks, there is surprisingly little diminution of genetic variation in the Maori. A complication in all these analyses which has both anthropological and forensic implications is that the term 'Maori' refers to a group of people who are themselves racially admixed. The degree of admixture within the Maori is discussed in Hamilton *et al.* (in prep.).

Likelihood tests of subdivision. The likelihood ratio tests (Table 5) lead us to the conclusion that it would be extremely unlikely to obtain the observed data from a single panmictic population. It should come as no surprise that data from five highly polymorphic VNTR loci produce this conclusion, and several other genetic markers produce similar measures of subdivision in other populations (Balaz, 1993). Despite the fact that significant heterogeneity can be detected with VNTR data, most human genetic variation, including VNTRs, is found within groups (Lewontin, 1982). Methods such as VNTR analysis allow detection of so many rare alleles that subpopulations can be distinguished. The challenge is to quantify the degree of subdivision, and to determine the best way to deal with population substructure in forensic applications (see Weir, 1994).

Mutation and population subdivision. The extraordinarily high rate of mutation of some VNTR loci is expected to reduce the degree of population subdivision, provided the mutation rate is almost as large as the migration rate. When the rates of migration and mutation are of the same order, an inverse relation between the rate of mutation and the degree of population subdivision is expected. In a sense, very high mutation rates erase the history of the population structure.

The results cited above are based on the classical island model, and this may not be the most appropriate model for the founding of New Zealand by the Maori. The island model specifies a group of subpopulations of constant size exchanging migrants at a constant rate. In reality, human populations in the Pacific can be characterized by a series of episodes of colonization and expansion. Such a population history could result in a positive correlation between mutation and G_{st}. With very high mutation rates, each island would get a new allele which then diverges over time. Founding events reduce the variation within populations, but mutation could maximize divergence between populations. Implicit in this argument is a model of mutation in which each mutation is to an allele not seen before (infinite alleles model), and VNTR loci probably violate this assumption. Because the molecular basis for VNTR variation is the variation in number of repeated copies of short sequence motifs, particular alleles can arise repeatedly by mutation as alleles gain and lose repeat copies. There is as least one case of two alleles which are the same size, and hence would be scored as identical in a Southern blot, but differ in sequence (Decorte, Marynen & Cassiman, 1994). Mutation can produce alleles of the same size, so extremely high mutation rates would result in either boundless increase in variance in copy number (allele size), or, because of the tendency of very high copy number alleles to lose copies, the distribution of allele sizes would settle to a steady state. If all subpopulations had the same pattern of mutation, they would arrive at the same steady distribution, and subdivision would be erased by mutation. We do not know where VNTR loci fall in this set of scenarios. Mutation rates of some VNTR loci exceed migration rates, but rather than appearing homogeneous across subpopulations, variation in these highly mutable loci exhibit significant population structure. It would appear that further knowledge of the mechanisms of mutational events that generate VNTR diversity may help to resolve this unexpected result.

Match probabilities and forensic implications. Consistent with other studies, a wide range of schemes for binning of VNTR data appears to make little difference to the conclusions about population subdivision (Weir, 1992a; Herrin, 1992). The data clearly allow rejection of the null hypothesis that the subpopulations are equivalent, and this raises the question of the most fair approach in estimating match probabilities for VNTR patterns in New Zealanders. Lander (1989) and Lewontin and Hartl (1991) pointed out that in a subdivided population, use of allele frequency estimates from the wrong ethnic group can result in large errors in match probability. Although they were correct in pointing out that the match probabilities can be changed by several orders of magnitude depending on the reference population used to estimate allele frequencies, Devlin and Risch (1992) and Weir (1992a, b) demonstrated that estimates of genotype frequency are correlated across ethnic groups. Identification of differences between major racial groups is not the most significant issue for forensics, because distinct databases are usually used for major racial groups. What is more troublesome is admixture or heterogeneity within the major groups. Despite order of magnitude differences in estimates of match probability, the range of match probabilities over the whole population is so large that exceedingly rare genotypes in one group are likely to be rare in another ethnic group. To this extent, estimates of match probabilities are actually quite robust to admixture among US Blacks, Caucasians and Hispanics. However, this does not imply that population substructure can be ignored in forensic applications, because frequency estimates may still differ by orders of magnitude when substructure is considered.

The magnitude of heterogeneity in match probability appears to be greater in New Zealand than in the U.S., in part because of the slight reduction in heterogeneity among Polynesians and Maori. Nevertheless, as Figure 5 shows, no five locus genotype is more frequent than 10^{-6}, so a random match is never a frequent event (Chakraborty & Kidd, 1991). Where there is still contention is in the method to arrive at a conservative estimate of match probability. The ceiling principle, which uses as estimates of allele frequencies the maximum among subpopulations, is generally conservative (Krane *et al.,* 1992), but it does not always yield a match probability higher than can be observed in a sample (Slimowitz & Cohen, 1993). Lange (1993) presents a lucid discussion of the issues, and we do echo his closing remark that, excluding close relatives, the identification of individuals is always made more accurate by simply testing additional VNTR loci.

Acknowledgements

The authors wish to acknowledge contributions from the following people: For assistance in blood sample collection: Dr. Graeme Woodfield, Auckland Regional Blood Centre; Dr. Ray Fong, Blood Transfusion Service, Wellington Hospital; Dr. Patricia Stapleton, DNA Diagnostics, Auckland; Mr. Sandor Milne, The Hepatitis Foundation, Whakatane. For providing data on estimates for VNTR locus mutation rates: Dr. Arthur Eisenberg, Department of Biology, Texas College of Osteopathic Medicine, Fort Worth, Texas and Dr. Patricia Stapleton. For providing the 'MacGel' software: S. Kessell, M. Gibbs and C. Millar, Molecular and Cellular Biology, Auckland University. For expanding the VNTR database: Ms. Debbie Monahan and Dr. Len Starling, ESR Forensic, Wellington Science Centre.

References

Anderson, A.J. & B.G. McFadgen, 1990. Prehistoric two-way voyaging between New Zealand and East Polynesia: Mayor Island obsidian on Raoul Island and possible Raoul Island obsidian in New Zealand. Archeology in Oceania 25:24–37.

Balazs, T.I., 1993. Population genetics of 14 ethnic groups using phenotypic data from VNTR loci. EXS. 67:193–210.

Budowle, B. & K.L. Monson, 1993. The forensic significance of various reference population databases for estimating the rarity of variable number of tandem repeat (VNTR) loci profiles. EXS.67:177–191.

Chakraborty, R. & L.T.I. Jin, 1992. Heterozygote deficiency, population substructure and their implications in DNA fingerprinting. Hum. Genet. 88:267–272.

Chakraborty, R. & K.K. Kidd, 1991. The utility of DNA typing in forensic work. Science 254:1735–1739.

Church, G.M. & W. Gilbert, 1984. Genomic Sequencing. Proc. Natl. Acad. Sci. USA. 81:1991–1995.

Cockerham, C.C. & B.S. Weir, 1993. Estimation of gene flow from F-statistics. Evolution 47:855–863.

Cohen, J.E., 1990. DNA fingerprinting for forensic identification: Potential effects on data interpretation of subpopulation heterogeneity and band number variability. Am. J. Hum. Genet. 46:358–368.

Crow, J.F. & K. Aoki, 1984. Group selection for a polygenic behavioural trait: estimating the degree of subdivision. Proc. Natl. Acad. Sci. USA 81:6073–6077.

Davidson, J.M., 1987. Maori origins, pp. 13–29 in The Prehistory of New Zealand, Longman Paul Limited, Auckland.

Decorte, R., R. Wu, P. Marynen & J-J. Cassiman, 1994. Identification of internal variation in the pseudoautosomal VNTR DXYS17, with nonrandom distribution of the alleles on the X and Y chromosome. Am. J. Hum. Genet. 54:506–515.

Devlin, B. & N. Risch, 1992. Ethnic differentiation at VNTR loci, with special reference to forensic applications. Am. J. Hum. Gen. 51:534–548

Devlin, B., N. Risch & K. Roeder, 1990. No excess of homozygosity at loci used for DNA fingerprinting. Science 249:1416–1420.

Feinberg, A.P. & B. Vogelstein, 1983. A technique for radiolabelling DNA restriction endonuclease fragments to high specific activity. Anal. Biochem. 132:6–13.

Felsenstein, J., 1989. PHYLIP – Phylogeny Inference Package (Version 3.2). Cladistics 5:164–166.

Flint, J., A.J. Boyce, J.J. Martinson & J.B. Clegg, 1989. Population bottlenecks in Polynesia revealed by minisatellites. Hum Genet. 83:257–263.

Hamilton, J.F., 1994. Multi-locus and single-locus DNA profiling in New Zealand. Ph.D. Thesis, Victoria University of Wellington, New Zealand.

Harding, R.M., A.J. Boyce & J.B. Clegg, 1992. The evolution of tandemly repetitive DNA: recombination rules. Genetics 132:847–859.

Hernandez, J.L. & B.S. Weir, 1989. A disequilibrium coefficient approach to Hardy-Weinberg testing. Biometrics 45:53–70.

Herrin, G. 1992. A comparison of models used for calculation of RFLP pattern frequencies. J. Forensic Sci. 37:1640–1651.

Hill, A.V.S., D.F. O'Shaughnessy & J.B. Clegg, 1989. The colonization of the Pacific: Some current hypotheses, pp. 246–285, in The Colonization of the Pacific: A Genetic Trail, edited by A.V.S. Hill and S.W. Serjeantson. Clarendon Press, Oxford.

Jeffreys, A.J., N.J. Royle, V. Wilson & Z. Wong, 1988. Spontaneous mutation rates to new length alleles at tandem-repetitive hypervariable loci in human DNA. Nature 332:278–281.

Krane, D.E., R.W. Allen, S.A. Sawyer, S.A. Petrov & D.L. Hartl, 1992. Genetic differences at four DNA typing loci in Finnish, Italian, and mixed Caucasian populations. Proc. Natl. Acad. Sci. USA 89:10583–10587.

Lander, E.S., 1989. DNA fingerprinting on trial. Nature 339:501–505.

Lange, K., 1993. Match probabilities in racially admixed populations. Am. J. Hum. Gen. 52:305–311.

Lewontin, R.C., 1982. Human Variety, pp. 1–13 in Human Diversity. Scientific American Books, W.H. Freeman and Company, San Francisco, CA.

Lewontin, R.C. & D.L. Hartl, 1991. Population genetics in forensic DNA typing. Science 254:1745–1750.

Manly, B.F.J., 1991. Randomization and Monte Carlo Methods in Biology. Chapman and Hall, New York.

Martinson, J.J., R.M. Harding, G. Philippon, F. Flye Saint Marie, J. Roux, A.J. Boyce & J.B. Clegg, 1993. Demographic reductions and genetic bottlenecks in humans: Minisatellite allele distribution in Oceania. Hum. Genet. 91:445–450.

Menotti-Raymond, M. & S. O'Brien, 1993. Dating the genetic bottleneck of the African cheetah. Proc. Natl. Acad. Sci. USA 90:3172–3176.

Nei, M., 1972. Genetic distance between populations. Am. Nat. 106:283–292.

Roe, A., 1993. Correlations and interactions in random walks and population genetics. Ph.D. Thesis, University of London.

Sambrook, J., E.F. Fritsch & T. Maniatis, 1989. In Molecular Cloning: A Laboratory Manual (2nd edition), Cold Spring Harbor Laboratory Press, Cold Spring Harbor, NY.

Serjeantson, S.W., 1989. HLA genes and antigens, pp. 120–173 in The Colonization of the Pacific: A Genetic Trail, edited by A.V.S. Hill and S.W. Serjeantson. Clarendon Press, Oxford.

Serjeantson, S.W. & A.V.S. Hill, 1989. The colonization of the Pacific: A genetic trail, pp. 286–294 in The Colonization of the Pacific: The Genetic Evidence, edited by A.V.S. Hill and S.W. Serjeantson. Clarendon Press, Oxford.

Shriver, M.D., L. Jin, R. Chakraborty & E. Boerwinkle, 1993. VNTR allele frequency distributions under the stepwise mutation model: a computer simulation approach. Genetics 134:983–993.

Slimowitz, J.R. & J.E. Cohen, 1993. Violations of the ceiling principle: Exact conditions and statistical evidence. Am. J. Hum. Gen. 53:314–323.

Smith, J.C., R. Anwar, J. Riley, D. Jenner, A.F. Markham & A.J. Jeffreys, 1990. Highly polymorphic minisatellite sequences: Allele frequencies and mutation rates for five locus-specific probes in a Caucasian population. J. Forensic Sci. Soc. 30:19–32.

Southern, E.M., 1975. Detection of specific sequences among DNA fragments separated by gel electrophoresis. J. Mol. Biol. 98:503–527.

Valdes, A.M., M. Slatkin & N.B. Freimer, 1993. Allele frequencies at microsatellite loci: the stepwise mutation model revisited. Genetics 133:737–749.

Weir, B.S., 1992a. Independence of VNTR alleles defined as fixed bins. Genetics 130:873–887.

Weir, B.S., 1992b. Independence of VNTR alleles defined as floating bins. Am. J. Hum. Gen. 51:992–997.

Weir, B.S., 1993. Independence tests for VNTR alleles defined as quantile bins. Am. J. Hum. Genet. 53:1107–1113.

Weir, B.S., 1994. The effects of inbreeding on forensic calculations. Ann. Rev. Genet. 28: 597–621.

Editor's comments

A side-benefit of the collection of DNA data from human populations is the light it may shed on human evolution. The authors discuss the colonization history of New Zealand in the light of such data. From a forensic viewpoint, too much should not be made of the differences between the major ethnic groups within New Zealand, as the forensic community in that country maintains separate databases for Caucasian, Maori and Pacific Islander (Buckleton *et al.*, 1987). It will be of interest in the future to examine subdivision within these groups, as opposed to within the country as a whole. The authors' comments on testing for independence will need to read along with the findings of Zaykin *et al.* and Maiste and Weir in this volume. The authors had not seen the Budowle *et al.* (1994) rebuttal to the paper of Krane *et al.* (1992).

B. S. Weir (ed.), Human Identification: The Use of DNA Markers, 51–53, 1995.

Conditioning on the number of bands in interpreting matches of multilocus DNA profiles

R.N. Curnow
Department of Applied Statistics, University of Reading, Reading, UK

Received 31 March 1994 Accepted 20 July 1994

Key words: DNA profiling, number of bands

Introduction and summary

There are clear and well-recognised problems in assessing the evidence provided by a match between a multi-locus DNA probe from a scene-of-crime sample and an identical profile obtained from a search of a data bank of profiles. Rarely, if ever, will the data bank be known to contain the profile of the criminal. Unless the scene-of-crime sample is unusual, a large data bank may contain a fortuitous matching profile. Thus, careful calculations similar to those used in assessing matches between individual DNA or protein sequences and DNA or protein sequences found by searching relevant data banks are needed if the occurrence of the match is to be used as evidence of involvement in the crime (Mott, Kirkwood & Curnow, 1989, 1990). Certainly, the match may assist the police in identifying potential culprits, but any use of the evidence of the match must be carefully scrutinised.

Even if the suspect is under investigation for reasons totally unconnected with his or her DNA profile, there can be problems of interpretation if the scene-of-crime profile is degraded or bands near the bottom edges of the gel are faint so that not all the bands are detectable with certainty. In this paper, we show that the calculation of the ratio of the likelihoods of the data if the suspect is guilty and if he is not guilty should be made conditional on the total number of bands in the suspect's profile. An identical argument applies to the interpretation of profiles in paternity disputes when the evidence is the possession by the putative father of all the child's bands not present in the mother's profile. The difficulties of a conditional argument when the other bands in the child's and mother's profiles are included are mentioned and an alternative approach based on decision rules and probabilities of decision errors are described.

Degradation of scene-of-crime samples

Let the number of bands in the scene-of-crime sample be j and in the suspect's profile be $i \geqslant j$. Also let N be the total possible number of bands detectable within the window of the electrophoretic gel. We shall make the usual, and well-supported, approximations that the frequency of each band in the relevant base population is the same, x, and that the occurrence of different bands are independent (Jeffreys, Turner & Debenham (1991)). Evett, Werrett and Smith (1989) modelled the degradation of the bands in the scene-of-crime sample by assuming that each present band had, independently, the same probability, m, of being detected. The probability that a random person with i bands would provide the scene-of-crime sample with j bands is

$$(mx)^j(1-mx)^{N-j},$$

whereas the probability of the sample if the suspect is the criminal is

$$m^j(1-m)^{i-j}.$$

The likelihood ratio, the probability of the scene-of-crime sample if the suspect is the criminal divided by the probability of the sample if he is not, is therefore

$$(1-m)^{i-j}/[x^j(1-mx)^{N-j}].$$

This likelihood ratio clearly, and correctly, depends on i, the number of bands in the suspect's profile. The possession of i bands has no direct relevance to the guilt, or otherwise, of the suspect and so the probabilities in the likelihood ratio are conditional on the number of bands in the suspect's profile. Evett, Werrett and Smith (1989) contains a discussion of the estimation of x and N and the effect of uncertainties in the value of m.

An alternative approach (Curnow, 1991) ignores the bands present in the suspect but not in the scene-of-crime sample and hence avoids assumptions about the probabilities of degradation. The probability that a random person includes j particular bands in his profile if he has i bands out of the total N bands is

$$\frac{(N-j)!}{(i-j)!(N-i)!} \Big/ \frac{N!}{i!(N-i)!}$$

The probability if the suspect is the criminal is clearly 1, so that the likelihood ratio is

$$\frac{N!}{i!(N-i)!} \Big/ \left(\frac{(N-j)!}{(i-j)!(N-i)!}\right)$$

Now we need to know only N, the total possible number of bands in a profile, in addition, of course, to i and j.

Table 1 shows the values of the likelihood ratios when the total possible number of bands is $N = 50$ and the number of bands detected in the scene-of-crime sample is $j = 15$. The number of bands in the suspect's profile, i, ranges from $i = 15$, the minimum possible, to $i = 50$, the maximum possible. The unconditional likelihood ratio usually quoted, x^{-j}, is the reciprocal of the average of the probabilities calculated above, averaged over the applicable events in the marginal distribution of i, which is binomial with $n = N$ and $p = x$, the band frequency. A value often quoted for x is $x = 0.14$. This gives the unconditional likelihood as $0.14^{-15} = 6.43 \times 10^{12}$. Clearly, from Table 1, the strength of the evidence of guilt will be much reduced for suspects with a large proportion of the possible bands in their profile. In such cases, the likelihood ratio should be calculated for the suspect's value of i.

Table 1. Likelihood ratios conditional on number of bands in suspect's profile.

i	Likelihood ratio
15	2.25×10^{12}
25	6.89×10^{5}
35	6.93×10^{2}
50	1

Likelihood ratios when the total possible number of bands is $N = 50$; the number of bands in the scene of crime sample is $j = 15$ and i is the number of bands in the suspect's profile.

Inferences in paternity cases

Most of the DNA profile evidence in paternity disputes is centred on the requirement, assuming no loss of bands and no new mutant bands in the child, that the putative father possess all the child's non-maternal bands, i.e. those not in the mother's profile. The same argument as applied above to the forensic situation shows that the likelihood ratio should be calculated conditionally on the number of bands in the putative father's profile. The likelihood ratio is the reciprocal of the probability that a random person unrelated to the child has the child's j non-maternal bands conditional on having a total of i bands out of the N possible bands in the profile. The previous formula for the likelihood ratio is still the relevant formula.

The calculation of the conditional likelihood ratio for all types of bands is more difficult. The probabilities, if the putative father is the father, that he has a band present in both mother and child and that he has a band not present in the child are different, and not certain, as is the case with the child's non-maternal bands. Hence the distribution of the number of bands possessed by the father given the profiles of the mother and child is the sum of three binomial distributions with three different p values, determined from the frequency of the bands in the population. The symmetry is lost and purely combinational arguments cannot be used. Curnow and Wheeler (1993) show how the probabilities of wrong decisions can be calculated given rules, based on the putative father's profile in relation to those of the mother and child, for deciding whether the putative father is or is not the father.

The decision-making approach does illustrate the dangers of relying on the evidence provided by a relatively small number of non-maternal bands in the child. In this situation the number of bands in both mother and child and in the putative father's profile may be useful in reducing the probability of saying the putative father is the father when he is not at little expense in increasing, from zero, the probability of saying he is not the father when he is. The dangers of wrong decisions when the putative father has a high proportion of all the possible bands in the profile is again clear.

Acknowledgement

I am grateful to a referee for helpful comments on the presentation of the results in this paper.

References

Curnow, R.N., 1991. DNA fingerprinting, pp 146–150 in The Use of Statistics in Forensic Science, edited by C.G.G. Aitken and D.A. Storey, Ellis Horwood, New York.

Curnow, R.N. & Wheeler, 1993. Probabilities of incorrect decisions in paternity cases using multi-locus DNA probes. J.R. Stat. Soc (A) 156: 207–223.

Evett, I.W., D.J. Werrett & A.F.M. Smith, 1989. Probabilistic analysis of DNA profiles. J. Forensic Science Soc. 29: 191–196.

Jeffreys, A.J., M. Turner & P. Debenham, 1991. The efficiency of multilocus DNA fingerprint probes for individualization and establishment of family relationships, determined from extensive casework. Am. J. Hum. Genet. 48: 824–840.

Mott, R.F., T.B.L. Kirkwood & R.N. Curnow, 1989. A test for the statistical significance of DNA sequence similarities for application in data bank searches. Comp. App. in the Biosciences 5: 123–131.

Mott, R.F., T.B.L. Kirkwood & R.N. Curnow, 1990. Test for the statistical significance of protein sequence similarities in databank searches. Protein Engineering 4: 149–154.

Editor's comments

The author discusses some of the complexities in dealing with multi-locus probes, where the genetic basis for each band may not be known.

B. S. Weir (ed.), Human Identification: The Use of DNA Markers, 55–67, 1995.

Match probability calculations for multi-locus DNA profiles

Peter Donnelly
School of Mathematical Sciences, Queen Mary and Westfield College, Mile End Road, London E1 4NS, UK
Current address: Departments of Statistics and Ecology and Evolution, University of Chicago, 5734 University Avenue, Chicago, IL 60637, USA

Received 16 May 1994 Accepted 28 July 1994

Abstract

The paper considers aspects of the match probability calculation for multi-locus DNA profiles and a related calculation which aims to assess the probability that a pair of profiles is concordant for the presence and absence of bands. It is suggested that levels of allelism and linkage for multi-locus profiles may be higher than reported in previous studies, and that comparison of bandsharing values between different studies is problematic. Our view is that the independence assumptions which underpin the calculations have not been established. The effect of ignoring (local) heterogeneities in band frequencies may be non-conservative. Concerns thus raised about the match probability calculation could be important in practical casework. The speculative nature of some aspects of the concordance probability calculation would seem to render it inappropriate for use in court.

Introduction

The multi-locus probe technique was the first to be used in DNA profiling for the purposes of human identification. For a review of this development, see for example, Jeffreys *et al.* (1991) and references therein; for details of the experimental technique, see for example Jeffreys, Turner and Debenham (1991). A multi-locus probe typically binds to alleles at a large (but unknown) number of loci. The resulting profiles, sometimes called 'DNA fingerprints', consist of a series of bands on an autoradiograph. Typically, profiles of different individuals will exhibit different numbers of bands, and there will be differences in the positions of these bands on the autoradiograph, corresponding to DNA fragments of differing electrophoretic mobilities. As an example, a recent study (Jeffreys, Turner & Debenham, 1991) of the two multi-locus probes 33.6 and 33.15 reported an average of about 17 bands in the scored region of the gel of each profile (for each probe) with a range, over the 3,404 individuals in the study, from about 5 to about 35 bands (again, for each probe).

Use of such probes for identification requires an assessment of how common particular profiles are in various populations. In criminal casework for example, if a suspect's profile is observed to match that from DNA known to belong to the true criminal, the strength of this evidence against the suspect will depend, among other things, on how likely an innocent individual would be to match that profile. The correct assessment of the strength of such evidence raises a number of subtle issues, see Balding and Donnelly (1995). In this paper we will concentrate on one central quantity, the 'match probability'. This is the probability that the profile of an individual from the relevant reference population will match the profile from the crime sample. Throughout, we will phrase the discussion in terms appropriate to a case in which the crime sample is known to have originated from the true criminal.

In some countries multi-locus probes were never used in criminal casework. In others, the technology has largely been superseded, for example by single locus probes. (Single locus probes detect alleles at a particular locus, in contrast to multi-locus probes which simultaneously detect alleles at many loci.) Nonetheless, a discussion of the statistical evaluation of multi-locus profile (MLP) evidence may be of more than historical interest. They may still arise in cases in which the crime sample was originally profiled using multi-locus probes (because the crime itself occurred when this was the standard technology) but for which a suspect has only recently been identified. A second area of relevance relates to the consideration on appeal of cases in which MLPs were originally used.

Various aspects of the match probability calculation for single locus profiles have recently been the subject of intense statistical scrutiny, both in the scientific literature and in the courts (Roeder, 1994, reviews many of these issues). Although the debate appears to be ongoing, this attention, and the associated empirical studies, have served to clarify a number of the issues involved. In contrast, relatively little attention appears to have been paid to the multi-locus profile calculation. Some of the issues are analogous to those arising in connection with single locus probes, while others are specific to the multi-locus context. By the time the statistical community became involved in the debate about the assessment of DNA evidence, multi-locus profiles were rarely presented to criminal courts. As a consequence, statistical aspects of the match probability calculation may not have been widely examined.

Throughout, our focus is on the use of MLPs in criminal forensic casework. As a consequence, we will not discuss other contexts in which they have proved valuable, such as human paternity cases or the assessment of relatedness in natural populations.

In the following section we introduce the model currently used (for example by the UK Forensic Science Service) to calculate the MLP match probability, and a related calculation which aims to derive the probability that two profiles will be concordant for the presence and absence of bands in the scored region of the profile. We will then assess existing data on band-sharing and the extent of allelism and linkage amongst bands in a MLP. The next section is concerned with the independence assumptions which underpin both calculations, while the penultimate section considers the effect of heterogeneity in band frequencies throughout the gel. The final section presents a number of conclusions.

It may be appropriate to draw attention to several issues that we will not address here. The strength of the evidence of matching DNA profiles depends on how common the profile in question is in the relevant population *and* on the probability that the suspect would match to the degree observed if indeed the suspect were the true criminal. It appears to be routine in casework to take this latter probability to be 1. Several aspects of the assessment of MLPs, including the subjectivity in recognising bands present in a given profile, and matches between profiles, and observations of discrepant bands in two profiles known to be from the same individual, suggest that such an assumption may be unwarranted. Further empirical studies on these aspects would be desirable.

In some cases the strength of the evidence is quantified via the likelihood ratio. The concerns just raised relate to the numerator of the likelihood ratio: inappropriately equating this to 1 will tend to overstate the strength of the DNA evidence against the defendant. The denominator is usually taken to be the match probability, although it should also include an assessment of the likelihood that laboratory error could lead to matching profiles when the suspect is not the source of the crime sample.

We have argued elsewhere (Balding & Donnelly, 1994, 1995) in connection with single locus probes that even in cases in which there is no direct evidence to incriminate close relatives of the suspect, their existence can have an important bearing on the interpretation of DNA evidence. Exactly the same arguments apply in the MLP context. Of course it would be preferable for all such relatives to be excluded, either by additional DNA testing, or by more traditional investigative or scientific means, as possible culprits. If this has not been done, it could be misleading simply to quote a match probability for unrelated individuals. Even if probabilities that specific relatives would match the suspect are given, it is important that their correct interpretation is understood. Suspects or their defense lawyers may be reluctant, for obvious reasons, to draw attention to the possibility that the true criminal is one of their close relatives.

The decision of the court in a particular case will depend on the DNA evidence and on all the other evidence presented at trial. The method for combining the DNA and non-DNA evidence, in a manner consistent with the laws of probability, is described in Balding and Donnelly (1995). One consequence of this is that in a particular trial, differences of one or two orders of magnitude in an already small match probability may mean the difference between conviction and acquittal, depending on the jury's assessment of the other evidence. In the ensuing discussion it should thus be remembered that concerns about aspects of the match probability calculation cannot be ignored simply on the grounds that they will not change the calculated value by much, and that it is extremely small in any case.

The models

We focus attention on the model for match probability calculation currently used by the U.K. Forensic Science Service (FSS). Suppose that a MLP is obtained from DNA left at the scene of the crime and from DNA obtained from the suspect, and that these profiles are declared to match. (This initial match/non-match decision is a subjective one on the part of the forensic scientist. In principle it uses more of the information in the profile than is used in the following method for calculating the match probability.)

The match probability calculation is based solely on the number of matching bands, of size greater than 4 kb, between the two profiles. We will denote this number of bands by n, and assume there are no discrepant or non-matching bands. The other relevant quantity is the so-called bandsharing value. In practice, the probability that a band in one profile will be matched by a band in the same position in the profile of an unrelated individual varies with the size of the band in question. Assessment of MLPs for forensic casework does not involve measuring the sizes of the DNA fragments which appear as bands on the autoradiograph (in contrast to standard practice for SLPs). Information about the probabilities of bands occurring in the same position in profiles from different individuals is gained from empirical studies of pairs of profiles from different individuals. These studies assess the proportion of times that bands in the profile from one individual in the pair have matching bands in the profile from the other individual. Several such studies have considered separate regions of the gel above 4 kb, for example 4–6 kb, 6–9 kb, and 9–23 kb. The practice of the FSS for the multi-locus probe 33.15 is to use 0.26 in place of the bandsharing value as this is larger than empirically observed bandsharing values in any of the relevant regions of the gel. The match probability, $\mathcal{M}$, is then calculated as if each of the n observed matches between suspect and crime bands were independent, each having probability 0.26:

$$\mathcal{M} = 0.26^n. \qquad (1)$$

The match probability calculation does not incorporate information about the shared absences of bands between two profiles. A method for doing so has been proposed. The calculation aims to assess the probability that two profiles from unrelated individuals will coincide in both the presence and absence of bands in the scored region of the gel. (For casework from the FSS this is typically the region about 4 kb; others have used the region above 3.5 kb.) The rationale behind the calculation given in Jeffreys, Turner and Debenham (1991, Table 5) is that the gel may be thought of as consisting of many positions, each of which either will be or will not be occupied by a band on a given profile. We denote the total number of such positions in the gel by $\mathcal{T}$ and the probability of a band being present in a particular position by p (our notation is different from that in Jeffreys, Turner and Debenham, 1991). It is assumed that the probability of a band being present is identical for each position in the gel and that the presence or absence of bands in different positions is independent. It is argued that the probability that in a given position both profiles have a band is p^2 and that the probability that both lack a band is $(1-p)^2$, so that the probability that both profiles agree in the presence and absence of bands is then given as

$$(p^2 + (1-p)^2)^{\mathcal{T}}. \qquad (2)$$

The preceding argument is sometimes phased in a slightly different way by referring to the total number of bands which could conceivably be present in a profile and multiplying together the probabilities that each such band is either present in both profiles or absent in both. As Jeffreys, Turner and Debenham (1991) note, some care is needed. If 'bands' are thought of synonymously with positions in the gel, then this is simply a restatement of the discussion above. If, on the other hand, 'bands' are taken to mean alleles of size bigger than 4 kb at loci to which the relevant multi-locus probe binds, then the argument is inaccurate because of the possibility that two matching bands may correspond to alleles from distinct loci which happen to have the same electrophoretic mobility.

In practice, $\mathcal{T}$ cannot be measured directly. Jeffreys, Turner and Debenham (1991), for example, estimate it by $\bar{N}/\hat{b}$ in which $\bar{N}$ denotes the average number of bands per individual in their study and $\hat{b}$ denotes their estimate of the bandsharing value. They evaluate the probability that both profiles agree in the presence and absence of bands, which we will call the *concordance probability* and denote by $\mathcal{C}$, twice, once using

$$(\hat{b}^2 + (1-\hat{b})^2)^{\bar{N}/\hat{b}}, \qquad (3)$$

and once using

$$(0.25^2 + (1-0.25)^2)^{\bar{N}/0.25}. \qquad (4)$$

Note that as defined, the match probability is a conditional probability and the concordance probability is unconditional. That is, the match probability relates to the probability that all the (n) bands present in a particular profile will be present in a second profile (from an unrelated individual). The concordance probability relates to the probability of identical patterns of bands and non-bands in two profiles (from unrelated individuals) before either profile has been observed. Conditional probabilities are the relevant ones in forensic casework. In connection with the conditional probability that a second profile would be concordant with a given profile having n bands in the scored region of the gel, the calculation analogous to (3) would be

$$\hat{b}^n(1-\hat{b})^{\bar{N}/\hat{b}-n}. \quad (5)$$

Empirical studies

As a prelude to discussing assumptions underlying the calculations described in the previous section, we will briefly review some findings from empirical studies related to those assumptions and to the numerical values used in the calculation.

Bandsharing estimates

It may be helpful to contrast the MLP calculation, via bandsharing, with the product rule calculation for a single locus profile (SLP). Consider a particular band in a SLP. The match probability calculation involves consulting a database to assess the frequency, for that particular probe, within a given population, of a band of the observed size (or of a size close enough to fall within the match guideline). Substantial efforts have been made in compiling databases for SLPs in order that this question may be addressed. In contrast, in the MLP setting, there is no direct information on how common the bands in question are in the relevant population. Empirical estimates of the bandsharing value can be thought of as an *average*, across many bands, of the proportion of times each band was shared in the pairs of individuals examined.

There are several estimates of bandsharing values in the literature. For definiteness, we confine attention to the multi-locus probe 33.15. Jeffreys, Wilson and Thein (1985), in a study of 20 unrelated British Caucasians, reported estimates of 0.27, 0.20, and 0.08, in the regions 4–6 kb, 6–10 kb and 10–20 kb respectively. Gill *et al.* (1987), in a study of 41 profiles, reported estimates of 0.26, 0.20, and 0.07 respectively for these same regions. An unpublished FSS study of nearly 500 individuals obtained similar estimates: 0.248, 0.201, 0.068, respectively, for the regions 4–6 kb, 6–9 kb, and 9–23 kb (M.J. Davie, personal communication). Based on a study of 1,702 pairs of Caucasian individuals, Jeffreys, Turner and Debenham (1991) report an estimate of bandsharing of 0.137 for the probe 33.15. This value applies to the entire region of the gel they considered (above 3.5 kb). Separate values for distinct subregions were not reported.

Several of these studies have been criticised elsewhere (Cohen, 1990) over their failure to use formal sampling methods in obtaining the profiles studied.

Jeffreys, Turner and Debenham (1991, page 837) refer to unpublished data on other population groups 'with as yet no evidence for significant shifts in the mean band-sharing frequency between different ethnic groups'. Published studies whose aim is a systematic investigation of possible differences between bandsharing values for different ethnic or racial groups would be helpful (of course, there have been considerable efforts in this direction in connection with SLPs).

While a direct comparison is not possible, we note the difference between the bandsharing estimate of 0.137 from Jeffreys, Turner and Debenham (1991) and those from the large unpublished FSS study. The size of each study makes it difficult to attribute this difference to sampling effects. The inclusion by Jeffreys, Turner and Debenham (1991) of the region 3.5–4 kb should have the effect of increasing observed bandsharing in their study (relative to the FSS study) because of the increased incidence of bands in this region.

Differences in the quality and quantity of DNA profiled, and perhaps in experimental conditions, in each study may account for some differences in observed bandsharing. The Jeffreys, Turner and Debenham (1991) study was based on paternity work for which good quality samples were presumably available, in contrast to the FSS study which included routine criminal casework. For example, Gill *et al.* (1987, page 40) note a smaller mean number of bands per profile (11) than Jeffreys, Wilson and Thein (1985) (15). In Jeffreys, Turner and Debenham (1991) this mean number is about 17.

Decisions about whether particular bands are present, and declarations of matches between bands, are subjective. These vary between scientists, and standard procedure for such decisions may differ between laboratories. This effect will also lead to differences between studies. It also occurs directly in criminal casework. For example, in one UK criminal case (**R-v-Bell**) four different FSS scientists examined the profiles in question. Their assessments of matching bands differed somewhat. At trial one of them gave evidence that there was a collection of 8 matching bands on which they were all agreed. A scientist from outside the FSS gave evidence on the same profiles that there were 'approximately 14' matching bands.

In view of these differences, it would seem inappropriate to rely on bandsharing estimates from one study in one laboratory in interpreting profiles obtained and assessed in another laboratory. In particular, an argument that a particular bandsharing value (for example 0.26) is necessarily conservative because it is larger than estimates from another study (for example 0.14) appears spurious.

In certain cases a bandsharing estimate for the whole of the region of the gel used may be unhelpful for other reasons. If most or all of the bands observed in the crime and suspect samples fall in, say, the 4–6 and 6–9 kb regions, the relevant bandsharing values are the ones for those regions. Comparison with a value for the whole gel, which is decreased substantially because of non-matching bands in the high molecular weight region, is not relevant if the profiles in question have no bands in that region, and may be misleading if they only have one or two such bands.

Allelism and linkage

Information about the extent to which pairs of bands are allelic (come from the same locus) or linked (correspond to DNA fragments near each other on the same chromosome) is contained in studies of the inheritance of the pair in sibships.

An initial study was made by Jeffreys *et al.* (1986). We concentrate here on the more extensive study by Jeffreys, Turner and Debenham (1991). These authors examined 36 large families of Asian (Indian subcontinent) origin. In each family, pairs of bands in each parent were examined to see if they segregated as if they were allelic (exactly one in each child) or as if they were linked (either both or neither in each child). The chance that a non-allelic (unlinked) pair will segregate fortuitiously as if they were allelic is $1/2^s$ where s is the size of the sibship. This is also the probability that a pair of unlinked (non-allelic) bands would segregate as if linked. Observed levels of apparently allelic or apparently linked pairs of bands can then be compared with the expected level due to fortuitious segregation.

Jeffreys, Turner and Debenham (1991, Table 3) present the results of their analysis. After allowance for background fortuitious segregation, they give the 'percentage of bands with allelic partners' ranging between 6.3% and 10% for the larger families. Similarly, their 'percentage of bands with linked partners' ranges between 1.5% and 6.4%. In the text they state that levels of allelism and linkage are low, 'with $< 10\%$ of bands having linked or allelic partners in a given parent'.

The 'percentage of bands with allelic partners' is calculated (A.J. Jeffreys, personal communication) as the number of allelic *pairs* of bands divided by the number of scored bands in the individual (and multiplied by 100 to give a percentage). Thus if 6 of 12 bands occurred as 3 allelic pairs, the percentage of bands with allelic partners would be calculated as 25% (not as 50%). The percentage of bands with linked partners is calculated in the same way, namely from the number of pairs of bands which are linked divided by the total number of bands (A.J. Jeffreys, personal communication).

With this definition of the percentages calculated, there appears to be an error in the formula given in the legend of Table 3 of Jeffreys, Turner and Debenham (1991) for the expected background level of pairs of bands which appear through fortuitious segregation to be allelic (linked). Their formula overstates by a factor of two the expected background. If the numerical values for the background given in the table were calculated from the formula given, the effect would be to underestimate the levels of allelism and linkage. This may also explain why observed levels of apparent allelism and apparent linkage for the smaller sibships were substantially smaller than the background calculated.

Jeffreys, Turner and Debenham (1991) note that mutation events or the electrophoretic comigration of bands may result in pairs of bands which are actually linked (allelic) not appearing to segregate as such, thus leading to underestimation of levels of allelism (linkage). The study reports an average number of 0.195 mutant bands per child for the probe 33.15 (this data relates to the Caucasian paternity assessments rather than the Asian sibships used to assess allelism and linkage). Consider a locus whose mutation probability is equal to the average value of 1.1% suggested by this data. For a pair of allelic bands in one parent at this locus, the chance that there would be at least one mutation to the inherited band in a sibship of 9 individuals is about 10%. A slightly different calculation shows that the probability that at least one mutation event would cause a linked pair (with mutation probability of 1.1% at each locus) to appear unlinked as a consequence of a mutation event in a sibship of 9 individuals is also about 10%. Mutation rates are not constant at the loci in question, and hence not necessarily equal to the average value across loci. It appears likely that most of the loci to which the probe binds will have mutation rates substantially smaller than this average value (Table 4 of Jeffreys, Turner and Debenham (1991) contains relevant information). For a locus with mutation probability of 0.1% for example the probability of a mutation event causing an allelic (linked) pair not to segregate as such would be 1% in a family of 9 individuals. We will argue below that two of the higher mutation rate loci are more likely to give rise to allelic pairs of bands. Variation in mutation rates, particularly in a way which is associated with the tendency to give rise to allelic or linked bands, makes quantification of the level of underestimation difficult. Nontheless, underestimation of both allelism and linkage due to mutations will be more substantial in the larger families. It also appears difficult to quantify the level of underestimation due to comigration.

Bands which are shared by both parents in the family are uninformative for allelism and linkage and so are not included in the analysis. This actually leads to another source of underestimation of levels of allelism and linkage. Write δ for the probability that a particular band in one parent is matched in the other parent. Then the probability that at least one of a pair of allelic bands will not be informative, so that the existence of the pair will not be observed, is about 2δ. Thus while the total number of bands included will be reduced, on average, by $1 - \delta$, the number of observed allelic pairs will be reduced by a factor of about $1 - 2\delta$, in both cases assuming constant probabilities of bandsharing throughout the gel. For the bandsharing value of 14%, this suggests an average level of underreporting of 15–20%. The effect of inhomogeneities through the gel in bandsharing probabilities seems more difficult to quantify, and the analysis would need to be changed if loci which contributed allelic bands exhibited higher or lower than average bandsharing values. A similar argument applies to assessed levels of linkage.

No assessment was made of the variability associated with reported estimates of allelism or linkage. One source of variability relates to the variation of the actual level of fortuitious allelism (linkage) around its expected value. Another arises from the fact that allelic (linked) pairs of bands can only be detected if they happen both to be present in one of the parents. Because of the larger number of families studied in the smaller sibship sizes, it might be argued that estimates from these families were less variable than those from the (smaller numbers of) larger sibships. That is, even though the expected background level of fortuitious cosegregation will be higher for the smaller families, the greater degree of replication here may make resulting estimates (after correction for background) more reliable than for those from the smaller number of larger families. The increase with sibship size of the effect of mutation in masking allelism and linkage is another reason for placing greater weight on the smaller families.

It is not clear whether the levels of background in Table 3 of Jeffreys, Turner and Debenham (1991) were calculated with the incorrect formula given in the paper. If they were, then calculated levels of linkage and allelism should each be around 10%. The above discussion then suggests that a figure of 12%–15% might be appropriate as an estimate of levels of allelism and that the same figures might give an appropriate estimate for the level of linkage. The Jeffreys, Turner and Debenham (1991) study would be informative on the extent to which bands which apparently occur in allelic pairs also occur in linked pairs. In the absence of this information, it would be conservative to assume that this does not occur often. This would result in a conclusion that on average 25–30% of bands would have allelic or linked partners in a typical profile. Recall that this means that 50–60% of the bands which occur in the profile are occurring as members of linked or allelic pairs.

A theoretical argument suggests that even this level of observed allelism is surprisingly low. The probe 33.15 detects the alleles from the two loci detected by the single locus probes MS1 and MS31 (Jeffreys, Turner & Debenham, 1991). Population data from experiments which use the same restriction enzyme (*Hin*f1) as in the multilocus study by Jeffreys, Turner and Debenham (1991) (e.g. Buffery *et al.*, 1991; Brinkman, Rand & Wiegand, 1991) shows that in Caucasians virtually all alleles for MS31 have length greater than 3.5 kb, while for MS1, roughly 2/3 of alleles are larger than 3.5 kb. (The rather smaller data set in Buffery *et al.* (1991) for Asians suggests that this is also approximately true for that group.) Thus, in their 33.15 profile, all (Caucasian or Asian) individuals will have at least one pair of allelic bands (the alleles from MS31) and about half will have at least two pairs (the additional pair being the alleles from MS1), *ignoring the many other loci which contribute to the profile*. In a parent with the average number of 17 bands these two loci alone would result in levels of allelism, calculated as described above, of at least 6% and often 12%. That is, these two loci explain virtually all of the observed level of allelism.

This requires highly skewed population distributions of allele length for almost all of the other loci to which the probe 33.15 binds. If we were to index these other loci, and write f_i for the population frequency of alleles of length at least 3.5 kb at the ith such locus, it would require $2\sum_i f_i$ to equal about 14, to explain the average number of bands observed, but $\sum_i f_i^2$ to be effectively zero, to ensure that individuals contain at most one allele from each locus. The VNTR alleles for which we do have reasonable information on population frequencies are not necessarily 'typical', having been chosen preferentially because of the variability exhibited. Nonetheless, the allelism data would thus suggest a very large number of loci (50–100) with the property that there are some, but no more than say 10% of, alleles in the population above 3.5 kb.

Independence

The match probability calculation (1) assumes that events involving matches of distinct bands are independent. Serious concerns about the justification of this independence assumption were raised in Cohen (1990). As that paper notes (page 364) 'without appropriate data, no amount of statistical theory can say whether different fragments are statistically independent or are statistically associated in a population'. The same applies to genetic theory. The only way to establish the independence assumption is by a direct assessment of whether particular collections of bands are possessed independently within the relevant population. No such study appears to have been undertaken. (Again this contrasts with the position on SLPs where considerable recent efforts have been directed at assessing some of the independence assumptions underlying the product rule used in that context.)

The independence issue relates to linkage disequilibrium in the relevant population. Linkage of the bands involved is neither necessary nor sufficient for linkage disequilibrium. The question of linkage is nonetheless indirectly relevant to the independence issue, as linked bands could well be in linkage disequilibrium within the population, and hence not be possessed independently.

The previous section suggested that linkage may be more common than had been supposed. From the figures given there, it might well be that a profile with ten bands contains one or two pairs of linked bands. If, as a consequence, it were felt prudent to ignore two of the bands in the match probability calculation, the effect would be to reduce the value calculated (via (1)) from about 1 in 700,000 to about 1 in 48,000. As noted earlier, a change of this magnitude could be important in a particular trial, not withstanding the fact that the new value (of 1 in 48,000) is still small.

For a pair of bands known to be allelic, the proper contribution to the match probability (assuming Hardy-Weinberg equilibrium) is $2q_1q_2$ where q_1 and q_2 are the frequencies in the relevant population of the bands in question. This will be smaller than the contribution 0.26^2 in (1) whenever $q_1q_2 < 0.0338$. This is plausible for many VNTR alleles, in the light of available population data and, for example, the theoretical argument given at the end of the section on allelism and linkage. In practice, however, for a multi-locus profile, it will not be known which bands are allelic. Even if this were known, the loci contributing the bands would not be known from the MLP itself without additional probing with specific single locus probes. Further, population allele frequency distributions for many of the loci involved are not available.

Various aspects of the data in Jeffreys, Turner and Debenham (1991) bear indirectly on the independence issue. One general concern is that it may not be sensitive to some departures from independence which might nonetheless be important in practice.

Their Table 2 notes underdispersion, relative to the model of independence and constant probability of band presence throughout the gel, in the histogram of numbers of bands present in the individuals scored. One possible cause of underdispersion is negative correlations in band presences, tending to make the independence assumption conservative. Another is variation with position in the gel of bandsharing values, which is known to occur. Data of this kind thus appear difficult to interpret in connection with the independence issue.

The bandsharing estimates for mother-father pairs in Jeffreys, Turner and Debenham (1991) are lower than those reported for arbitrary pairs of individuals in other studies. In the presence of population structure (one possible cause of non-independence), mother-father pairs would be expected to be from the same subpopulations, and hence more genetically similar than arbitrary pairs. Earlier remarks about the dangers of comparing bandsharing values from different studies apply here. The sampling scheme by which the pairs were selected also bears on the generality of such an inference. In any event, Caucasian data is not directly relevant to population structuring in other ethnic groups.

The above discussion relates to the independence, within the population, of bands present in a profile. Effects which lead to positive correlations in band presences are also likely to lead to positive correlations in band absences. The rationale behind the concordance probability calculation assumes that such correlations do not exist. In addition, the combined presences and absences of bands are assumed to be mutually independent. This is false. Allelism will lead to positive correlations between the presence of bands in one position and their absence in others. A limit to the total number of possible bands (or to the number of loci detected by the probe) would have the same effect. Each of these applies to some degree in the context of MLPs.

It is difficult to assess the extent to which calculations of the concordance probability along the lines of (3) could thus overstate the rarity of concordant profiles. For reasons analogous to those given above in connection with the match probability calculation, the limited information which bears indirectly on the issue does not appear especially helpful.

Heterogeneity of band frequencies

The match probability and concordance probability calculations assume that bandsharing values are constant throughout the gel. This is an approximation. In this section we consider theoretically the consequences of this assumption, by considering a slightly less simplified model which allows for local heterogeneity in band frequencies.

Bandsharing is known to vary between different regions of the gel. We consider the effect of heterogeneity within one such region (for example the region from 4–6 kb). Suppose the region of the gel in question consists of T 'positions' or subregions, labelled $1, 2, \ldots, T$. (It is unrealistic to assume that there is a discrete collection of positions in the gel within which bands must be located. We do so to facilitate the discussion. The central points would also apply to a more complicated model in which the gel is continuous.) We will assume for the moment that the total number of positions, T, is known exactly (this is generalised below). Write x_i, $i = 1, 2, \ldots, T$, for the relative frequency in the population with which a band will be present in position i in a particular profile. Assume that if the x_i's were known, the presence or absence of bands in different positions within a profile would be independent and further that the presence or absence of bands would be independent between profiles.

The frequencies x_i are not known with certainty, although studies of bandsharing bear on them indirectly. A natural approach then is to treat them as unknown and to encapsulate our uncertainty about them in the form of a probability distribution for the vector $(x_1, x_2, \ldots, x_T)$. (Note that while each x_i is a frequency, so $0 \leqslant x_i \leqslant 1$, there is no requirement that $x_1 + \cdots + x_T = 1$.) We will thus think of $(x_1, x_2, \ldots, x_T)$ as being random, where the "randomness" is over our uncertainty (given available information) about their actual values in the relevant population.

Throughout, we will assume that $(x_1, x_2, \ldots, x_T)$ is exchangeable. Formally, this is a statement that if we relabelled the positions, our uncertainty about the new vector $(x_1, x_2, \ldots, x_T)$ would not change. In view of the fact that we are focussing on a subregion of the gel the exchangeability assumption seems reasonable, at least approximately: there seems no reason to expect, for example, that the values of x_3 and x_9 will be systematically different from x_8 and x_2.

Cohen (1990) considered the effect of heterogeneity in the x_i's (although he did not introduce our notation) when the x_i's are known exactly. He argued that in this case it was conservative to replace each x_i by the mean of the x_i's and to treat them as being constant. Effectively the same point is made in Jeffreys, Turner and Debenham (1991). (Neither of these papers states explicitly that the x_i's are assumed known, but as the discussion below shows, their argument is not in general valid without that assumption.) The assumption that the x_i's are known exactly, that is, that there is no uncertainty about their values, seems inappropriate for MLPs. None of the published data bear directly on the x_i's.

Now, consider a particular position, i, in two profiles from different individuals.

$$
\begin{aligned}
&\text{P(band at } i \text{ in second profile} | \text{band at } i \text{ in first profile)} \\
&= \frac{\text{P(band at } i \text{ in both profiles)}}{\text{P(band at } i \text{ in first profile)}} \\
&= \frac{\mathrm{E}(x_i^2)}{\mathrm{E}(x_i)}. \qquad (6)
\end{aligned}
$$

The exchangeability assumption ensures that the right hand side of (6) is the same for each i, and we *define* this quantity to be the bandsharing value, denoted by

b:

$$b \equiv \frac{\mathrm{E}(x_i^2)}{\mathrm{E}(x_i)}. \tag{7}$$

The probability of a band in position i in a particular profile is just $\mathrm{E}(x_i)$. Again the exchangeability assumption ensures that this probability is the same for each i. We call this probability the *band probability* and denote it by p:

$$p \equiv \mathrm{E}(x_i). \tag{8}$$

The first point to note is that the bandsharing value is different from the band probability. In fact (Jensen's inequality)

$$b \geqslant p, \tag{9}$$

with equality only in the special case in which all the x_i's take the same value which is known exactly. One view on the intuition behind the inequality (9) is that the information that there is a band in position i in one profile should cause us to believe in larger values for x_i than before we had this information. The probability that there is a band at i in another profile, given knowledge of the band there in the first profile, is thus larger than it would have been without the information about the first profile.

We can rewrite (7) as

$$b = p\left(1 + \frac{\mathrm{Var}(x_i)}{p^2}\right),$$

so that the difference between the bandsharing value b and the band probability p depends on the uncertainty about x_i, as measured by its coefficient of variation.

Now suppose we compare profiles for M pairs of individuals. Write

$$\chi_{ji}^{(1)} = \begin{cases} 1 & \text{if there is a band in position } i \text{ of the} \\ & \text{first individual of the } j\text{th pair} \\ 0 & \text{otherwise} \end{cases}$$

and

$$\chi_{ji}^{(2)} = \begin{cases} 1 & \text{if there is a band in position } i \text{ of the} \\ & \text{second individual of the } j\text{th pair} \\ 0 & \text{otherwise.} \end{cases}$$

Now write $N_j^{(1)}$ and $N_j^{(2)}$ for the number of bands observed in the first (respectively second) member of the jth pair and S_j for the number of shared bands in the jth pair:

$$N_j^{(1)} = \sum_{i=1}^{T} \chi_{ji}^{(1)}, \quad N_j^{(2)} = \sum_{i=1}^{T} \chi_{ji}^{(2)},$$

$$S_j = \sum_{i=1}^{T} \chi_{ji}^{(1)} \chi_{ji}^{(2)}.$$

Further, write

$$\bar{N}^{(1)} = M^{-1} \sum_{j=1}^{M} N_j^{(1)}, \quad \bar{N}^{(2)} = M^{-1} \sum_{j=1}^{M} N_j^{(2)},$$

$$\bar{S} = M^{-1} \sum_{j=1}^{M} S_j,$$

for the empirical averages of the numbers of bands and numbers of shared bands in each pair.

Now, using exchangeability of the x_i,

$$\mathrm{E}(\bar{S}) = \mathrm{E}(S_j) = T\mathrm{E}(x_i^2),$$

and

$$\mathrm{E}(\bar{N}^{(1)}) = \mathrm{E}(\bar{N}^{(2)}) = \mathrm{E}(N_j^{(1)}) = \mathrm{E}(N_j^{(2)}) = T\mathrm{E}(x_i).$$

Thus

$$\hat{b} \equiv \frac{\bar{S}}{\frac{1}{2}(\bar{N}^{(1)} + \bar{N}^{(2)})}$$

is a natural estimator of the bandsharing value b. (Another natural estimator is the empirical average of the bandsharing per pair: $M^{-1} \sum_{j=1}^{M} S_j / \frac{1}{2}(N_j^{(1)} + N_j^{(2)})$.) Jeffreys, Turner and Debenham (1991, Table 1) use the estimator $\hat{b}$ for bandsharing. Previous authors are not explicit about which estimator they use. In some cases there appears to be confusion between the estimates and the probability which is being estimated.

What is the correct calculation of the contribution to the match probability within the region of the gel of interest for our simple model? Suppose matches are observed at locations $\{i_1, \ldots, i_l\} \equiv L$. Then assuming that the suspect is innocent and unrelated to the criminal,

$$\begin{aligned} \mathcal{P} &= \mathrm{P}(\text{bands in positions in } L \text{ in suspect} \mid \\ &\qquad \text{these bands in criminal}) \\ &= \frac{\mathrm{E}(\prod_{i \in L} x_i^2)}{\mathrm{E}(\prod_{i \in L} x_i)}. \end{aligned} \tag{10}$$

We argued above that our uncertainty about the x_i's should be encapsulated in an exchangeable probability distribution. The simplest such distribution would involve independence of the x_i's. In this case,

$$\mathcal{P} = \frac{\mathrm{E}(\prod_{i \in L} x_i^2)}{\mathrm{E}(\prod_{i \in L} x_i)} = \left(\frac{\mathrm{E}(x_i^2)}{\mathrm{E}(x_i)}\right)^l = b^l. \tag{11}$$

That is, in the case in which the x_i's are assumed independent, the contribution to the match probability for this region of the gel will be given by the appropriate power of the bandsharing value, as suggested by the formula (1).

It might be argued, however, that independence is inappropriate in this context. Variability in gels and experimental conditions means that in repetitions of the process, a fragment of the same length may find itself in slightly different (perhaps neighbouring in our model) positions. In addition, plausible beliefs about the mutation mechanism at the loci in question, and population genetics arguments, will result in a tendency for relatively smooth population distributions of allele length at each locus (as is typical at VNTR loci studied) so that knowledge that an allele of a particular length is relatively common (uncommon) will increase the likelihood that alleles of 'neighboring' lengths will also be relatively common (uncommon). Each of these effects will lead to positive correlations, at least among x_i's for nearby positions. There will be a (multinomial type) negative correlation effect, due to limits in total numbers of fragments in a population: at each particular locus, knowledge that alleles in position i are relatively common in the population will tend to make all other lengths less common simply because some of the alleles which might have been of this length are now known to be in position i. While the positive correlation effects relate to nearby locations on the gel, this negative correlation effect, which may be weaker, extends throughout the region in question.

In general, the correct contribution, given at (10), will be different from the factor b^l suggested by the calculation (1). A general assessment of (10) is impossible. Its value will depend on the distribution of the x_i's. It is, however, natural to conjecture that if this involves positive correlation amongst the unknown x_i's, then the correct calculation (10) will be *larger* than the 'naive' calculation b^l, in which case multiplication of a bandsharing value may lead to a non-conservative error.

We now consider the effect of heterogeneities in band frequencies on concordance probabilities. We continue with the model used above, and in particular, we continue to assume here that were the frequencies x_i known, presences and absences of bands within and between profiles are independent.

The probability that two profiles agree in the presence and absence of bands (in the region of the gel under consideration) is then

$$E\left(\prod_{i=1}^{T}(x_i^2 + (1-x_i)^2)\right). \tag{12}$$

Relatively little direct information is available about T, the total number of positions (in that region of the gel). The correct approach then is to encapsulate our uncertainty about T in a probability distribution (in general, T will not be independent of the x_i).

Direct comparison of (12) and (3) seems impossible in general. In particular, it is far from clear that the two expressions should give similar values. We can, however, make several observations. Consider first T, the total number of positions. Under the exchangeability assumption, if T were fixed, $\mathrm{E}(\bar{N}) = Tp$ where $p \equiv \mathrm{E}(x_i)$ is the bandsharing probability. Thus, $\bar{N}/p$ is a natural (point) estimator of T. (In the current context, $\bar{N}$ would refer to the average number of bands per profile in the region of the gel under consideration.) Since $p \leqslant b$, use of $\bar{N}/b$ will tend to underestimate T. When T is unknown, we might well have $\mathrm{E}(T) < \bar{N}/\hat{b}$. However, Jensen's inequality ensures that for non-random a,

$$\mathrm{E}(a^T) \geqslant a^{\mathrm{E}(T)},$$

so that the effect of using an underestimate of T in the exponent of (3) in place of the proper calculation is unclear (in general it will depend on the amount of uncertainty about T).

We now suppose that T were actually known, in order to assess the factors in the product (12). If the x_i are independent, then (12) reduces to

$$\begin{aligned}(\mathrm{E}(x_i^2) + \mathrm{E}((1-x_i)^2))^T &= (1-2p(1-b))^T \\ &\geqslant (1-2b(1-b))^T,\end{aligned}$$

which would be the contribution from (3) (this inequality is a non-conservative direction).

Thus, while a direct comparison is impossible, there are several aspects of the calculation (3) which may be non-conservative in the presence of uncertainty about the x_i's. These include the effect of positive correlations in uncertainty about the unknown x_i's, the inappropriate equating of the bandsharing value and the band probability, and the assumption that T is equal to its estimated value, rather than allowing for the uncertainty involved. Many of these comments also apply to conditional concordance probabilities, as would those raised in connection with the match probability.

In summary, the argument by earlier authors that ignoring heterogeneity in the gel leads to a conservative error appears inappropriate, since it relies on the population frequencies of bands in different positions being known. We have derived expressions, in the context of a model which still relies on the usual independence assumptions, for match probabilities and concordance probabilities in the presence of heterogeneity (the details of which are not assumed known). Evaluation of the difference between these and the usual match and concordance probability calculations does not seem straightforward in the absence of considerably more information about the extent of actual heterogeneity. Nonetheless, there are aspects of the comparison which suggest that the error may be non-conservative.

It might be hoped that the use in practice of a value (0.26) larger than observed levels of bandsharing would compensate for any such non-conservativeness in the match probability calculation. We argued earlier (see *Bandsharing estimates*) that a comparison between the value 0.26 used in casework and that of 0.14 obtained in Jeffreys, Turner and Debenham (1991) is inappropriate. For the region of the gel from 4–6 kb, the appropriate comparison would appear to be between 0.26 and the value 0.248 obtained in the large FSS study. It is impossible to be definitive, but it is far from clear that this would provide adequate allowance for heterogeneity in this region. The "slack" between 0.26 and observed (FSS) values in other regions is greater, but where the overall balance between this slack and possible non-conservativeness due to heterogeneity lies appears to us to be unclear. Assessment of the effect of heterogeneity on concordance probabilities, and comparison with existing calculations, is still more complicated.

Conclusions

Justification of the independence assumption underlying the match probability calculation for multi-locus profiles requires a study of the joint possession of subsets of bands within appropriate populations. In the absence of such studies, our view is that the independence assumptions cannot be considered to have been established.

Jeffreys, Turner and Debenham (1991) argued in their summary that 'the level of allelism and linkage between different hypervariable DNA fragments scored with these probes [33.6 and 33.15] is . . . low, implying substantial statistical independence of DNA fragments'. It would be unfortunate if the conclusion 'substantial statistical independence' were used to justify the product-rule nature of the MLP match probability calculation. In casework, the important issue is the extent to which use of the independence assumption may lead to an inappropriate overstatement of the rarity of the profile. In particular trials, differences of one or two orders of magnitude can be important. The quantitative difference between 'substantial statistical independence' and independence is unclear, and it seems unlikely that the authors meant their wording to have a precise quantitative interpretation.

We have suggested that levels of allelism and linkage may be rather higher than had been reported, with perhaps half the bands in a profile belonging to allelic or linked pairs. Were this to be the case, the premise of the above statement appears questionable. In any case, as the authors themselves note, a conclusion of statistical independence does not follow from a premise of no linkage (or allelism).

The effect in criminal casework, using one multi-locus probe and fewer scored bands, of this increased level of linkage may be more significant than in the paternity casework described in Jeffreys, Turner and Debenham (1991) in which two probes were used with an average total of 35 bands. It might be thought prudent to assume that linked bands may be in linkage disequilibrium in the population. As Jeffreys, Turner and Debenham (1991) suggest, one possible remedy is simply to ignore in the match probability calculation a number of bands equal to (or perhaps slightly larger than) the expected number of linked pairs of bands in the profile. For the levels of linkage suggested here, in a profile from criminal casework with 8–12 bands, this would increase the calculated match probability by a factor of perhaps 15 or 50. In a profile with 8 bands originally, the calculated match probability would change from about 1 in 48,000 to about 1 in 3,000 (with two bands ignored). For a profile with 12 bands originally, it would change from about 1 in 10 million to about 1 in 700,000 (with two bands ignored) or 1 in 200,000 (ignoring three bands).

The match probability calculation makes the simplifying assumption that the frequency of bands in different positions of the gel is constant. This frequency is known to decrease approximately with increasing fragment length. We have examined the effect (under the usual independence assumptions) of heterogeneity in this frequency within regions of the gel for which band sizes are comparable. The magnitude of any error made in assuming constant frequency seems difficult to assess in the absence of detailed information about actual variability in frequencies. We have shown that if population relative frequencies (treated as being random in view of our uncertainty about their actual values) are independent, then the correct calculation is indeed given by the appropriate power of the bandsharing value. We have argued that this independence assumption may be inappropriate, and that because of correlation effects the resulting error may be non-conservative.

The assessment of MLPs is a subjective one, and the number of scored bands differs between scientists. In addition, there may be differences between laboratories and between samples obtained for different purposes. As a consequence, it seems inappropriate to compare bandsharing values between studies.

The correct comparison for the value of 0.26 used in casework would appear to be with 0.248, 0.201, and 0.068 for the regions 4–6 kb, 6–9 kb, and 9–23 kb, obtained in a large FSS study. It might be argued that the use of the conservative bandsharing value of 0.26 in current practice for criminal casework compensates for possible non-independence and heterogeneity effects. Such an argument appears, at best, difficult to substantiate, in view of the lack of direct information on the actual extent of possible non-independence and heterogeneity. Our view is that the position is unclear.

The match probability calculation does not take account of the joint absences of bands, nor of other aspects of the two profiles, such as the region below 4 kb, and similarities or differences in patterns of band intensities. It is more plausible, but still not established, that were these to be quantified the result would be smaller than the match probability calculated from (1). (Note, however, that in some cases in which some of these other aspects of the profiles, for example band intensities, have been scrutinized, experts have differed on whether or not they support or question the hypothesis that the defendant is the source of the crime sample.)

The concordance probability calculation represents one method for quantifying some of the other information in matching profiles. There is a difference between its use in the scientific literature in order to present an approximation for the probability of concordant profiles, and its use in evidence as a numerical summary of the rarity of matching profiles. It is not clear that those who introduced the model would advocate its presentation in court. Our view is that this would be inappropriate. The independence assumptions which underpin it are known to be violated. Knowledge about the total number of 'positions' in the gel is limited. The error introduced by the simplifying assumption of constant population frequency for bands in different positions is difficult to assess, but could be non-conservative.

Acknowledgements

It is a pleasure to acknowledge helpful discussions with Drs. D.J. Balding, J.F.Y. Brookfield, M.W. Bruford, and R.A. Nichols, and to thank the referees of an earlier draft of the paper for their comments. I am also grateful to Mr. M.J. Davie for permission to refer to unpublished work and to Professor A.J. Jeffreys for clarification of issues relating to his analysis of multilocus profile data. This work was supported in part by SERC grants GR/F 98727, GR/G 11101, and B/AF 1255. Part of the work was completed while the author was visiting the Isaac Newton Institute for Mathematical Sciences, and the support and hospitality of the Institute is gratefully acknowledged.

References

Balding, D. & P. Donnelly, 1994. How convincing is DNA evidence. Nature 368: 285–286.

Balding, D. & P. Donnelly, 1995. Inference in forensic identification. J. Roy. Stat. Soc. A, In press.

Brinkman, B., S. Rand & P. Wiegand, 1991. Population and family data of RFLP's using selected single- and multi-locus systems. Legal Medicine 104: 81–86, 179.

Buffery, C., F. Burridge, M. Greenhalgh, S. Jones & G. Willot, 1991. Allele frequency distributions of four variable number tandem repeat (VNTR) loci in the London area. For. Sci. Int. 52: 53–64.

Cohen, J., 1990. DNA fingerprinting for forensic identification: potential effects on data interpretation of subpopulation heterogeneity and band number variability. Am. J. Hum. Genet. 46: 358–368.

Gill, P., J. Lygo, S. Fowler & D. Werrett, 1987. An evaluation of DNA fingerprinting for forensic purposes. Electrophoresis 8: 38–44.

Jeffreys, A., N. Royle, I. Patel, J. Armour, A. MacLeod, A. Collick, I. Gray, R. Neumann, M. Gibbs, M. Crosier, M. Hill, E. Signer & D. Monckton, 1991. Principles and recent advances in human DNA fingerprinting, in DNA Fingerprinting: Approaches and Applications, T. Burke, G. Dolf, A. Jeffreys & R. Wolff (eds.) Birkhauser Verlag, Basel, pp. 1–19.

Jeffreys, A., M. Turner & P. Debenham, 1991. The efficiency of multilocus DNA fingerprint probes for individualization and establishment of family relationships, determined from extensive casework. Am. J. Hum. Genet. 48: 824–840.

Jeffreys, A., V. Wilson & S. Thein, 1985. Individual-specific 'fingerprints' of human DNA. Nature 316: 76–79.

Jeffreys, A., V. Wilson, S. Thein, D. Weatherall & B. Ponder, 1986. DNA "fingerprints" and segregation analysis of multiple markers in human pedigrees. Am. J. Hum. Genet. 39: 11–24.

Roeder, K., 1994. DNA fingerprinting: a review of the controversy. Stat. Sci. 9: 222–278.

Editor's comments

The author continues the investigation of multilocus probes begun introduced by Curnow in this volume. He contrasts bandsharing estimates from Jeffreys *et al.* (1991) and the UK Forensic Science Service. Much of the difference is attributable to the greater resolving power of Jeffreys' gels (J.S. Buckleton, personal communication). Tests for independence of multilocus probe bands, and studies of the joint possession of subsets of bands within populations, were conducted by the UK Forensic Science Service, but not published (J.S. Buckleton, personal communication). In spite of the complexities of multilocus probe profiles, it is interesting to note that, just as when several single-locus probes are used, the same profiles are not found for unrelated people in large populations.

B. S. Weir (ed.), Human Identification: The Use of DNA Markers, 69–87, 1995.

Population genetics of short tandem repeat (STR) loci

Peter Gill & Ian Evett
Service Development, The Forensic Science Service, Priory House, Gooch Street North, Birmingham, B5 6QQ, UK

Received 25 August 1994 Accepted 5 October 1994

Key words: F-statistics, genetic distance, populations, short tandem repeat (STR)

Abstract

To investigate the population genetics of short tandem repeat (STR) polymorphisms in human populations, we have studied the allele frequency distributions of four STR loci (HUMTH01, HUMVWA31, HUMF13A1 and HUMFES) in 16 different population surveys which can be categorised within three broadly defined ethnic groups: Caucasian, Asian (Indian subcontinent), and African (Afro-Caribbean and US black). We have observed that allele frequency distributions of populations within ethnic groups are similar; consequently, genetic distances are an order of magnitude lower than between ethnic groups. Inbreeding coefficients (F-statistics) and calculations of the number of mean heterozygous loci per individual, along with estimates of variance, did not suggest that the populations were substructured. This included a study of an immigrant Asian population known to comprise at least three different sub-groups. Finally, an indication of the discriminating power is given by calculation of likelihood ratios (LR) of each individual tested across all four loci. Approximately 70% of Caucasians give an LR of greater than 10,000; the test is even more discriminating in Afro-Caribbeans – approximately 90% of tests are greater than 10,000.

Introduction

The analysis of polymorphic short tandem repeat (STR) sequences (Weber & May, 1989; Edwards *et al.*, 1991) by automated fluorescence (Ziegle *et al.*, 1992; Kimpton *et al.*, 1993; Fregeau & Fourney, 1993) is becoming increasingly important in genetic applications, particularly for gene mapping and identification purposes. For forensic applications, there are certain advantages in using STR analysis. Because the loci are comparatively small (<200 bp), they can be detected even when DNA is highly degraded; recently Gill *et al.* (1994) demonstrated that STRs could be reliably analysed from 70 year-old bones. Although there are fewer alleles present, compared to high molecular weight (>2 kB) hypervariable DNA, with automated fluorescence, alleles can be easily distinguished by incorporating an internal standard size marker in each electrophoretic track to enable accurate measurement of a DNA fragment to within 1 bp (Kimpton *et al.*, 1993). Multiplexing (i.e. carrying out several amplifications of loci in a one-tube reaction) is feasible with automated fluorescence. Furthermore, alleles of different loci can be detected even if they overlap in size, because the primers can be prepared with different dye-labels. Dinucleotide repeat sequences are not used because stutters cannot be avoided and this renders interpretation difficult in forensic applications where there may be mixtures of body fluids present. Hence we have considered only trimeric and tetrameric repeating sequences, in order to avoid this problem. Accordingly, we have developed a quadruplex consisting of the loci HUMTH01 (Edwards *et al.*, 1991; Polymeropolous *et al.*, 1991a), HUMVWA31 (Kimpton, Walton & Gill, 1992), HUMFES (Polymeropolous *et al.*, 1991b), and HUMF13A1 (Polymeropolous *et al.*, 1991c) for use in routine casework throughout the UK. All four loci have discrete alleles which can unambiguously be identified (Kimpton *et al.*, 1993); the utility and robustness of the system in forensic casework is described elsewhere (Kimpton *et al.*, 1994; Lygo *et al.*, 1994).

The purpose of this paper is to report on comparisons in frequency distributions between 16 different databases: seven Caucasian, five Indian Asian, and four Afro-Caribbean/US black. The comparisons are presented graphically by ethnic group and they are supplemented by measures of genetic distance. In general, distances within ethnic groups are small. Wright's F statistics are also presented for each of the three ethnic groups, demonstrating that F_{ST} values are of the order of 0.001.

Materials and methods

Populations

Samples were collected from various locations (Table 1), subdivided into ethnic groups as follows:

Caucasian
a) UK (1): A general Caucasian population derived from the UK – data collected by the Forensic Science Service, UK.
b) UK (2): A general Caucasian population derived from the UK – data collected by the Metropolitan Police Forensic Science Laboratory, London.
c) UK (Derbyshire): A local population collected from the county of Derbyshire and including 140 samples from Strathclyde.
d) US (1): Caucasian population (Foster City) – data from J. Robertson (ABI), USA.
e) US (2): Caucasian population (US Army) – data from D. Fisher (AFDIL), USA.
f) Germany: Caucasian population – data from M. Prinz and Hou Yiping (Cologne, Germany).
g) Sweden: data from S. Holgerson (Linkoping, Sweden).

Asian
a) UK: immigrant population of mixed ethnic origin from Birmingham.
b) Bangladeshi: population (from M. Webb, Cellmark Diagnostics, Abingdon, UK).
c) Hindu (Punjabi): population (from S. Mastana, Loughborough University, UK).
d) Sikh (Punjabi): population (from S. Mastana, Loughborough University, UK).
e) Chinese: data from M. Prinz (Cologne).

Black/Afro-Caribbean
a) UK, Afro-Caribbean (1): collected by the Forensic Science Service, UK.
b) UK, Afro-Caribbean (2): collected by the Metropolitan Police Forensic Science Laboratory, UK.
c) US, black (US Army): data from D. Fisher (AFDIL), Washington, USA.
d) Barbadian: collected by the Metropolitan Police Forensic Science Laboratory, UK.

Data derived from HUMVWA, HUMTH01, HUMF13A1 and HUMFES were analysed for all populations (Chinese and German databases were not analysed for HUMF13A1).

Preparation and analysis of DNA is described by Kimpton *et al.*, (1993). Nomenclature of STRs follows the recommendations of the DNA commission of the International Society for Forensic Haemogenetics (1992).

Statistical analysis

F statistics:
The following definitions apply:
Wright's F_{IT}: the overall inbreeding coefficient, i.e. the correlation of genes within individuals.
Wright's F_{ST}: the coancestry, i.e. the correlation of genes between individuals within populations.
Wright's F_{IS}: the average within population inbreeding coefficient, i.e. the correlation of genes within individuals within populations.

The methods of Reynolds, Weir and Cockerham (1983) and Weir (1990) were used to estimate F statistics. In addition, under a drift model, F_{ST} provides a natural measure of genetic distance, $D = -\ln(1 - F_{ST})$, that is proportional to time since divergence of the populations being compared (Reynolds, Weir & Cockerham, 1983).

Gene diversity was calculated by:

$$H = 1 - \sum p_i^2$$

where p_i = the allele frequency of the ith allele.

A test of random association of alleles at different loci was carried out by measuring the variance of the number of heterozygous loci per individual (Brown, Feldman & Nevo, 1980; Chakraborty, 1984); a test criterion for non-random association is given by the

Table 1. Summary of populations analysed for up to four STR loci (HUMVWA31, HUMTH01, HUMF13A1 and HUMFES). For each population the sample size; k, the number of different kinds of alleles observed; gene diversity and observed heterozygosity is recorded.

Ethnic group	Population	Locus	Sample size	number of alleles (k)	Gene diversity	observed heterozygosity
Caucasian	(UK1)	VWA	423	9	0.81	0.79
		TH01	423	5	0.78	0.77
		F13A1	423	12	0.73	0.72
		FES	423	6	0.69	0.70
Caucasian	(UK2)	VWA	245	7	0.80	0.78
		TH01	247	6	0.77	0.76
		F13A1	252	11	0.73	0.65
		FES	252	6	0.68	0.68
Caucasian	(UK) Derbyshire	VWA	724	8	0.81	0.82
		TH01	724	6	0.77	0.79
		F13A1	724	12	0.73	0.73
		FES	724	6	0.67	0.65
Caucasian	(US1)	VWA	80	7	0.81	0.84
		TH01	80	5	0.75	0.78
		F13A1	80	10	0.78	0.83
		FES	80	6	0.69	0.76
Caucasian	(US2)	VWA	103	7	0.80	0.83
		TH01	100	5	0.76	0.72
		F13A1	103	9	0.74	0.80
		FES	103	5	0.64	0.62
Caucasian	Sweden	VWA	52	9	0.82	0.77
		TH01	54	5	0.75	0.74
		F13A1	49	7	0.71	0.71
		FES	51	6	0.67	0.75
Caucasian	German	VWA	122	8	0.80	0.84
		TH01	122	6	0.78	0.74
		FES	123	6	0.68	0.72
Asian	Birmingham	VWA	175	8	0.80	0.81
		TH01	175	5	0.79	0.86
		F13A1	175	11	0.77	0.75
		FES	175	8	0.72	0.74
Asian	Punjabi Hindu	VWA	52	8	0.80	0.87
		TH01	52	6	0.73	0.65
		F13A1	52	9	0.75	0.77
		FES	52	6	0.69	0.58
Asian	Punjabi Sikh	VWA	42	7	0.79	0.71
		TH01	42	5	0.76	0.74
		F13A1	42	11	0.81	0.81
		FES	42	5	0.65	0.64
Asian	Bangladeshi	VWA	128	7	0.81	0.77
		TH01	128	5	0.78	0.75
		F13A1	128	12	0.78	0.83
		FES	128	6	0.71	0.67
Chinese		VWA	121	8	0.79	0.80
		TH01	121	5	0.68	0.70
		FES	121	7	0.67	0.69

Table 1. Continued

Ethnic group	Population	Locus	Sample size	number of alleles (k)	Gene diversity	observed heterozygosity
African	UK Afro-Caribbean (1)	VWA	195	10	0.82	0.80
		TH01	195	6	0.75	0.78
		F13A1	195	13	0.80	0.79
		FES	195	7	0.76	0.80
African	UK Afro-Caribbean (2)	VWA	175	9	0.81	0.77
		TH01	177	6	0.74	0.66
		F13A1	177	14	0.79	0.79
		FES	177	6	0.78	0.81
African	US African	VWA	41	7	0.81	0.85
		TH01	42	6	0.73	0.71
		F13A1	41	11	0.81	0.90
		FES	42	6	0.76	0.76
African	Barbados	VWA	178	9	0.81	0.81
		TH01	189	6	0.72	0.73
		F13A1	173	14	0.81	0.80
		FES	178	7	0.74	0.73

95% confidence interval, which was calculated for each population group.

Results

A compilation of sample sizes, gene diversity and number of alleles at each locus is given in Table 1. A compilation of allele frequencies for each of the populations is given in the appendix.

Allele frequencies

Allele frequencies were calculated from the genotypes collected from 16 different populations (Appendix). In all populations HUMVWA and HUMFES showed unimodal distributions, the highest frequencies occurring mid-way in the allelic range (allele 17 for HUMVWA and allele 11 for HUMFES). HUMF13A1 had the largest number of different alleles; the most common was identified as allele 5. Characteristically, those in the range of 8–17 repeats (Appendix) were rare in all populations examined. Alleles at the HUMTH01 locus were distributed bimodally; all alleles with the exception of a rare 11 repeat had frequencies greater than 5%. The most common alleles were different in all ethnic groups at this locus.

Genetic distances and F statistics

Distances (Table 2) between closely related ethnic groups such as Caucasians were low (≤ 0.002); similar values were obtained within the Asian group and the African group. Comparisons between ethnic groups revealed much greater divergences which were always an order of magnitude greater than within ethnic group comparisons.

F-statistics were estimated for each of the three major ethnic groups which included all of the populations listed in Table 1 with the exception of the Chinese population. F_{ST} was estimated between 0.002–0.003 and is the same order of magnitude as listed by Morton (1992); F_{IT} (the relationship between genes within individuals) is more variable, ranging between 0.005–0.011 and F_{IS} (the degree of inbreeding within populations) is between 0.004–0.008 (Table 3). This suggests that there is little difference between each value of F_{ST}, F_{IT} and F_{IS} between each ethnic group.

Non-random association among genetic loci

Mixtures of populations with different allele frequencies can produce non-random association of alleles at two or more unlinked loci or between alleles within one locus. Brown, Feldman and Nevo (1980) showed that the expectations of mean and variance of the number of heterozygous loci per individual can be calculated as functions of locus-specific heterozygosities

Table 2. Genetic distances ($-\ln(1 - F_{ST})$) calculated between populations. Populations are divided into three broad ethnic groups – Caucasian, Asian and Afro-Caribbean.

		Caucasian					Asian Indian Subcontinent				African			
Ethnic type	Population	UK (2)	Derbyshire	US(1)	US(2)	Sweden	UK	Hindu	Sikh	Bangladeshi	Afro-Caribbean (1)	Afro-Caribbean (2)	US black	Barbadian
White Caucasian	UK (1)	-0.0001	0.0001	0.0006	0.0029	0.0001	0.0114	0.0101	0.0157	0.0117	0.0163	0.0248	0.0163	0.0222
	UK (2)		0.0001	0.0002	0.0021	0.0001	0.0107	0.01	0.0139	0.0112	0.0151	0.0235	0.0161	0.0212
	UK Derbyshire			0.0006	0.0014	-0.0001	0.0111	0.009	0.0141	0.0116	0.0163	0.0247	0.0168	0.0215
	US(1)				0.0012	-0.00004	0.0091	0.0051	0.0123	0.0094	0.0132	0.0216	0.0135	0.019
	US(2)					-0.0003	0.0119	0.0068	0.0125	0.013	0.0161	0.0247	0.0177	0.0203
	Sweden						0.0138	0.0099	0.017	0.0127	0.0179	0.0274	0.0177	0.0229
Asian Indian Subcontinent	UK Asian (FSS)							0.0021	0.0015	0.0002	0.0056	0.0105	0.0122	0.0117
	Hindu								0.0015	0.0023	0.0119	0.0203	0.0183	0.0184
	Sikh									0.0012	0.0133	0.0189	0.022	0.0208
	Bangladeshi										0.0096	0.0155	0.0159	0.0163
African	Afro-Caribbean (1)											0.001	0.0006	0.0009
	Afro-Caribbean (2)												0.0008	0.0009
	US black													-0.0004

Table 3. F-Statistics calculated for three different ethnic groups comprising populations listed in Table 1.

Caucasian	F_{ST}	0.003
	F_{IT}	0.011
	F_{IS}	0.008
Asian	F_{ST}	0.002
(Indian Subcontinent)	F_{IT}	0.01
	F_{IS}	0.008
Afro-Caribbean	F_{ST}	0.002
US Black	F_{IT}	0.005
	F_{IS}	0.004

under the assumption of random association of alleles. In addition, 95% confidence intervals were calculated from the standard deviation of the variance of the number of heterozygous loci expected per individual (Chakraborty, 1981; 1984; Deka, Chakraborty & Ferrell, 1991). The observed variance was compared to the upper confidence interval. The results of these tests are summarised in Table 4.

The mean number of heterozygous loci in the quadruplex per individual varied between 2.87–3.2 for Asian and Caucasians (Table 4). The African populations had the highest mean heterozygosity per individual (3.03–3.56). There was no evidence to suggest that any of the samples have come from substructured populations, as variances were within the expected confidence intervals.

Likelihood ratio and evidential value

When a match is found between a crime sample and a suspect in forensic casework, it is necessary to derive an indicator of the evidential value. This is done by calculating the probability of a match given two competing hypotheses which are generally of the kind:

The crime sample was left by the defendant
The crime sample was left by an unknown person, unrelated to the defendant

The ratio of the two probabilities is called the likelihood ratio (LR).

When the typing system is essentially discrete as with STRs, and under the assumption of independence, the LR is simply the inverse of the relative frequency of the observed genotype in the relevant population. If the suspect and perpetrator of a crime are relatives then their genotypic frequencies are not independent and a different analysis, using the four gene theory described by Weir (1994), is appropriate. Furthermore, in the absence of information about a particular population sub-group, the same theory can be used to estimate conditional genotype frequencies from a database comprising samples derived from a general population of the ethnic group of interest. Four-gene measures are necessary even for low-level relatedness imposed by finite population size, even though the effect is trivial. In practice, F_{ST} can be used to estimate genotype frequency using the two gene theory proposed by Balding and Nichols (1994) although this ignores higher order measures accommodated by Weir (1994); the method is nevertheless a good approximation (B.S. Weir, pers. comm.). It follows that realistic estimates of F_{ST} are needed to make realistic predictions on the effect of population substructuring to forensic calculations.

Heterozygosity is one indicator of the forensic value of a typing system, because it is a measure of its discriminating power (Jones, 1972). Another is provided by the distribution of the LR to be expected in such cases for each of the three main ethnic groups. The horizontal axis of Figures 1 and 2 is scaled in terms of log (10) of the LR and the vertical scale is the percentage of those cases (in which the suspect is truly the offender) in which the log LR exceeds the value indicated on the horizontal axis by the appropriate curve. Thus, for example in Figure 1 it can be seen that in approximately 70% of cases involving Caucasians, the LR will exceed 10,000, whereas 90% of Afro-Caribbeans will exceed this figure; the Asian group appears intermediate. By contrast, comparison of populations within an ethnic category, e.g. Asians (Indian subcontinent), showed no obvious differences in the distributions of LRs (Fig. 2).

This experiment is incomplete without a consideration of the number of duplicates or chance matches observed between individuals within a population database. In Table 5, the match probabilities of each of the three ethnic groups are combined across four loci to give an overall match probability for the STR quadruplex. The number of matches between individuals (who are known to be different) were counted by pairwise comparisons as follows: the first individual is compared to the second, then the third and so on until the end of the database, then the second individual is compared to the third, fourth and fifth, continuing as before. Thus a database of n individuals comprises $n(n-1)/2$ comparisons. If individuals

Table 4. Mean number of heterozygous loci observed in each individual (out of four loci) and the variance. Expected distributions, variance and the confidence interval of the variance were calculated directly from heterozygosities using the theory of Chakraborty (1981). For all populations four loci were analysed; with the exception of the German and Chinese populations where only three were analysed.

Ethnic Group	Population	Mean no. of heterozygous loci observed	Observed Variance	95% confidence interval of variance	
				lower	upper
Caucasian	UK (1)	2.98	0.76	0.66	0.85
	UK (2)	2.89	0.73	0.67	0.92
	Derbyshire (UK)	3.10	0.72	0.67	0.81
	US (1)	3.20	0.67	0.44	0.84
	US (2)	2.97	0.70	0.54	0.93
	Sweden	2.91	0.92	0.46	1.07
	Germany	2.30	0.50	0.40	0.66
Asian	Birmingham	3.15	0.64	0.52	0.79
	Hindu	2.87	0.71	0.49	1.04
	Sikh	2.90	0.92	0.47	1.09
	Bangladeshi	3.02	0.67	0.56	0.90
	Chinese	2.20	0.56	0.45	0.71
African	Afro-Caribbean (1)	3.56	0.52	0.52	0.79
	Afro-Caribbean (2)	3.03	0.72	0.58	0.87
	US African	3.25	0.55	0.33	0.86
	Barbados	3.10	0.62	0.56	0.85

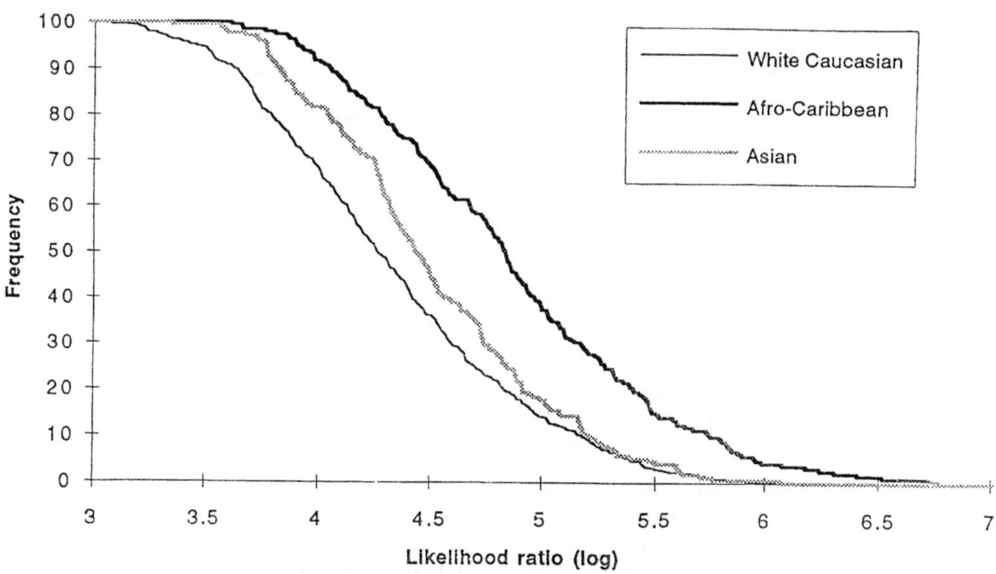

Fig. 1. Comparison of likelihood ratios between different ethnic groups. A likelihood ratio was calculated for each sample within the database. The percentage of samples exceeding any given value shown on the X axis is plotted, e.g. 70% of Caucasians give an LR > 10^4. The three ethnic groups compared are Afro-Caribbean (1), Caucasian (UK1) and Asian (Birmingham).

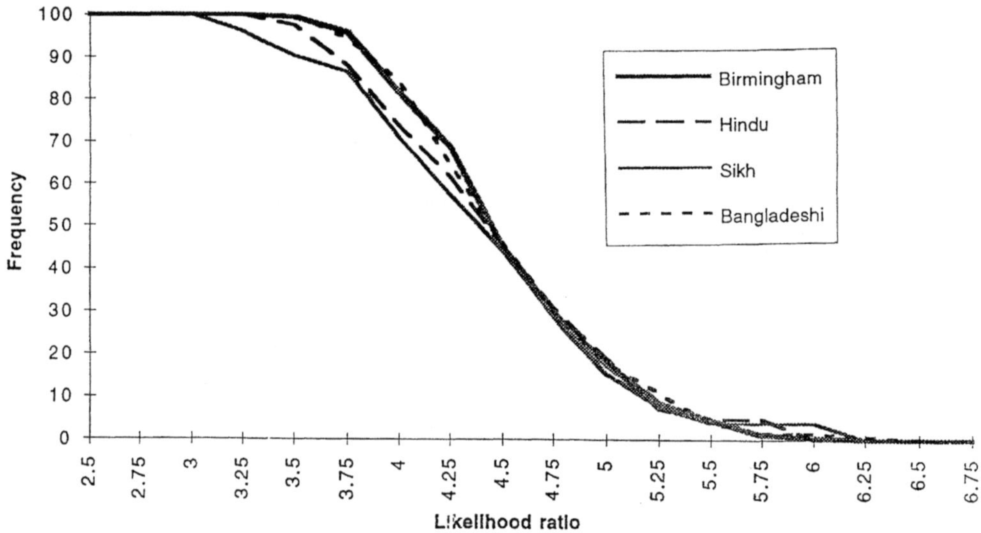

Fig. 2. Comparison of likelihood ratios of populations within the Asian ethnic group. A likelihood ratio was calculated for each sample within the database. The percentage of samples exceeding any given value shown on the X axis is plotted, e.g. 70% of samples give an LR > 10^4. The four different populations compared are Birmingham, Punjabi Hindu, Punjabi Sikh and Bangladeshi.

Table 5. Calculation of match probabilities across four loci for three ethnic groups and determination of expected v. observed chance matches by carrying out pairwise comparisons of population data.

	Match probability				Combined match	Number of duplicates in the database		
	HUMVWA31	HUMTH01	HUMF13A1	HUMFES	probability	expected	observed	number of pairwise comparisons
White Caucasian (UK1)	0.06	0.09	0.12	0.16	0.0001	9.14	13	89253
Asian (Birmingham)	0.08	0.08	0.09	0.13	0.00004	0.56	0	15225
Afro-Caribbean (1)	0.06	0.10	0.06	0.10	0.00007	1.38	0	18915

were found to match at four loci, further biochemical analysis was carried out using two additional STRs (data not shown); in every case, individuals were different when a fifth STR locus was considered. In Table 5 the number of expected matches derived from the match probability estimate is compared to the number of matches actually observed. Slightly more matches than expected were observed in the Caucasian group; in the remaining two groups no matches were observed. The power of this analysis is dependent upon the size of the database used, but nevertheless serves to demonstrate that the frequency of expected versus observed chance matches at four loci are similar to each other. A similar analysis demonstrating the efficacy of pairwise comparisons was carried out by Evett, Scranage and Pinchin (1993) using RFLP databases.

Discussion

Although differences were observed between ethnic groups (Asian (Indian subcontinent); Caucasian and African), allele frequency distributions were similar within respective ethnic groups. Gene diversities were greatest in Afro-Caribbean and US African populations; these populations originate from ancestral

Table 6. Comparison of mean number of heterozygous loci observed in each individual and the variance estimate for 14 loci (following the method of Chakraborty, 1981) using data described by Kimpton *et al.* (1993). Results from three different UK ethnic groups were analysed (55 Caucasians (UK1); 50 Afro-Caribbean (UK1) and 50 Asians (Birmingham)).

	Caucasian		Afro-Caribbean		Asian (Birmingham)	
	Expected	Observed	Expected	Observed	Expected	Observed
Mean no. of heterozygous loci	10.49	10.45	10.78	10.78	10.82	10.82
Variance	2.49	2.99	2.32	2.95	2.27	1.95
Lower confidence interval	1.57		1.42		1.41	
Upper confidence interval	3.39		3.21		3.14	

African populations and have high genetic admixture from other populations including Caucasians (Urquhart & Gill, 1993). Significant heterogeneity was not demonstrated within any of the populations studied. The Birmingham Asian population consists of individuals derived from different ethnic sub-groups (Haskey, 1991) including 'Indian,' Bangladeshis and Pakistanis in the approximate proportions (5:4:1); but precise details of the origin of these samples cannot be ascertained because of the considerable difficulty of obtaining this information from immigrant populations within the UK. Examination of genetic distances (Table 2) suggests a close affinity to Bangladeshis.

Calculations of the distribution of the number of heterozygous loci (Chakraborty, 1981) are more powerful when more loci are used. Inevitably, this means that fewer individuals can be analysed in view of the labor intensive nature of experimental analysis. Kimpton *et al.* (1993) recently compared 52 Asians from Birmingham with 55 Caucasians and 50 Afro-Caribbeans at 14 different STR loci. The mean numbers of heterozygous loci per individual (Table 6) were no different from each other, regardless of the ethnic group studied. Variances of the number of heterozygotes per individual were all within their expected 95% confidence intervals (i.e. there was no indication of excess homozygosity within any group).

Heterozygosity studies are complemented by calculation of F-statistics and genetic distances. Whereas Nichols and Balding (1991) and Balding and Nichols (1994) advocate the routine use of relatively high F_{ST} values (0.05), Morton (1992) suggests that nearly all forensic populations are less than 0.01, while Brookfield (1991) states that 0.05 is 'absurdly high' and 0.01–0.02 is more reasonable. Morton, Collins and Balazs (1993) also criticise use of 0.05 on the grounds that this level of inbreeding has only been observed in a few isolates and is unrealistic. Balding and Nichols (1994; page 129) also make an unconditional generalisation that there *a priori* reasons to suppose that the suspect (if innocent) is drawn from the same sub-population as the criminal. In contrast, Morton (1992) described three conditional models: 1) the suspect (if innocent) and perpetrator are drawn independently and randomly from a major population (either from different or the same sub-populations); 2) the suspect (if innocent) and perpetrator are related to the suspect as closely as a spouse would be (i.e. from the same sub-population); 3) the suspect (if innocent) and perpetrator are related. If marriages are not consanguineous then the first condition will usually be satisfied, especially in industrialised countries where mobility of individuals is commonplace and isolated or inbred communities are rare. Morton (1992) also suggests F_{IT} (approximately 1% in this study) should be substituted for F_{ST} in exceptional cases where the suspect (if innocent) and perpetrator are drawn from the same local (i.e. isolated) population. In addition, Morton, Collins and Balazs (1993) showed that F_{ST} calculated from hypervariable loci is an order of magnitude lower than F_{ST} derived from traditional blood groups and isozymes; this is because the higher mutation rate of the former sometimes exceeds 1% per gamete (Jeffreys *et al.*, 1988), resulting in increased gene diversity in populations, acting in the opposite direction to genetic drift. A consideration of mutation rates of STRs is therefore relevant to assessing the effect of kinship on gene diversity. Weber and Wong (1993) estimated the average mutation rate of 28 different STR loci to be approx 1.2×10^{-3} which compares favorably with Edwards *et al.*'s (1992) estimate of $< 4.3 \times 10^{-3}$; consequently, mutation rates of STR loci appear to be intermediate between hypervariable loci and blood groups. A realistic F_{ST} for all of the ethnic groups (which includes immigrant populations) in this study (Table 3) is 0.003. The effect of F_{ST} on genotype frequencies is greatest when alleles are rare. The number of alleles per STR locus in a population is much lower than alleles of hypervariable high molecular weight (VNTR) loci. This means that allele frequencies are correspondingly higher, and can reach 50%. An allele frequency of 5% would give an LR of 200 per heterozygous locus; given an F_{ST} of 0.003 this would reduce the LR to 178. Allele frequencies of 10% lead to an LR of 50, which is reduced to 47 (calculated using the formula of Balding & Nichols, 1994).

Within Asian populations, cousin marriages are more common than within Afro-Caribbeans and Caucasians, hence increased kinship would be expected in the former. However, this is not reflected in the F-statistics (Table 3); all three ethnic groups were similar. Because allele frequencies and gene diversities were similar within ethnic groups (Appendix) this means that there is little overall difference to the frequency estimate when results are combined across the four loci examined. Comparing the alternative use of two databases from an Asian sub-group, e.g. Sikh v. Bangladeshi, to determine the frequency of chance association of a given DNA profile does not give results which differ to any great extent (Figure 2). Larger differences were found between Afro-Caribbeans and Caucasians (Appendix) – comparison of a Caucasian with an Afro-Caribbean database could, for example, give a likelihood ratio an order of magnitude different (Fig. 1).

These results contradict the conclusions of Wall *et al.* (1993) who suggested significant variation between closely related ethnic groups (e.g. Gujarati and Pakistani). One of the loci used was HUMF13A1. However, the results at this locus have since been shown to be inaccurate (B. Parkin, pers. comm.). In addition, the variation observed at the CD4 locus in this study has yet to be verified.

The data presented in this paper show the effects of population substructuring to be minimal in the forensic context and that F_{ST} values in the region of .002–.003 are appropriate for most cases. The alternative suggestion by Balding and Nichols (1994; page 133) that 'migrant groups tend to be genetically distinct and differentiated from each other and a value of 5% may be reasonable' is not supported by this study.

Finally, a full analysis to test independence within and between loci has been carried out on the data presented in this paper and will be the subject of a separate publication.

Acknowledgements

The authors are grateful to the many individuals who have contributed either samples or data for analysis, including D. Fisher, M. Greenhalgh, S. Holgerson, S. Mastana, M. Prinz, J. Robertson and M. Webb. In addition, we are grateful to B.S. Weir, R. Chakraborty and N.E. Morton for helpful discussions and advice on the statistical analyses presented in this paper.

References

Balding, D.J. & R.A. Nichols, 1994. DNA profile match probability calculation: how to allow for population stratification, relatedness, database selection and single bands. Forens. Sci. Int. 64: 124–140.

Brookfield, J., 1991. The effect of population subdivision on estimates of the likelihood ratio in criminal cases using single-locus DNA probes. Heredity 69: 97–100.

Brown, A.H.D., M.W. Feldman & E. Nevo, 1980. Multilocus structure of natural populations of *Hordeum sponataneum*. Genetics 96: 523–536.

Cavalli-Sforza, L.L. & A. Piazza, 1993. Human Genomic Diversity in Europe: A summary of recent research & prospects for the future: Eur. J. Hum. Gen. 1: 3–18.

Chakraborty, R., 1981. The distribution of the number of heterozygous loci in an individual in natural populations. Genetics 98: 461–466.

Chakraborty, R., 1984. Detection of nonrandom association of alleles from the distribution of the number of heterozygous loci in a sample. Genetics 108: 719–731.

Chakraborty, R., R. Deka, L. Jin & R.E. Ferrell, 1992. Allele sharing at six VNTR loci and Genetic Distances among three ethnically defined human populations. Am. J. of Hum. Biology 4: 387–397.

Deka, R., R. Chakraborty & R.E. Ferrell, 1991. A population genetic study of six VNTR loci in three ethnically defined populations. Genomics 11: 83–92.

DNA recommendations. 1992 report concernng recommendations of the DNA Commission of the International Society for Forensic Haemogenetics relating to the use of PCR based polymorphisms. Int. J. Leg. Med. 105: 63–64.

Edwards, A., A. Civitello, H.A. Hammond & C.T. Caskey, 1991. DNA typing and genetic mapping with trimeric and tetrameric tandem repeats. Am. J. Hum. Genet. 49: 746–756.

Edwards, A., H.A. Hammond, L. Jin, C.T. Caskey & R. Chakraborty, 1992. Genetic variation at five trimeric and tetrameric tandem repeat loci in four human population groups. Genomics 12: 627–631.

Evett, I.W., J. Scranage & R. Pinchin, 1993. An illustration of the advantages of efficient statistical methods for RFLP analysis in forensic science. Am. J. Hum. Genet. 52: 498–505.

Fregeau, C.J. & R.M. Fourney, 1993. DNA typing with fluorescently tagged short tandem repeats: a sensitive and accurate approach to human identification. Biotechniques 15: 100–119.

Gill, P., P.L. Ivanov, C. Kimpton, R. Piercy, N. Benson, G. Tully, I Evett, E. Hagelberg & K. Sullivan, 1994. Identification of the remains of the Romanov family by DNA analysis. Nature Genetics 6: 130–135.

Haskey, J., 1991. The ethnic minority populations resident in private households – estimates by county and metropolitan district of England and Wales. Population Trends 63, HMSO: 22-35.

Jones, D.A., 1972. Blood samples: Probability of discrimination. J. Forens. Sci. Soc. 12: 355–359.

Jeffreys, A.J., N.J. Rouyle, V. Wilson & Z. Wong, 1988. Spontaneous mutation rates to new length alleles at tandem-repetitive hypervariable loci in human DNA. Nature 332: 278–281.

Kimpton, C.P., A. Walton & P. Gill, 1992. A further tetranucleotide repeat polymorphism in the vWF gene. Hum. Mol. Genet. 1: 287.

Kimpton, C.P., P. Gill, A. Walton, A. Urquhart, E.S. Millican & M. Adams, 1993. Automated DNA profiling employing 'multiplex' amplification of short tandem repeat loci. PCR Methods and Applications 3: 13–22.

Kimpton, C.P., D. Fisher, S. Watson, M. Adams, A. Urquhart, J.E. Lygo & P. Gill, 1994. Evaluation of an automated DNA profiling system employing multiplex amplification of four tetrameric STR loci. Int. J. Leg. Med. (in press).

Lygo, J.E., P.E. Johnson, D.J. Holdaway, S. Woodroffe, J.P. Whitaker, T.M. Clayton, C.P. Kimpton & P. Gill, 1994. The validation of short tandem repeat (STR) loci for the use in forensic casework. Int. J. Leg. Med. (in press).

Morton, M.E., 1992. Genetic structure of forensic populations. Proc. Natl. Acad. Sci. USA 89: 2556–2560.

Morton, N.E., A. Collins & I. Balazs, 1993. Kinship bioassay on hypervariable loci in Blacks and Caucasians. Proc. Natl. Acad. Sci. USA 90: 1892–1899.

Nichols, R.A. & D.J. Balding, 1991. Effects of population structure on DNA fingerprint analysis in forensic science. Heredity 66: 297–302.

Polymeropoulos, M.H., D.S. Rath, H. Xiao & C.R. Merril, 1991a. Tetranucleotide repeat polymorphism at the human tyrosine hydrolase gene (TH). Nucleic Acids Res. 19: 3753.

Polymeropoulos, M.H., D.S. Rath, H. Xiao & C.R. Merril, 1991b. Tetranucleotide repeat polymorphism at the human c-fes/fps proto-oncogene (FES). Nucleic Acids Res. 19: 4018.

Polymeropoulos, M.H., D.S. Rath, H. Xiao & C.R. Merril, 1991c. Tetranucleotide repeat polymorphism at the human coagulation factor XIII A subunit gene (F13A1). Nucleic Acids Res. 19: 4036.

Reynolds, J., B.S. Weir & C.C. Cockerham, 1983. Estimation of the coancestry coefficient: Basis for a short term genetic distance. Genetics 105: 767–779.

Urquhart, A. & P. Gill, 1993. Tandem-repeat internal mapping (TRIM) of the Involucrin Gene: repeat number and repeat-pattern polymorphism within a coding region in human populations. Am. J. Hum. Genet. 53: 279–286.

Wall, W.J., R. Williamson, M. Petrou, D. Papaioannou & B.H. Parkin, 1993. Variation of short tandem repeats within and between populations. Hum. Mol. Genet. 2: 1023–1029.

Weber, J.L. & P.E. May, 1989. Abundant class of human DNA polymorphism which can be typed using the polymerase chain reaction. Am. J. Hum. Genet. 44: 388–396.

Weber, J.I. & C. Wong, 1993. Mutation of human short tandem repeats. Hum. Mol. Genetics 2: 1123–1128.

Weir, B.S., 1990. Intraspecific variation, pp. 373–410 Molecular Systematics and Evolution, published by Sinauer Associates, Inc, Sunderland, Massachusetts, USA.

Weir, B.S., 1994. The effects of inbreeding on forensic calculations. Annual Review of Genetics 28: 597–621.

Ziegle, J.S., Y. Su, P. Kevin, L.N. Corcoran, P.E. Mayrand, L.B. Hoff, L.J. McBride, M.N. Kronick & S.R. Diehl, 1992. Application of automated DNA sizing technology for genotyping microsatellite loci. Genomics 14: 1026–1031.

Editor's comments

The authors present data generated by the move from VNTR to STR loci for human identification. The data they present for samples within major racial groupings address some of the concerns about population substructuring discussed by Balding and Nichols in this volume.

Appendix

Comparison between Caucasian (UK1) Afro-Caribbean (1) and Asian Birmingham populations (Figs. 1–4).

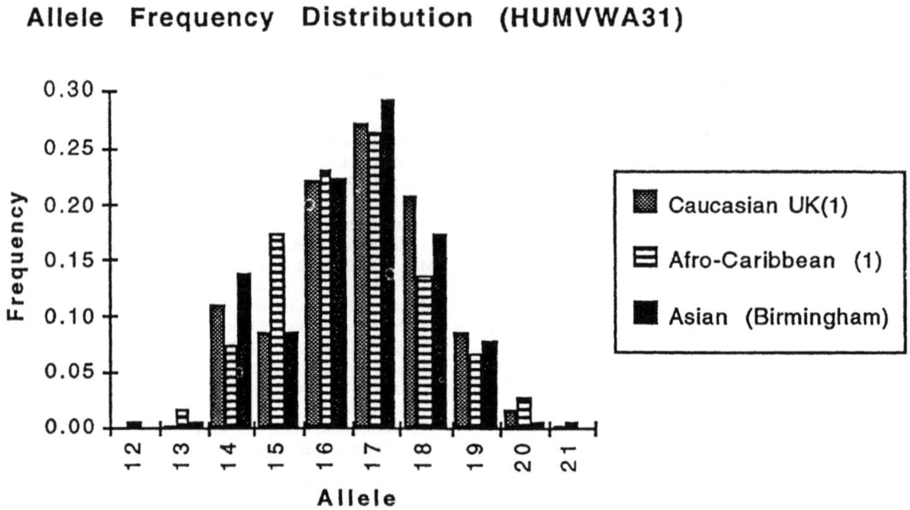

Fig. 1. HUMVWA31.

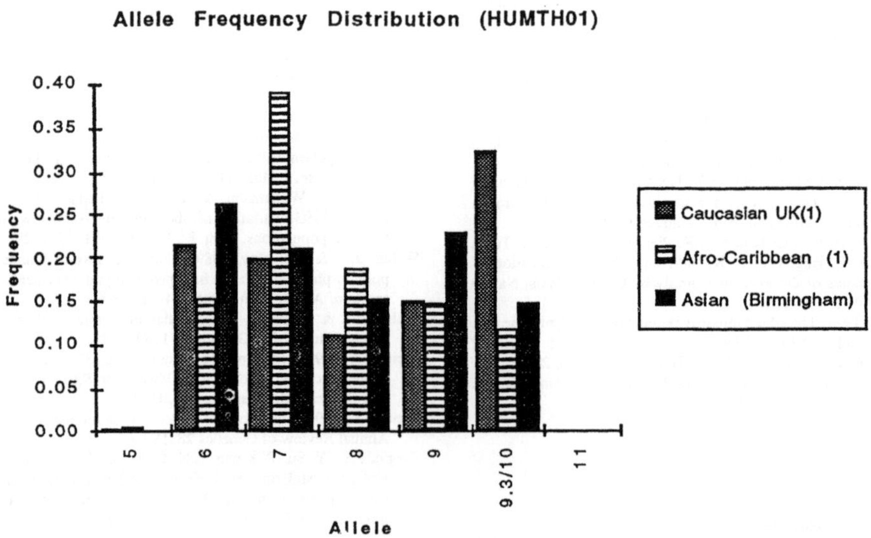

Fig. 2. HUMTH01.

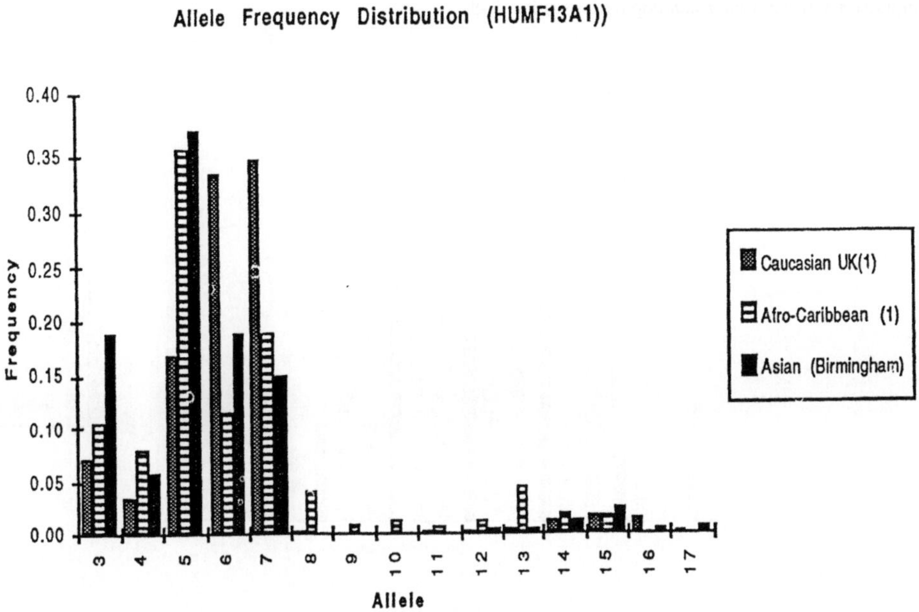

Fig. 3. HUMF13A1.

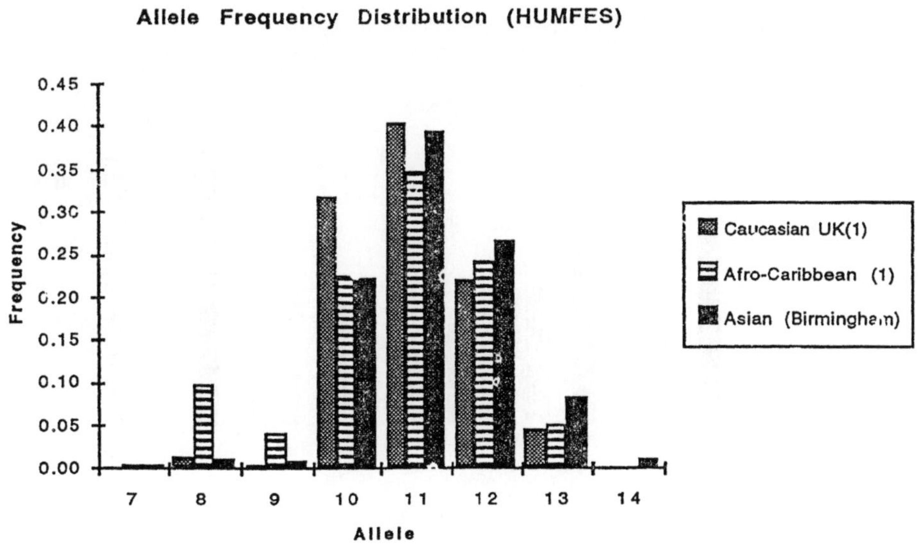

Fig. 4. HUMFES.

Comparison between Caucasian populations (Figs. 5–8).

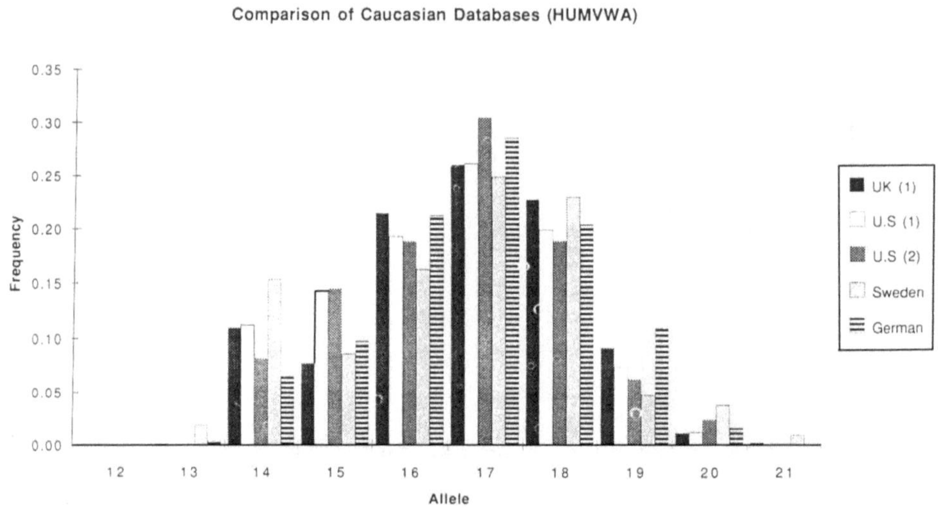

Fig. 5. HUMVWA31.

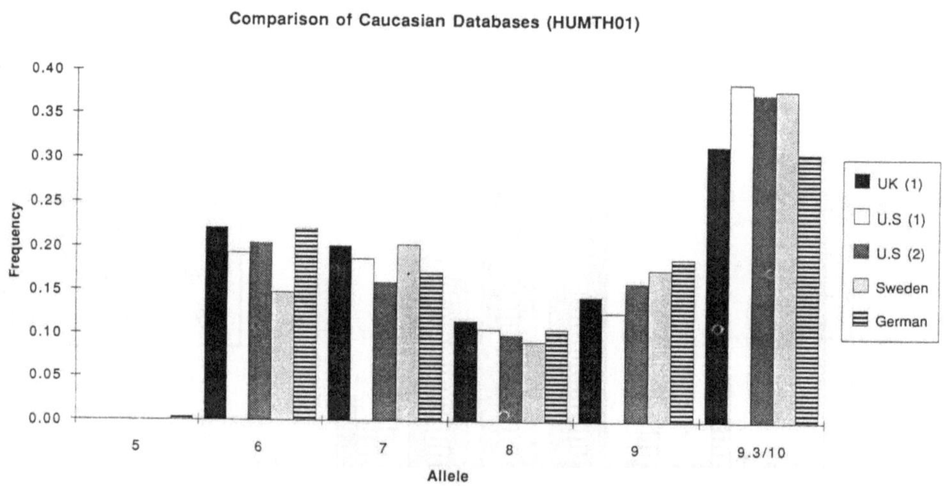

Fig. 6. HUMTH01.

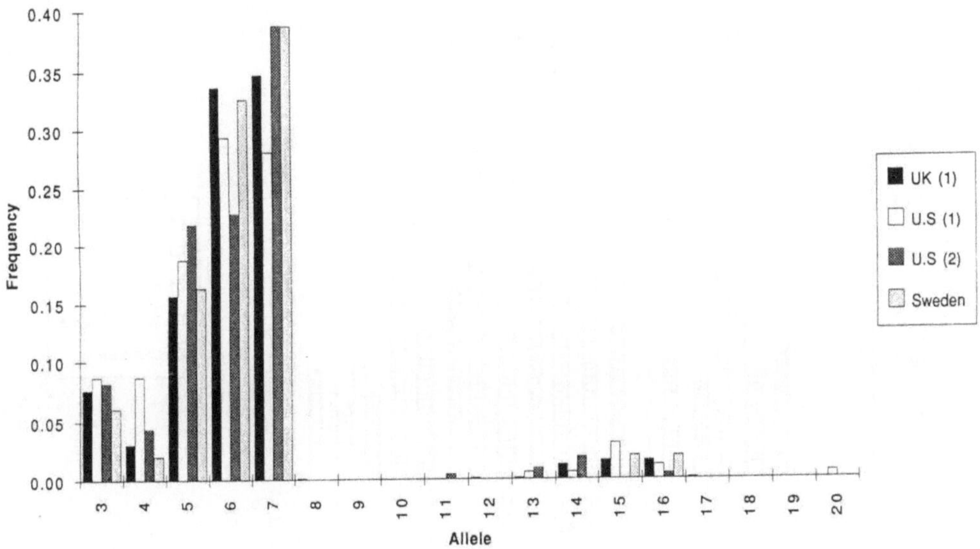

Fig. 7. HUMF13A1.

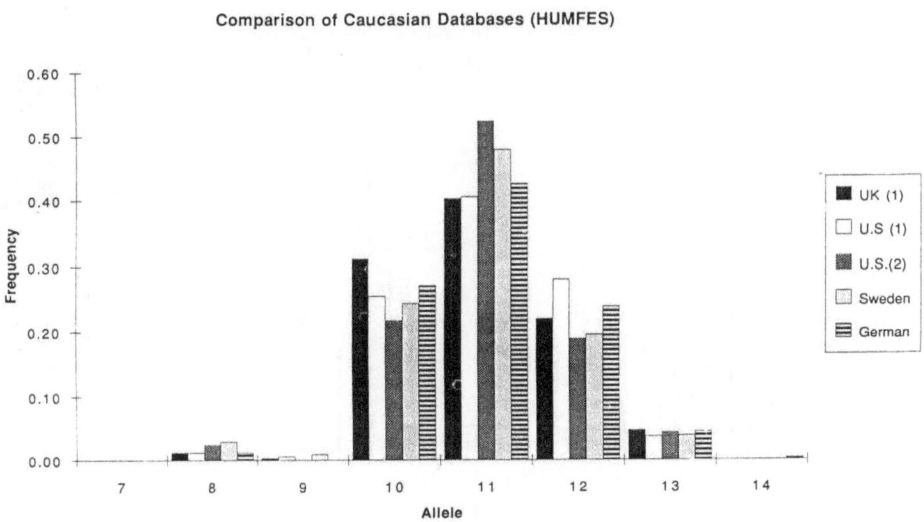

Fig. 8. HUMFES.

Comparison between Asian populations (Figs. 9–12).

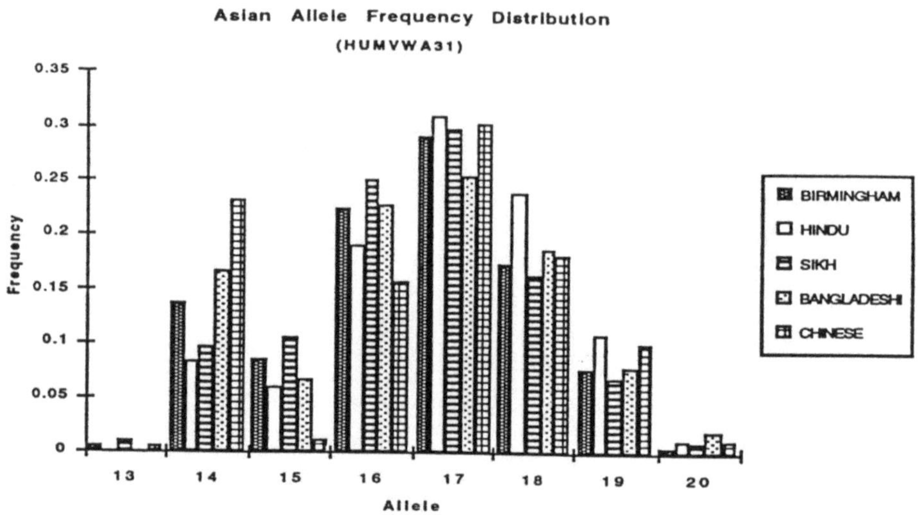

Fig. 9. HUMVWA31.

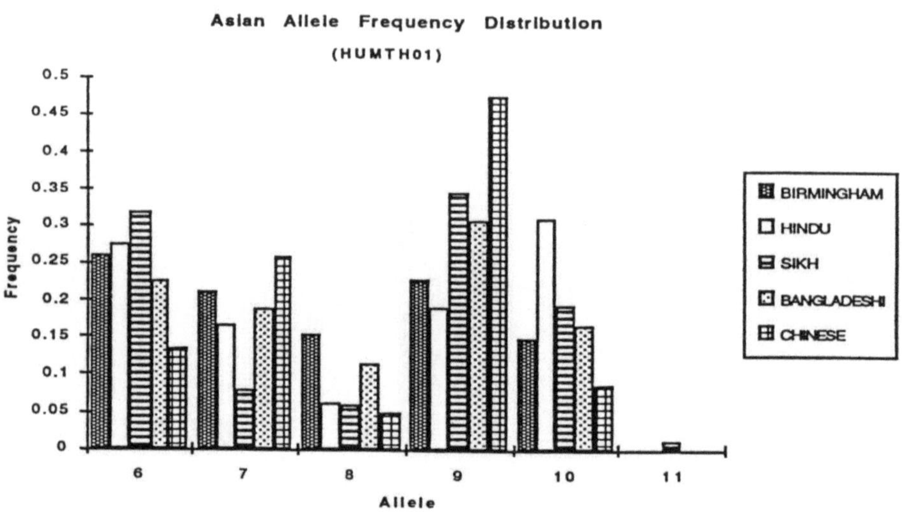

Fig. 10. HUMTH01.

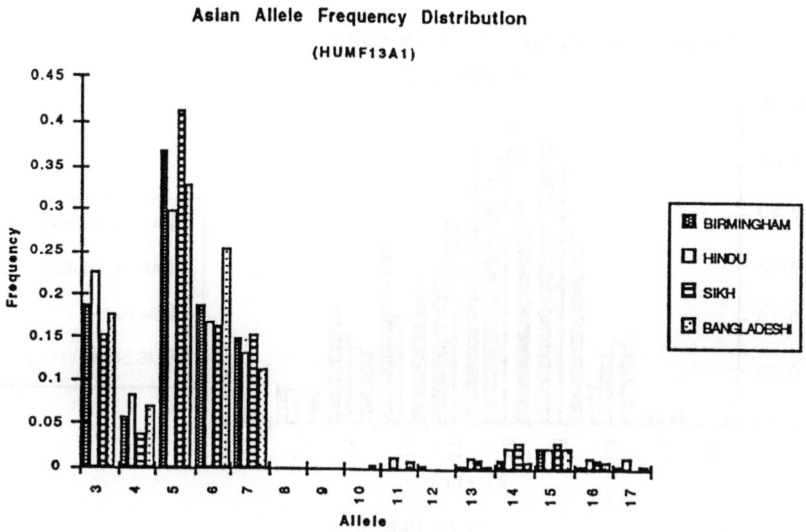

Fig. 11. HUMF13A1.

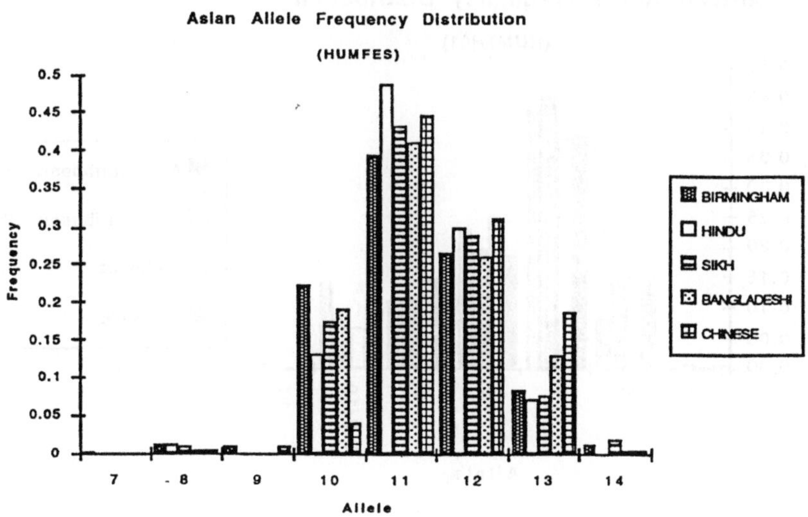

Fig. 12. HUMFES.

Comparison between Afro-Caribbean and US black populations (Figs. 13–16).

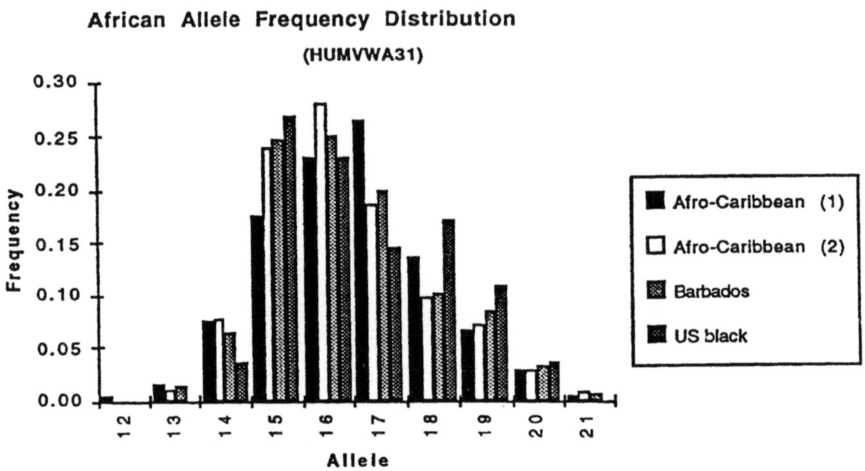

Fig. 13. HUMVWA31.

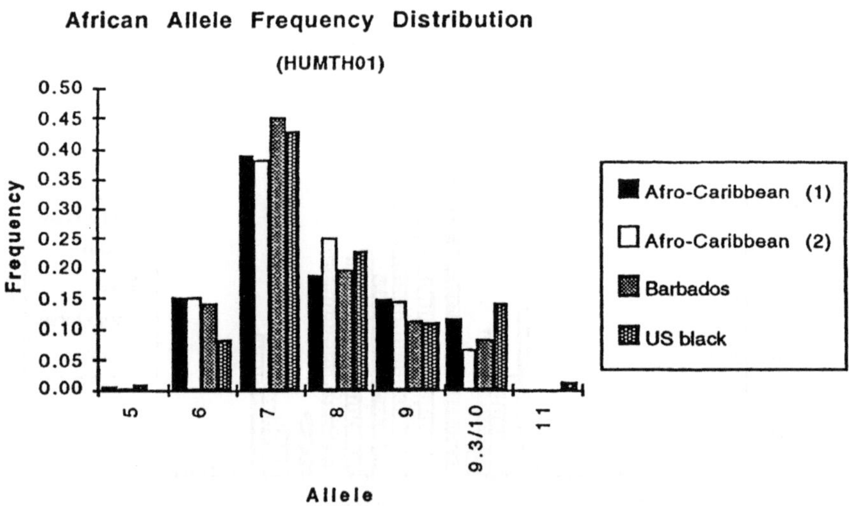

Fig. 14. HUMTH01.

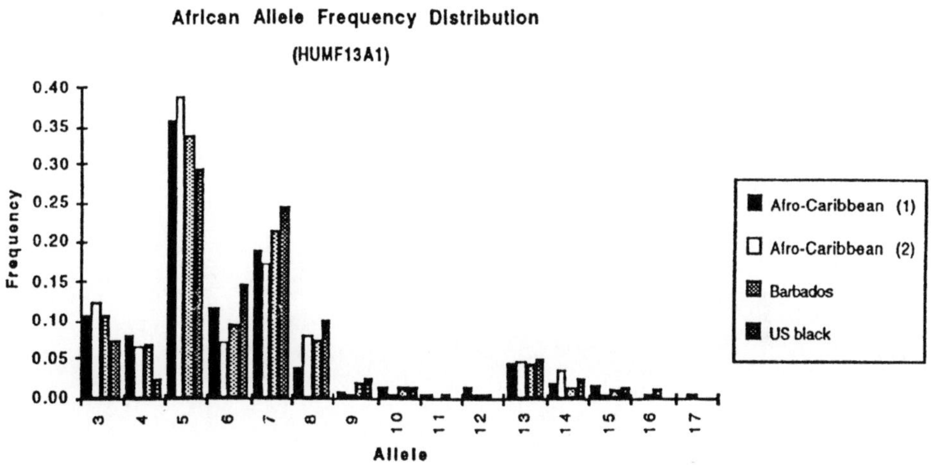

Fig. 15. HUMF13A1.

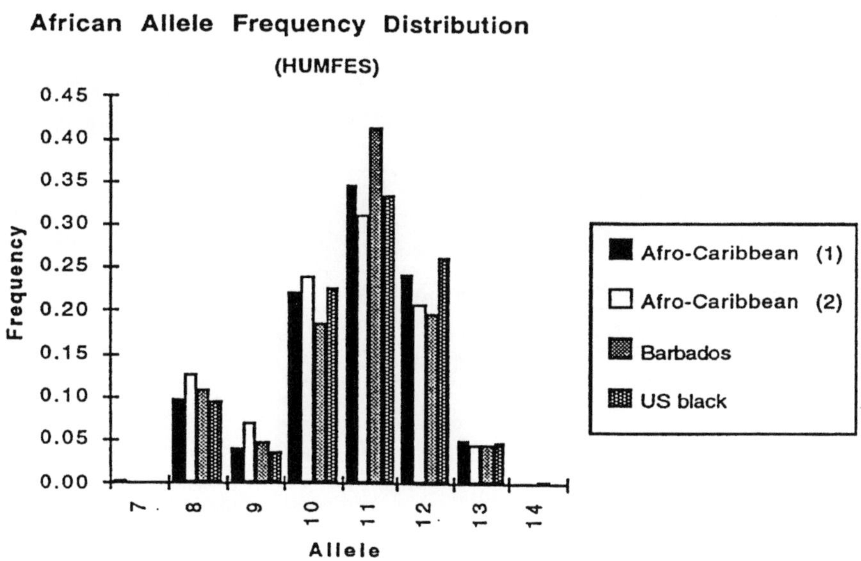

Fig. 16. HUMFES.

B. S. Weir (ed.), Human Identification: The Use of DNA Markers, 89–98, 1995.

Assessing probability of paternity and the product rule in DNA systems

David W. Gjertson[1,2] & Jeffrey W. Morris[2]*
[1] *UCLA Tissue Typing Laboratory, 950 Veteran Avenue, Los Angeles, CA 90024, USA*
[2] *Long Beach Genetics, 2384 East Pacifica Place, Rancho Dominguez, CA 90220, USA*
* *Author for correspondence: Long Beach Genetics, 2384 East Pacifica Place, Rancho Dominguez, CA 90220, USA*

Received 20 April 1994 Accepted 18 July 1994

Key words: DNA, probability of paternity, product rule

Abstract

The genetic resolution of paternity disputes begins with an intricate detection of inherited traits and finishes with a statistical inference (the probability of paternity, W). Notwithstanding some initial fanfare, statistical inference is a necessary component of DNA-based paternity tests because band patterns may be rare but not yet unique, and even rare events in a vacuum are meaningless. The genetic match must be combined with other evidence for relevancy, thus a Bayesian approach is preferred when computing W. This paper reviews the standard model used to compute W and discusses the model's various properties and assumptions. The standard model is extended to include DNA systems in which alleles are operationally continuous due to measurement error. This extension avoids problems associated with 'matched/non-matched' binned decisions. After outlining the model assumptions for a single DNA system, particular attention is given to the product rule – the procedure of multiplying intermediate probabilities across genetic loci to form a combined W. An empirical alternative to the product rule is also assessed and correlated with standard procedures.

Introduction

Any human trait that is fully expressed at birth, remains constant throughout life (i.e. is unaffected by age, disease, or other environmental conditions) and is under strict genetic control can be used to 'resolve' parenthood. Even before Mendel defined the gene, racial and other anthropological measures (e.g., body build) had been used to exclude paternity. For example, Archer (1810) reports observations that a white woman may conceive twins, one of which is white, the other mulatto, through intercourse with a white male and a black male. In general, however, unpredictable transmission and overt display (the allegation itself could be based on the trait) make such physical characteristics unreliable for resolving parentage disputes. Landsteiner's discovery of ABO agglutinogens started the modern incorporation of blood traits into the process of determining fatherhood. Today, a laboratory's repertoire of tests includes surface markers on red and white blood cells, red cell enzymes and serum proteins, as well as DNA polymorphisms.

Historically, likelihoods for genetic markers in pedigree and parentage studies were derived using discrete data methodologies employing multinomial distributions and Mendel's laws. In contrast to standard trait identification (e.g., ABO blood grouping), detection of restriction fragment length polymorphisms (RFLP), variable number of tandem repeats (VNTR) and other DNA loci by gel electrophoresis introduces a potentially sizable, continuous measurement error into an otherwise simple Mendelian system. Gjertson *et al.* (1988) formally incorporated allele measurement error into the Essen-Möller version of the probability of paternity (W). Other important discussions on the biostatistical evaluation of DNA parentage studies include: 1) adaptation of mixture lane experiments (Morris, Sanda & Glassberg, 1989); 2) estimation of VNTR distributions and probabilities in forensic identification and paternity cases (Devlin, Risch &

Roeder, 1991, 1992); 3) extended models for correlated measurement errors (Berry, 1991; Berry, Evett & Pinchin, 1992); and 4) multilocus minisatellite probe techniques (Evett, Werrett & Buckleton, 1989; Jeffreys, Turner & Debenham, 1991). Missing from the literature is a thorough review of the various assumptions underlying calculations in paternity cases. The main purpose of this report is to examine those assumptions.

In particular, several reports (Lander, 1989; Cohen, Lynch & Taylor, 1991; Lewontin & Hartl, 1991) have questioned one assumption – the *product rule* – the procedure of multiplying proportions of matching alleles across DNA loci to form a combined proportion of phenotype matches. Generally, application of the product rule hinges on the absence of any significant population substructure arising through inbreeding of group members within some defined population. Numerous position papers (e.g. Devlin, Risch & Roeder, 1990; Chakraborty & Kidd, 1991; Morton, 1992; Weir, 1992) have addressed this issue of population heterogeneity. (Further, several authors – Balding, Chakraborty, Clark, and Morton – have contributed chapters to this volume concerning the issue of population substructure.)

In forensic identification cases, four or more DNA loci are often tested, requiring an assumption of no significant effect of departure from eight-fold (two bands per locus), or greater, independence of bands. Directly validating this degree of independence which yields frequencies of one in millions or billions is practically impossible given the size of existing data bases. However, only two or three DNA systems are usually necessary to resolve disputed paternity suits under current evidentiary statutes in most of the United States (Kaye & Kanwischer, 1988). The task of directly validating two- or three-way independence (which usually leads to frequencies of only one in a few hundred) is possible given the size of several ethnic data bases. Thus, we also wish to investigate the validity of the product rule for disputed paternity following a suggestion of C. Brenner (pers. comm.) that product-rule free probabilities can be empirically obtained from large multilocus data bases.

The probability of paternity (*W*)

The standard model for W

Essen-Möller (1938) and his mathematical colleague Quensel (Essen-Möller & Quensel, 1939) devised a formula (generally known as the Essen-Möller formula) for the putative father-mother-child cases which enabled serological findings to be expressed numerically as a probability of paternity. They arrived at the relation $W = X/(X+Y)$ in which the sum of the probabilities of the hypotheses 'paternity' (X) and 'nonpaternity' (Y) equals one. A somewhat different approach was suggested by Gurtler (1956) to report the ratio $PI = X/Y$ as the paternity index with large values suggesting fatherhood.

Twenty-three years after Essen-Möller published his formula, Ihm (1961) showed that the formula can be derived from a straightforward application of Bayes' theorem using the Bayes' postulate of equal *a priori* probabilities of paternity. Define x, y and z as the observed phenotypes of the mother (MO), putative father (PF) and child (CH), respectively. Let $\alpha = (\alpha_1, \alpha_2)$ and $\beta = (\beta_1, \beta_2)$ be the true maternal and paternal genotypes. Further, let $\theta = (\theta_1, \theta_2)$ be the transmitted maternal (θ_1) and paternal (θ_2) haplotypes of the child. Define k to be the status indicator where $k = 1$ if the putative father is the true father of the child and $k = 0$ if the putative father is unrelated to the child. As defined, $W = P(k = 1|x, y, z)$. By Bayes' theorem,

$$W = \frac{P(x, y, z|k = 1)\pi_1}{P(x, y, z|k = 1)\pi_1 + P(x, y, z|k = 0)\pi_0};$$

in which $\pi_1 = P(k = 1)$ and $\pi_0 = P(k = 0)$ are the *a priori* probabilities for and against paternity. The probabilities of the phenotypes given status, $P(x, y, z|k)$, can be written as

$$P(x, y, z|k) = P(z|x, y, k)P(x, y|k).$$

Since the probabilities of the phenotypes of the mother and putative father are independent of status, $P(x, y|k)$ equals $P(x, y)$ and cancels from the expression for W. Let P_k equal $P(z|x, y, k)$. By intermediate conditioning,

$$P_k = \int_{\Theta} P(z|\Theta) d \left[\int_{\alpha,\beta} P(\Theta|\alpha, \beta) dH_k(\alpha, \beta|x, y) \right],$$

where dH_k represents the posterior distribution for maternal and paternal genotypes. Alternatively, since

θ, α, β are discrete,

$$P_k = \sum_{\Theta} f(z|\Theta)\left[\sum_{\alpha,\beta} A(\Theta|\alpha,\beta)P(\alpha,\beta|x,y,k)\right],$$

where $f(z|\theta)$ is the density function for the distribution of z given θ, A is a matrix of transmission probabilities and $P(\alpha,\beta|x,y,k)$ is the posterior probability of the genotypes.

First, it is assumed there is no ambiguity in the assignment of alleles from serological data. This implies that a genotype produces only one observable phenotype. Thus, $f(z|\theta) = 1$ if phenotype conforms to genotype and 0 otherwise. Elements of the A matrix are obtained by

$$P(\Theta|\alpha,\beta) = \begin{cases} 1 & \text{if mating type conforms,} \\ & \alpha_1 = \alpha_2, \beta_1 = \beta_2, \\ \frac{1}{2} & \text{if mating type conforms,} \\ & \alpha_1 = \alpha_2, \beta_1 \neq \beta_2 \text{ or} \\ & \alpha_1 \neq \alpha_2, \beta_1 = \beta_2, \\ \frac{1}{4} & \text{if mating type conforms,} \\ & \alpha_1 \neq \alpha_2, \beta_1 \neq \beta_2, \\ 0 & \text{if mating type nonconforming.} \end{cases}$$

The posterior probabilities $P(\alpha,\beta|x,y,k)$ can be obtained via the joint distributions of genotypes and phenotypes.

$$P(\alpha,\beta|x,y,k=1) = \\ = \frac{f(x|\alpha)f(y|\beta)dG(\alpha)dG(\beta)}{\int_\alpha\int_\beta f(x|\alpha)f(y|\beta)dG(\alpha)dG(\beta)},$$

and

$$P(\alpha,\beta|x,y,k=0) = \frac{f(x|\alpha)dG(\alpha)}{\int_\alpha f(x|\alpha)dG(\alpha)}dG(\beta),$$

where f is defined as above and $G(.)$ is the distribution function for α or β. It is not necessary that α and β share a common distribution, although it is assumed that the functional form is similar. However, the number and frequency of alleles may differ if for example mother and likely fathers are from different racial groups. Notice that when $k = 0$, the putative father (y) is not the true father (by definition); if he is also not related to the true father, then $P(\alpha,\beta|x,y,k=0) = P(\alpha,\beta|x,k=0)$. Finally, assuming equal *a priori* probabilities,

$$W = \frac{P_1}{P_1 + P_0}.$$

Also note that the paternity index $PI = P_1/P_0$. Since P_1 was derived from $P(x,y,z|k=1)$ and P_0 was derived from $P(x,y,z|k=0)$, PI is truly a likelihood ratio (Baur *et al.*, 1986), that is, the ratio of the likelihood of the observations conditioned on two mutually exclusive hypotheses.

The extended model for W to include DNA systems

For DNA polymorphisms absent mutation and silent genes, allelic uncertainty arises solely from determining length of enzyme-cleaved fragments from mobility in gel electrophoresis. Inherent in electrophoresis systems is a resolution limitation that causes bands to have shape and width, thus producing a continuous measurement error. To compensate for this uncertainty, one must create either statistically-based matching criteria or modeling assumptions, for incorporation into W. Following are highlights of the latter process as briefly summarized from Gjertson *et al.* (1988).

Actual true gene size plus measurement error constitutes the underlying model for an observed DNA allele (log scale) where the errors are assumed to be normally distributed with mean zero and variance σ^2. Symbolically, $x = \alpha + \epsilon_1$, $y = \beta + \epsilon_2$, and $z = \theta + \epsilon_3$, where ϵ's are distributed $N(0, \sigma^2)$.

A conditional component of P_k is H_k, the posterior distribution for α and β, given x and y. In order to derive H_k, a prior distribution (G) for α and β must be defined. To be consistent with the error measurement model, the prior distribution of, say, α' is represented as a mixture of a form,

$$dG(\alpha') = \frac{1}{\sigma_g^2}\left[\sum_i g_i\phi\{(\alpha_1' - \alpha_i)/\sigma_g\}\right] \times \left[\sum_j g_j\phi\{(\alpha_2' - \alpha_j)/\sigma_g\}\right],$$

where α_i and α_j $(i, j = 1, \ldots, T)$ are the empirically observed alleles, g = empirical allele frequency, σ_g is error associated with empirical determination, and ϕ represents the standard normal density function. We assumed constant variance (σ_g^2) on the log scale and independence of alleles which are thought to exhibit variability by rearrangement of internal DNA segments of constant or varying length (Wyman & White, 1980). Parameter estimates for such distributions are obtained by maximum likelihood via the EM algorithm. In addition, we 'smoothed' (reduced the number of mixture components) G, concluding that W values were relatively insensitive to changing distribution parameters as long as the number of components was large ($\geqslant 20$) (data not shown). Finally, we chose T such that $\hat{\sigma}_g \approx \hat{\sigma}/\sqrt{T}$ (analogous to sampling variability of the mean of T *iid* random variables).

Under these models, appropriate expressions for P_1 and P_0 are

$$P_1 = \frac{1}{16\pi\sigma^2} \times\{[\exp(-(z_1-x_1)^2/4\sigma^2) + \exp(-(z_1-x_2)^2/4\sigma^2)] \times[\exp(-(z_2-y_1)^2/4\sigma^2) + \exp(-(z_2-y_2)^2/4\sigma^2)] \times[\exp(-(z_1-y_1)^2/4\sigma^2) + \exp(-(z_1-y_2)^2/4\sigma^2)] \times[\exp(-(z_2-x_1)^2/4\sigma^2) + \exp(-(z_2-x_2)^2/4\sigma^2)]\},$$

and

$$P_0 = \frac{1}{4\sqrt{2}\pi\sigma(\sigma_g^2+\sigma^2)^{1/2}} \times\sum_t g_t\{[[\exp(-(z_1-x_1)^2/4\sigma^2) + \exp(-(z_1-x_2)^2/4\sigma^2)] \times\exp(-(z_2-\beta_t)^2/2(\sigma_g^2+\sigma^2))] + [\exp(-(z_1-\beta_t)^2/2(\sigma_g^2+\sigma^2)) \times[\exp(-(z_2-x_1)^2/4\sigma^2) + \exp(-(z_2-x_2)^2/4\sigma^2)]]\},$$

where $g_1, \ldots, g_T$ are the allele frequencies for the alleles $\beta_1, \ldots, \beta_T$. Heuristically, P_1 is a genetic constant (a quantity proportional to $\frac{1}{4}$ since DNA polymorphisms are single locus codominant systems) multiplied by terms representing the relative size differences due to measurement error between DNA fragments of the putative father, mother and child. Likewise, P_0 is a weighted sum of the mixture distribution over possible DNA components in which each term of the sum involves a genetic constant (proportional to the mixture component's weight, g_t) multiplied by quantities representing distances between measured bands for mother and child, and the mean of the particular Gaussian component.

Strictly speaking, P_1 and P_0 are density functions. However, since W depends only on the ratio P_1/P_0, the mean value theorem for integrals applies, thereby demonstrating W's appropriateness. Equations for P_1 and P_0 asymptotically approach the standard Essen-Möller calculation as σ and σ_g go to zero.

Properties and assumptions surrounding *W*

The 'tester' (laboratory) of a paternity dispute has only to perform procedures as accurately as possible in accordance with scientifically accepted principles to produce phenotypic data on the trio. The judge of the dispute wants to make a decision, based in part on these data. Accordingly, any measure which summarizes paternity should be examined in terms of decision theory. To use this theory, one must define a loss function that contrasts the true states of nature and possible actions. Here, loss represents the detriment to the putative father. (Note that loss could also represent a child's loss, the mother's loss or society's loss.) If he is falsely excluded, he loses the benefits derived from a father/child relationship (loss A); if he is falsely included, the putative father loses money in the form of child support payments he must make (loss B). Ihm (1981) showed that the Bayes' test (the rule which minimizes the average expected loss with respect to some prior distribution for k) has the form:

$$\text{reject } H_0 \text{ if } W > A/(A+B).$$

A whole subclass of Bayes' decision rules has been defined, depending on the cutoff value. Furthermore, it can be shown that these Bayes' rules form a complete class (Ferguson, 1967), implying that a search for the most desirable test can be restricted to this set. For a given prior distribution, one can select the best rule by specifying A and B. For example, $A/(A+B)$ can be set at 99%, consistent with criteria specified by the California Family Code, Section 7555. Thus, if the tested man mismatches or W is greater than 99% based on an initial battery of tests, the case is concluded and results are issued. Otherwise, further genetic tests are executed until either the criterion of Section 7555 is reached or all available genetic tests are exhausted. Attainment of this California threshold, $P_1/P_0 \geqslant 100$, transfers the burden of proof to the party denying paternity, and testing to this level results in settlement of more than 99% of cases without need for trial.

Several assumptions are necessary to compute W. These assumptions can be categorized as fundamental, specific, empirical and changeable. Fundamental claims imply correctness in laws of genetics and mathematics. Specific assumptions include (1) undisputed maternity, (2) accurately identified subjects, and

(3) true lab results. In practice, empirical observation determines $G(.)$; however, the error in sampling a population is usually ignored when computing W. Selvin (1983) has examined methods and effects of propagating this error into the standard calculation for W. Large uncertainties in gene and haplotype frequencies can produce substantial variation in W. By the invariant principle of maximum likelihood estimators, W values with optimal large sample properties (e.g., consistency) can be achieved using maximum likelihood estimates for gene and haplotype frequencies. Randomness of mating (where possible fathers are not related to the mother or to each other), race, and prior probability comprise the three changeable assumptions.

As a general rule, mating is random (Morris & Gjertson, 1994), but methods have been developed to handle exceptional cases. Goldgar and Thompson (1988) suggested an interesting variation of theme using Bayesian interval estimation of genetic relationships. (Also see an analogous report by Morris *et al.*, 1988.) In the Goldgar/Thompson method, k (now named the 'coefficient of relationship') is permitted to assume values from zero through one. For example, $k = 1$ is a direct relationship (i.e., the putative father is the father of the child), whereas $k = 0.5$ indicates a second-degree relationship such as the putative father being the brother of the father of the child. Bayesian methods for estimating kinship are derived.

Race is intended to define a population in the formulation of W. Gene frequencies may vary markedly from one race group to the next. When using DNA polymorphisms absent mutation and silent alleles, only the race of the alternative father influences W, as demonstrated by formulas for P_1 and P_0. In practice, a subject's race is assigned by interview and the alternative father's race is equated with that of the putative father such that transmission probabilities are computed for that racial group. Our casework provides some justification for this practice because of the strong racial concordance among the tested men when a mother names two or more alleged fathers for a child. More importantly, the court can weigh the physical characteristics of the child to challenge and change the race of the alternative father. In many cases, it is a useful practice to tabulate W stratified on a variety of race assumptions.

With respect to prior probability, the 'tester' lacks prior information except for the plaintiff's (usually the state or mother) accusation of paternity, and the defendant's (usually the tested man) denial of fatherhood. By convention and for illustrative purposes, π_1 is usually set at one-half. Should the case proceed to trial, the tier of fact needs to establish the prior probability. As with race, justifying the assignment of prior probability may be difficult, e.g., an adopted putative father with a fluctuating sperm count, or a mother who is a prostitute having intercourse with an unknown number of men. In such cases, it is appropriate to test sensitivity (robustness) in W by calculation under a variety of prior probability assumptions. Thus, graphs of W versus π_1 are informative and useful in the courtroom. Monitoring the distribution of W can justify certain prior probabilities for empirical validation studies (Mickey, Gjertson & Terasaki, 1986; Morris, 1989). For any given test battery, dividing the rate at which tested men are excluded by the rate at which nonfathers are excluded produces estimates of the mean prior probability of nonpaternity (π_0). Such estimates typically range from 0.2 to 0.3. (See Potthoff & Whittinghill, 1965; Hummel, Kundinger & Carl, 1981; Baur, Kittner & Wehner, 1981, for additional arguments in the assignment of prior probabilities.)

The 'DNA' extension of W incorporates new modeling assumptions, including additional parameters which must be empirically estimated. Modeling assumptions are (1) normal errors in measuring DNA alleles (log scale), (2) a normal mixture distribution for DNA alleles, and (3) a smoothing distribution (a bivariate normal for ordered responses) which replaces the mixture in the derivation of P_1 and in maternal components of P_0 (Gjertson *et al.*, 1988). Regarding the first assumption, a Gaussian function agrees with other published reports on measurement of DNA lengths by gel electrophoresis (Agard, Steinberg & Stroud, 1981; Elder *et al.*, 1983).

Since empirical study determines $G(.)$, smearing frequencies of observed DNA alleles with Gaussian weights is justifiable and consistent with the model. In our original article, σ_g, a parameter of $G(.)$, was set equal to σ. Here, the formula for P_0 preserves the distinction. Through consensus among several studies, the variability associated with the mean for an allele in the empirical distribution (σ_g) would be less than the variability about a single measurement (σ), analogous to sampling variability of the mean of independent random variables. Devlin, Risch and Roeder (1991) have developed empirical distributions explict for VNTR loci.

Finally, the exact form for the smoothing distribution is quite arbitrary with respect to P_k so long as its variance is much greater than σ^2. In Bayesian terms, sharp likelihoods will dominate posterior distributions derived using diffuse priors (see Box & Tiao, 1973). For example, suppose α is uniformly distributed; the formula for P_1 is still correct given a large range of alleles compared to σ^2.

The product rule

As mentioned previously, forensic use of the product rule in DNA blood stain cases has created great controversy. In paternity, no such controversy has ensued, although if multiple, independent genetic markers are used in the diagnosis, the resulting probability of paternity W_N (at $\pi_1 = \frac{1}{2}$) is given by a variation of the product rule,

$$W_N = \frac{\prod^N P_{1_n}}{\prod^N P_{1_n} + \prod^N P_{0_n}},$$

where $n = 1, \ldots, N$ indexes the individual genetic systems. (Alternatively, the genetic systems can be viewed as a series of independent sequential tests where the prior probability of paternity for system n in the series is equal to W for system $n-1$, and the prior at $n = 1$ is defined as π_1.) The ability to formulate W_N without the product rule is desired for two reasons. First, this ability would enable validation studies, and, second, a case specific, product-rule free W_N would eliminate objection by skeptics of validation studies.

We have examined a method (gamete count) to determine W_N across genetic loci without using the product rule (Morris & Gjertson, 1993). The method requires that genotype be directly determined from phenotype, so practical limits need be set on inherited bands. Based on empirical data, bands differing by more than 3% of length were considered not matched. By direct count, determine the proportion of individuals in the multilocus phenotype data base that match at all loci tested. For each of the randomly matching phenotypes, the relative 'fit' to the mother-child pair compared to the phenotype of the tested man is determined The average relative 'fit' gives the proper correction term which, when multiplied by the nonexclusion proportion, yields PI_N and therefore W_N. For example, suppose in a paternity case tested at three VNTR loci, only one of the tested man's eight equiprobable gametes matches the paternal marker at each locus, yielding 1/8 as the numerator of the 'fit' term. If the 43 of 6,146 matching nonfathers consist of 38 who match as well as the tested man and five who are apparent homozygotes at one of the three loci, the 'fit' denominator term would be $(1/43)(38 \times 1/8 + 5 \times 2/8)$, yielding a 'gamete count' PI_N of $6146/48 = 128$.

Comparisons of gamete count and product rule PI_N's for 358 2-locus (TBQ7/D10S28 and EFD52/D17S26) cases and 276 3-locus (YNH24/D2S44, TBQ7/D10S28 and EFD52/D17S26) cases (restriction enzyme: Hae III) are shown in Figures 1 and 2, respectively. PI_N's were obtained as described above from 2-locus (3-locus) data bases of sizes 6,146 (2,343) for Caucasians, 4,148 (1,601) for Hispanics, and 2,432 (748) for Blacks. One match count was added to two (0.5%) 2-locus cases and 73 (26%) 3-locus cases with zero direct observed matches. Testing was performed sequentially as described above, so that all 2- and 3-locus product rule PI_N's $\geqslant 100$.

The clouds of 2-locus gamete and product-rule PI_N points (Fig. 1) are closely packed about an $\sim 45°$ line regardless of race. All 95% confidence intervals for slope and intercept of the regression lines include one and zero, respectively. Squared correlation coefficients indicate that gamete count PI_N's explain 80% or better of the amount of variability in product rule PI_N's, the remaining 20% or so being divided between sampling variation and possible deviation from the product rule.

Also, strong positive correlations exist between gamete and product-rule 3-locus PI_N's among Caucasians and Hispanic cases (Fig. 2). Again, 95% confidence limits for slope and intercept equalled 0.96 to 1.24 and -1.38 to 0.33 for Caucasians and 0.81 to 1.08 and -0.52 to 1.15 for Hispanics, respectively. For Blacks, however, the 3-locus data base is substantially smaller (748) and the test battery more polymorphic (i.e. 3-locus product rule PI_N's are shifted to the right relative to those of other races) so that 56% of the 41 cases resulted in zero direct observation matches. Consequently, we could not test the relationship between Black gamete and product-rule PI_N's. Even though further study is indicated, each of the 41 3-locus PI_N's were associated with at most four directly observed matches out of 748, providing confidence for legal decisions based on product rule $PI_N \geqslant 100$.

Conclusions

In this report, we reviewed standard formulas for W and extended W for DNA systems with continuous

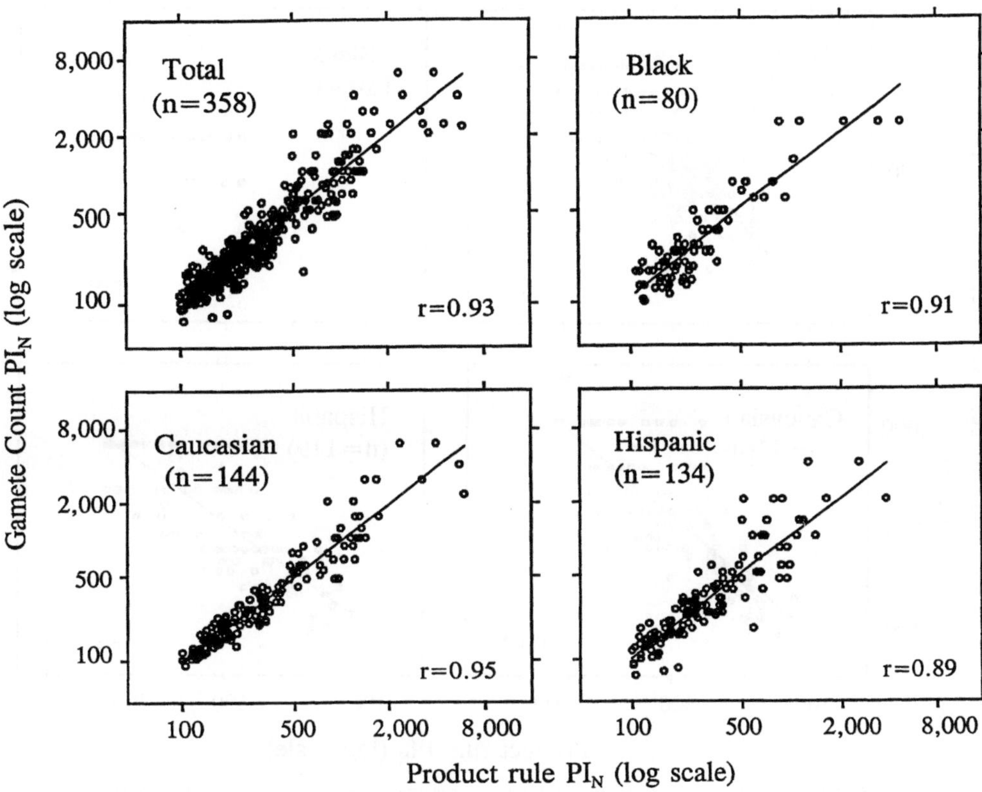

Fig. 1. Correlations of 2-locus (TBQ7 & EFD52) gamete count and product rule PI_N – total and stratified by race.

measurement errors. We discussed assumptions surrounding these models and presented supportive evidence for product rule PI's using three DNA probes among three racial groups. For completeness, we note that the formulas do not assume correlated measurement errors among an individual's two bands within a single probe. Positive correlations are typical in many DNA laboratories, usually as a result of band shifting (Berry, Evett & Pinchin, 1992). Under band shifting, if one band is measured too low, the other band in the same sample will also be measured too low. In the present study, not only were all bands translated according to bracketing DNA standards (Gjertson *et al.*, 1990) to remove uniform band shifts, but measured lengths represented an average of two technicians's digitizations. In the past, both practices have tended to dilute correlations among measurably heterozygous individuals (e.g., r values ranging from 0.2 to 0.5 compared with 0.9 reported by Berry, Evett & Pinchin, 1992). Here, a third practice was used whereby a putative father/child coelectrophoresis lane was analyzed as a separate experiment to remove the effects of residual correlations. Due to Dykes (1987), the motivation behind coelectrophoresis of DNA from the putative father and child is that, commonly, 'DNA is DNA'. If the putative father is the true father of the child, then, regardless of gel impurities or band shifting, the paternal DNA fragments (one from each of the putative father and child) should co-migrate as one band. Otherwise, if the putative father is a nonfather and the DNA test excludes him, then the child's paternal band should be distinguishable from bands of the putative father.

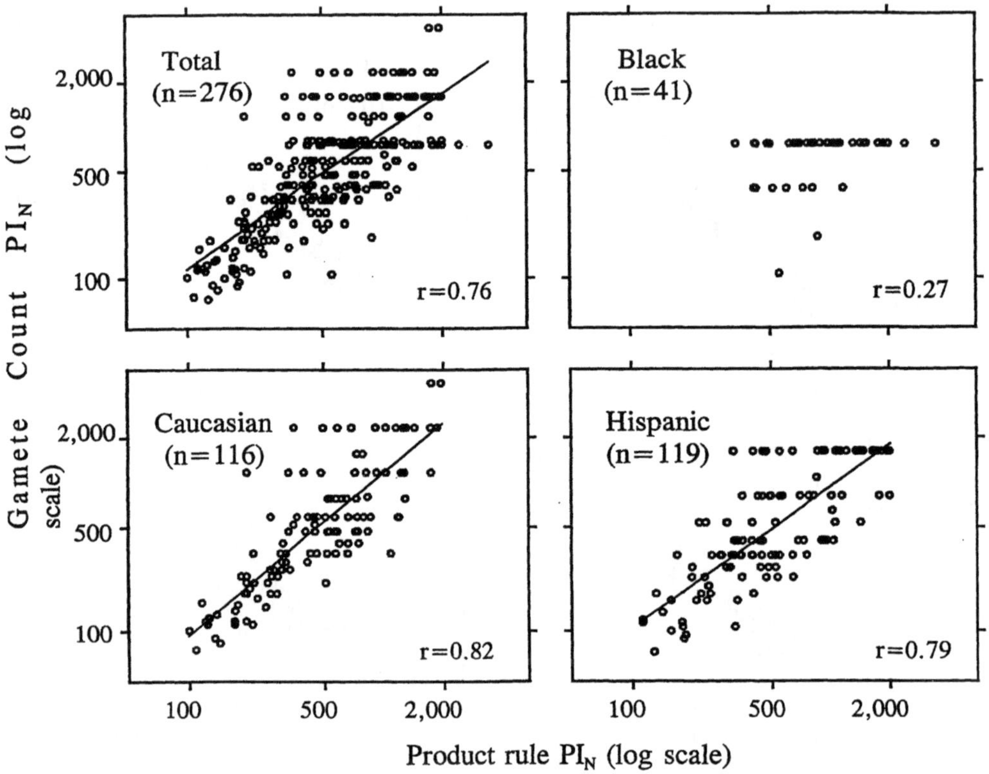

Fig. 2. Correlations of 3-locus (YNH24, TBQ7 & EFD52) gamete count and product rule PI_N – total and stratified by race.

None of our W models have directly included mutation or silent allele mechanisms. A single point mutation substituting one DNA base pair may completely eliminate an enzyme restriction site, producing a 'rogue' gene capable of passing to offspring. On the other hand, insertion/deletion anomalies may produce only small disturbances in DNA fragment sizes. Martin (1983) and Mickey *et al.* (1983) have examined effects of silent genes and mutational crossovers in the statistical evaluation of blood group findings in paternity testing. Silent alleles may be easily accounted for by incorporation into the distribution function for alleles. Allowance for mutation can be incorporated into W by altering elements of the matrix of transmission probabilities of maternal and paternal genes to offspring. In practice, mutations are rare (American Association of Blood Banks, AABB, Standards require that DNA loci used for paternity testing have low mutation rates: typically they are less than 0.1%) and much more likely to falsely exclude true fathers than to falsely include nonfathers. Thus, explicitly accounting for mutations when a match is observed is unnecessary. Further, AABB Standards require at least two DNA-loci non-matches in order to report a case exclusion, so that false exclusions due mutation would require two or more rare events.

References

Agard, D.A., R.A. Steinberg & R.M. Stroud, 1981. Quantitative analysis of electrophoretograms: a mathematical approach to super-resolution. Anal. Biochem. 111: 257–268.

Archer, F., 1810. Facts illustrating a disease peculiar to the female children of Negro slaves. Med. Reposit. 1: 319–323.

Baur, M.P., C. Rittner & H.D. Wehner, 1981. The prior probability parameter in paternity testing. Its relevance and estimation by maximum likelihood, pp. 389–392 in Lectures of the Ninth International Congress of the Society for Forensic Hemogenetics. Bern.

Baur, M.P., R.C. Elston, H. Gurtler, K. Henningsen, K. Hummel, H. Matsumoto, W. Mayr, J.W. Morris, L. Niejenhuis, H. Polesky, D. Salmon, J. Valentin & R. Walker, 1986. No fallacies in the formulation of the paternity index. Am. J. Hum. Genet. 39: 528–536.

Berry, D.A., 1991. Inferences using DNA profiling in forensic identification and paternity cases. Stat. Science 6: 175–205.

Berry, D.A., I.W. Evett & R. Pinchin, 1992. Statistical inference in crime investigations using deoxyribonucleic acid profiling. Appl. Statist. 41: 499–531.

Box, G.E.P. & G.C. Tiao, 1973. Bayesian inference in statistical analysis. Addison-Wesley, Reading, MA.

Chakraborty, R. & K. Kidd, 1991. The utility of DNA typing in forensic work. Science 254: 1735–1739.

Cohen, J.E., M. Lynch & C.E. Taylor, 1991. Forensic DNA tests and Hardy-Weinberg equilibrium. Science 253: 1037–1038.

Devlin, B., N. Risch & Roeder, 1990. No excess of homozygosity at loci used for DNA fingerprinting. Science 249: 1416–1420.

Devlin, B., N. Risch & K. Roeder, 1991. Estimation of allele frequencies for VNTR loci. Am. J. Hum. Genet. 48: 662–676.

Devlin, B., N. Risch & K. Roeder, 1992. Forensic inference from DNA fingerprints. J. Am. Stat. Assoc. 87: 337–350.

Dykes, D.D., 1987. Parentage testing using restriction fragment length polymorphisms (RFLPs), pp. 59–85 in Clinical Application of Genetic Engineering, edited by L.C. Lasky and J.M. Edwards-Moulds. American Association of Blood Banks, Arlington, Virginia.

Elder, J.K., A. Amos, E.M. Southern & G.A. Shippey, 1983. Measurement of DNA length by gel electrophoresis. I. Improved accuracy of mobility measurements using a digital microdensitometer and computer processing. Anal. Biochem. 128: 223–226.

Essen-Möller, E., 1938. Die Beweiskraft der Ahnlichkeit im Vaterschaftsnachweis – theoretische Grundlagen. Mitt. Anthrop. Ges. (Wien) 68: 9–53.

Essen-Möller, E. & C.E. Quensel, 1939. Zur Theories des vaterschaftsnachweises Aufgrund von Ahnlichkeitsbefunden. Z. Ges. Gerichtl. Med. 31: 70–96.

Evett, I.W., D.J. Werrett & J.S. Buckleton, 1989. Paternity calculations from DNA multilocus profiles. J. Forens. Sci. Soc. 29: 249–254.

Ferguson, T.S., 1967. Mathematical statistics, a decision theoretic approach. Academic Press, Inc., New York.

Gjertson, D.W., M.R. Mickey, J. Hopfield, T. Takenouchi & P.I. Terasaki, 1988. Calculation of probability of paternity using DNA sequences. Am. J. Hum. Genet. 43: 860–869.

Gjertson, D.W., J. Hopfield, P.A. Lachenbruch, M.R. Mickey, T. Sublett, C. Yuge & P.I. Terasaki, 1990. Measurement error in determination of band size for highly polymorphic single-locus DNA markers, pp. 3–11 in Advances in Forensic Haemogenetics 3, edited by H.F. Polesky & W.R. Mayr. Springer-Verlag, Heidelberg.

Goldgar, D.E. & E.A. Thompson, 1988. Bayesian interval estimation of genetic relationships: application to paternity testing. Am. J. Hum. Genet. 42: 135–142.

Gurtler, H., 1956. Principles of blood group statistical evaluation of paternity cases at the University Institute of Forensic Medicine Copenhagen. Acta. Med. Leg. Soc. (Liege) 9: 83–94.

Hummel, K., O. Kundinger & A. Carl, 1981. The realistic prior probability from blood group findings for cases involving one or more men. Part II. Determining the realistic prior probability in one-man cases (forensic cases) in Freiburg, Munich, East Berlin, Austria, Switzerland, Denmark, and Sweden, pp. 81–87 in Biomathematical Evidence of Paternity, edited by K. Hummel & J. Gerchow. Springer-Verlag, Berlin.

Ihm, P., 1961. Die mathematischen Grundlagen, vor allem fur die statistische Auswertung des serologischen und antropologischen Gutachtens, pp. 128–145 in Die Medizinische Vaterschaftsbegutachtung mit biostatistischem Beweis, edited by K. Hummel. Fischer, Stuttgart.

Ihm, P., 1981. The problem of paternity in the light of decision theory, pp. 53–68 in Biomathematical Evidence of Paternity, edited by K. Hummel & J. Gerchow. Springer-Verlag, Berlin.

Jeffreys, A.J., M. Turner & P. Debenham, 1991. The efficiency of multilocus DNA fingerprint probes for individualization and establishment of family relationships, determined from extensive casework. Am. J. Hum. Genet. 48: 824–840.

Kaye, D.H. & R. Kanwischer, 1988. Admissibility of genetic testing in paternity litigation: a survey of state statutes. Fam. Law Q. 22: 109–116.

Lander, E.S., 1989. DNA fingerprinting on trial. Nature 339: 501–505.

Lewontin, R.C. & D.L. Hartl, 1991. Population genetics in forensic DNA typing. Science 254: 1745–1750.

Martin, W., 1983. Consideration of 'silent genes' in the statistical evaluation of blood group findings in paternity testing, pp. 245–257 in Inclusion Probabilities in Parentage Testing, edited by R.H. Walker. American Association of Blood Banks, Arlington, Virginia.

Mickey, M.R., J. Tiwari, J. Bond, D. Gjertson & P.I. Terasaki, 1983. Paternity probability calculations for mixed races, pp. 325–347 in Inclusion Probabilities in Parentage Testing, edited by R.H. Walker. American Association of Blood Banks, Arlington, Virginia.

Mickey, M.R., D.W. Gjertson & P.I. Terasaki, 1986. Empirical validation of the Essen-Möller probability of paternity. Am. J. Hum. Genet. 39: 123–132.

Morris, J.W., 1989. Experimental validation of paternity probability. (Letter to the Editor.) Transfusion 29: 281.

Morris, J.W., R.A. Garber, J. d'Autremont & C.H. Brenner, 1988. The avuncular index and the incest index, pp. 607–611 in Advances in Forensic Haemogenetics 1. Springer-Verlag, Berlin.

Morris, J.W. & D.W. Gjertson, 1993. The paternity index, population heterogeneity, and the product rule, in press in Advances in Forensic Haemogenetics 5. Springer-Verlag, Berlin.

Morris, J.W. & D.W. Gjertson, 1994. Population genetics issues in disputed parentage, pp. 63–66 in Proceedings from The Fourth International Symposium on Human Identification 1993. Promega Corporation, Madison, Wisconsin.

Morris, J.W., A.I. Sanda & J. Glassberg, 1989. Biostatistical evaluation of evidence from continuous allele frequency distribution deoxyribonucleic acid (DNA) probes in reference to disputed paternity and disputed identity. J. Forens. Sci. 34: 1311–1317.

Morton, N.E., 1992. Genetic structure of forensic populations. Proc. Natl. Acad. Sci. USA 89: 2556–2560.

Potthoff, R.F. & M. Whittinghill, 1965. Maximum-likelihood estimation of the proportion of nonpaternity. Am. J. Hum. Genet. 17: 480–494.

Selvin, S., 1983. Some statistical properties of the paternity ratio, pp. 77–88 in Inclusion Probabilities in Parentage Testing, edited by R.H. Walker. American Association of Blood Banks, Arlington, Virginia.

Weir, B.S., 1992. Population genetics in the forensic DNA debate. Proc. Natl. Acad. Sci. USA 89: 11654–11659.

Wyman, A.R. & R. White, 1980. A hihgly polymorphic locus in human DNA. Proc. Natl. Acad. Sci. 77: 6754–6758.

Editor's comments

The authors present a carefully reasoned account of Bayesian inference in the paternity context. Their presentation of an alternative to the product rule for Paternity Index values offers a means of avoiding objections to the rule.

B. S. Weir (ed.), Human Identification: The Use of DNA Markers, 99–105, 1995.

The forensic debut of the NRC's DNA report: population structure, ceiling frequencies and the need for numbers

D.H. Kaye
College of Law, Arizona State University, Box 877906, Tempe AZ 85287-7906, USA

Received 16 July 1994 Accepted 10 August 1994

Key words: forensic DNA, ceiling frequencies, NRC report, courts

Abstract

This paper reviews judicial opinions that have discussed the April 1992 recommendations of a committee of the U.S. National Research Council concerning the statistics of forensic DNA profiles obtained with single-locus VNTR probes. It observes that a few courts have held 'ceiling frequencies' (as opposed to less 'conservative' estimates) admissible, but that the implications of the scientific criticisms of the ceiling procedures have yet to be addressed adequately in court opinions. It urges courts to distinguish between policy judgments and scientific assessments in both the NRC report and the scientific literature, and to defer less to the former than to the latter.

Introduction

In deciding whether the results of DNA tests are admissible in criminal cases, courts in the United States have relied on the testimony of scientists about the state of the science, on the scientific literature, and on other judicial decisions and opinions. These inputs – and the resulting caselaw – have not been static. Since 1988, a fusillade of objections to the admission of DNA tests has been raised in court. These include questions about the possible effects of contaminants on forensic samples, the use of radioactive isotopes in exposing X-ray film, the addition of ethidium bromide, *ad hoc* corrections for band shifting, the records of laboratories on proficiency tests, the criteria and procedures for deciding whether two VNTR fragments 'match', the size and sources of databases used to assess the significance of matching bands, and the method of calculating the frequency of matching DNA patterns within a reference population. Three phases in the judicial responses to DNA evidence can be discerned: an initial period of enthusiastic acceptance of the evidence, a damping counterreaction, and the current period of mixed outcomes and opinions expressing grave concern over certain aspects of DNA evidence. (For reviews of these legal developments, see Kaye, 1993; Thompson, 1993).

In April 1992, a committee of the National Research Council (NRC, 1992) entered this free fire zone. This paper examines the courts' use of the NRC report in their opinions. I confine the discussion to one issue – the statistics of DNA 'matches' with single-locus VNTR probes – for the committee's views on this topic have been the most controversial. I do not purport to show that the report has *caused* the courts to reach any particular results; it is possible, and often probable, that the outcomes of cases in which the report is cited would have been the same had no report been written. Nor do I assess the contribution the report has made to clarifying the issues, resolving doubts about other matters, and prompting reforms in forensic work. I merely seek to describe how one major part of the report has been treated in opinions and to offer some suggestions to avoid its being misunderstood or misused.

Proving the existence of a controversy

In some jurisdictions, the lack of general acceptance within the scientific community precludes admission of scientific evidence. In others, it weighs against admissibility but is not invariably fatal (see, for example, McCormick, 1992, § 203). Pretrial hearings lasting months and generating thousands of pages of testimony have addressed the question of scientific consensus, and courts frequently cite the NRC report to support a finding that a major scientific controversy is raging over the proper method for ascertaining the frequency of a 'match'.

Prior to the NRC report, the standard practice was to estimate multi-locus frequencies by multiplying single-locus frequencies, which in turn, were obtained by multiplying the two 'allele' frequencies at each locus. The allele frequencies were derived from histograms of fragment weights within a database. Under the protocol developed by the Federal Bureau of Investigation, the histograms involved very large class intervals to ensure that the single-locus frequencies would not be underestimated. The similarly sized fragments falling into each large 'bin', which are analogous to an allele, may be called a 'binelle'; the set of binelles, which is analogous to a genotype, may be called a 'binotype' (compare Devlin, Risch & Roeder 1992). The computations, in other words, presupposed Hardy-Weinberg equilibrium and linkage equilibrium: if the estimated binelle frequencies of a heterozygote at the jth locus are p_j and q_j, then the estimated binotype frequency P for a match at n loci is

$$P = 2^n \prod_{j=1}^{n} p_j q_j$$

In court, the most potent criticism of this simple calculation[1] has been the possibility of population structure – the presence of subgroups, with varying binelle frequencies, that tend to mate among themselves. One exposition of this criticism often cited in judicial opinions is Lewontin and Hartl (1991). The critics argue that, until direct studies of subpopulations are completed, there is no way to be certain that the departures from equilibrium do not make the standard frequency estimates too small (or for that matter, too large) by several orders of magnitude. Other population geneticists and statisticians maintain that direct studies of subpopulations are unnecessary and unlikely to be productive (see, for example, Devlin, Risch & Roeder, 1994; Roeder, 1994).

This debate is not easy for the courts to penetrate. In *People v. Pizarro*, 12 Cal. Rptr. 436, 456 (Ct. App. 1992), for instance, a California court of appeals quoted at length from early scientific publications, and lamented:

> The difficulty is, where does this place us? It places us in the middle of the conflict as to whether or not the basic theory of population genetics involving broad racial and ethnic groups as opposed to the argument of substructure has any general acceptance in the relevant scientific community – a conflict which we cannot resolve on the present record.

The NRC report can only have reinforced the perception of dissension. If there was any doubt left in early 1992 that prominent scientists were divided over the adequacy of the forensic computations, the NRC committee put it to rest. Starting with People v. Barney, 10 Cal. Rptr. 2d 731, 741 (Ct. App. 1992), court after court has noted the committee's report of 'considerable dispute' and a 'substantial controversy'. A New Mexico appellate court was especially impressed:

> The report discusses the debate over the need for subpopulation databases, and concludes that they indeed are necessary. This report is indicative of the absence of general acceptance. There is not just one author trying to make a point, but rather a group of people that has reached a consensus in rejecting one aspect of the current methods of forensic use of DNA evidence.

State v. Anderson, 853 P.2d 135, 146 (N.M. Ct. App. 1993).

Indeed, a few judges have reacted to the NRC report like sharks scenting blood. Despite the report's endorsement of the principles underlying VNTR studies, these judges perceive vulnerability without pinpointing its location or extent. In *State v. Futch*, 860 P.2d 264 (Or. Ct. App. 1993), prosecution and defense experts presented binotype frequencies ranging from 10^{-10} to 10^{-5}, and a majority of the appellate court concluded that such disagreements were matters for the jury to sort out. But a dissenting judge, citing only the first page of the report's summary, insisted that:

> The National Research Council, an equally august body [as the Office of Technology Assessment], reports that important questions have been raised about the reliability and validity of forensic RFLP. . . . [T]he technique has not yet achieved general acceptance.

[1] Some courts thinks that P can be computed even more simply, as the product of all the binelle frequencies. State v. Cauthron, 846 P.2d 502, 513 (Wash. 1993).

In short, the most common use of the NRC report has been to support the perception of a schism within the field of population genetics on the validity of using equation (1) with binelle frequencies estimated from racial databases. However, the disagreements may be more subtle than both the NRC report and the court opinions suggest. The NRC committee did not come down in favor of either side in the debate on the forensic implications of population structure. Assuming arguendo that population structure is a problem, it offered a compromise. In jumping to a solution without fully examining the problem, the report does not address all the implications of its convenient assumption that the racial populations are sharply structured. In many cases, the reference population in which to estimate the binotype frequency is a general population, like Caucasians or African-Americans, rather than any special ethnic subpopulations. As Chakraborty *et al.* (1992) and Kaye (1993) have observed, even if binelle frequencies differ among the subpopulations, using equation (1) with binelle frequencies estimated without regard to substructure typically results in overestimates of the true general population frequency.[2] Since the appropriate reference population in most cases is a general population, the controversy over possible departures from equilibrium is less troubling than it might first appear.[3]

Hitting the ceiling

The *Anderson* court's claim that the NRC committee 'reached a consensus in rejecting' the use of equation (1) with binelle frequencies estimated from racial databases flies in the face of the report itself and news accounts of the committee and its deliberations. The *Barney* court was far closer to the truth when it acknowledged (10 Cal. Rptr. 2d at 741) that '[t]he report does not, however, choose sides in the debate'; it merely 'assume[s] for the sake of discussion that population substructure may exist'. Unable to agree that population structure leads to dramatic errors in estimating VNTR binotype frequencies, the committee 'decided to assume that population substructure might exist' and to propound one particularly 'conservative' solution to this hypothesized problem – the notorious 'ceiling principle' (NRC committee, 1992, pp. 80, 94).

The ceiling procedure seeks one plausible upper bound on the binotype frequency – both in the population and any of its subpopulations. Like the standard method, it uses equation (1); however, instead of taking actual binelle frequencies from the appropriate reference population, it substitutes for each such binelle frequency in (1) the largest frequency seen in *any* subpopulation or population. (This is merely the rough idea – the method as presented in the NRC report is more involved; even so, it retains ambiguities that have been exploited in litigation. For criticisms of the general approach and the NRC committee's implementation, see Devlin, Risch and Roeder (1993, 1994); Morton (1992); Weir (1992, 1993a, 1993b). For rejoinders, see Lander (1993); Lempert (1993).)

Courts have responded to the proposal to use ceiling frequencies in several ways. Some have cited the proposal as a basis for excluding the standard calculations. *Commonwealth v. Lanigan*, 596 N.E.2d 311, 316 (Mass. 1992), is one of the first such cases. There, the Supreme Judicial Court of Massachusetts noted that the prosecution's failure to follow the 'considered, conservative' prescription of the NRC committee 'underscored the wisdom of . . . excluding the test evidence'.

Other courts have intimated that ceiling frequencies may be admissible. On the basis of the NRC report and other materials, the Colorado Supreme Court in *Fishback v. People*, 851 P.2d 884 (Colo. 1993), took notice of the 'substantial controversy' about population structure but sidestepped the issue on the theory that since the controversy materialized well after the trial court admitted standard calculations, that court did not err. At the same time, the Supreme Court went out of its way to suggest that 'this dispute may be rendered essentially moot if [the 'ceiling principle' is] determined to be generally accepted'.

An appellate court in Illinois, agreeing with a lower court that general scientific acceptance of the standard method was lacking, nevertheless directed the trial court to consider admitting ceiling estimates. The appellate court pointedly explained in *People v. Watson*, 629 N.E. 2d 634 (Ill. App. 1994), that '[a]t least in our view, the NRC Report, which was not previously available to the trial court, suggests that the DNA evidence should be admitted on the basis of this more conservative probability calculation for which the requisite consensus may now exist'. The same disposition, accompanied by a virtually identical observation

---

[2] This result assumes that the database from which binelle frequencies are taken is representative of the population and its subpopulations and that each subpopulation is in equilibrium.

[3] For a legal argument for admissibility based on this idea, see Kaye (1993).

concerning the NRC report, can be found in an earlier case, *United States v. Porter*, 618 A.2d 629 (D.C. App. 1992).

Vermont's Supreme Court responded to a conviction in *State v. Vandebogart*, 616 A.2d 483, 494 (Vt. 1992), involving DNA evidence similarly. It too remanded the case to the trial court, which it said 'must conduct a hearing in order to determine whether the NRC's recommended ceiling principle has gained general acceptance in the relevant scientific community . . '.

So, too, some Justices of the Massachusetts Supreme Judicial Court are contemplating the admissibility of ceiling frequencies. In *Commonwealth v. Daggett*, 622 N.E.2d 272, 277 (Mass. 1993), a plurality opinion stated that '[i]f the NRC's conclusion regarding the ability of the 'ceiling frequency' approach to correct for the errors that might result from the use of a general population database reflects that of the general scientific community, a numerical assessment concerning the likelihood that a given DNA match is false which relies on a general population database could be admissible if the 'ceiling frequency' method of calculation is employed by the expert'.

The Supreme Court of Washington was less cautious in its endorsement of ceiling frequencies. The court was so 'encouraged' by the NRC report that it commented in *State v. Cauthron*, 846 P.2d 502 (Wash. 1993), that '[a]lthough we lack the scientific expertise to either assess or explain the methodology, its adoption by the Committee indicates sufficient acceptance within the scientific community . . .'.

On remand, the trial court in *Cauthron* was so impressed with these remarks that it decided that the issue of general acceptance of the interim ceiling principle was settled. Consequently, it ordered a hearing on whether Cellmark adhered to this procedure, and ruled that 'the statistical evidence in a DNA case is no longer subject to a *Frye* hearing if it is founded on the NRC approved Interim Ceiling Principle'. State v. Cauthron, No. 88-1-01253-3 (Super. Ct. Snohomish County Aug. 20, 1993) (Report of Proceedings).

Still, such pronouncements are dicta – they are not essential parts of the opinions and are not strictly binding in later cases. A handful of appellate courts, however, have held ceiling frequencies admissible. In *State v. Alt*, 504 N.W.2d 38 (Minn. Ct. App. 1993), the Minnesota Court of Appeals held that 'the statistical frequencies of individual loci should be admitted in this case if calculated according the NRC modified ceiling principle'. Since the defendant did not dispute the general acceptance of the NRC's ceiling procedure, the court of appeals reversed the exclusion of such numbers, but expressly left open the possibility that defendants might demonstrate the lack of general acceptance in later cases.

Despite a state statute purporting to make probability calculations of genotype frequencies admissible in criminal cases, *Alt* stopped short of admitting the multilocus binotype frequency because of several idiosyncratic decisions of the Minnesota Supreme Court forbidding testimony about the product of probabilities of independent events. In *State v. Bloom*, 516 N.W. 2d 159 (1994), however, the Supreme Court abruptly decided that this rule need not apply to DNA evidence. In previous opinions, the same court had fashioned a rule excluding the probability that is of most interest to the jury for fear that jurors would misuse even a properly computed joint probability; there was no suggestion in those cases that the probabilities, which involved traditional genetic markers and microscopic comparisons of hair fibers, were inaccurate.[4] Nevertheless, the *Bloom* court distinguished its earlier decisions on the basis of an affidavit from Daniel Hartl. As an expert for the prosecution, Hartl expressed his 'belief that the [exclusionary] rule was correct at a time when genetic typing could be performed only with blood groups and other types of genetic systems that are not highly polymorphic', since these systems produce 'a long chain of multiplication of probabilities yielding a small, and unjustified, match probability for the whole set of genes'. In the case at bar, Hartl was prepared to testify that a nine-locus match found from a search of a database of convicted sexual offenders constituted 'overwhelming evidence that, to a reasonable degree of scientific certainty, the DNA from the victim's vaginal swab came from [defendant], to the exclusion of all others', and that using the 'interim ceiling method', there was a 1 in 634,687 chance of a random match across the five loci first tested. Evidently persuaded that such calculations for a DNA binotype are less of a problem because of 'the very conservative nature of the probability figures obtained using the NRC's approach', the court concluded that 'a DNA exception to the rule against admission of quantitative, statistical probability evidence . . . is justified'. It held that 'any properly qualified prosecution or defense expert may . . . give an opinion as to random match probability

[4] In actuality, the hair comparison frequencies are difficult to defend, but for reasons that have nothing to do with any independence assumptions. See Panel on Statistical Assessments as Evidence in the Courts (1989, pp. 64–65).

using the NRC's approach to computing that statistic', and it approved as well of certain forms of qualitative characterizations of the meaning of a 'match', although its treatment of this issue seems garbled.[5]

The only other definitive adoption of the ceiling procedure to date may have come in *Springfield v. State*, 860 P.2d 435 (Wyo. 1993). There, ceiling figures not only were held admissible in their own right, but they seem to have swept in more standard computations on their coattails. In 1988, a knife-wielding, masked man forced his way into a woman's home and raped her. In a trial four years later, the state introduced the testimony of an FBI agent that the binotype frequency of the rapist was 1/250 thousand among Native Americans (using the FBI's Native American database) and much less among African-Americans (1/150 million), Caucasians (1/250 million), and Hispanics (1/25 million). It also offered testimony about the NRC report, including ceiling frequencies of 1/17 million for African-Americans and 1/221 thousand for Native Americans. The Wyoming Supreme Court upheld the admission of these estimates, on the ground that they 'will assist the trier of fact and that the . . . ceiling principle has provided the most conservative statistical estimate as recommended by the NRC report'.

These opinions fail to mention the sometimes strident criticisms of the ceiling method (noted above) by some population geneticists and statisticians. The post-NRC report literature is just beginning to diffuse into the caselaw. In *People v. Wallace*, 17 Cal. Rptr. 2d 721 (Ct. App. 1993), the author of the *Barney* opinion cautioned the scientific community:

> While we are in no position to choose sides in this ongoing debate, we note that its persistence threatens the admissibility of an extremely important forensic tool. This is no time for purist insistence that DNA evidence should be admitted on one's own terms or not at all. . . . Our hope is that the key players in this debate will . . . agree to a compromise on statistical calculation. Otherwise they risk preventing any general acceptance at all, thus precluding the admissibility of DNA analysis evidence.

[5] The court insisted that 'the expert [should not] be allowed to say that defendant is the source to the exclusion of all others or to express an opinion as to the strength of the evidence. But should a properly qualified expert, assuming adequate foundation, be allowed to express an opinion that, to a reasonable scientific certainty, the defendant is (or is not) the source? We believe so'.

This risk materialized in *State v. DeFroe*, No. 92-1-03699-8 (Super Ct. King County June 23, 1993), and *State v. Hollis*, No. 92-104603-9 (Super Ct. King County June 23, 1993) (Findings of Fact and Conclusions of Law). The trial court ruled that even with ceiling frequencies, DNA evidence was not generally accepted and therefore inadmissible. After reviewing post-NRC articles on the ceiling principle and affidavits and testimony from many expert witnesses, the court found the 'interim-ceiling principle' to be 'a statistical technique without scientific basis, contrived by compromise or pressure from the law enforcement community . . . '[6]

The view that the dispute over the ceiling principle demonstrates a fatal lack of general scientific acceptance springs from a failure to appreciate the nature and purpose of the ceiling procedure. As Lempert (1993) makes clear, the proposal was not devised for scientific applications and cannot be defended on strictly scientific grounds. It is intended to supply numbers that are useful in court – hardly a scientific concern. It involves debatable policy judgments that these numbers should be conservative and that they should be as conservative as the ceiling binotypes are. Scientists have been drawn into this debate over policy matters, and their views may well be valuable, but courts have little reason to defer to their opinions. The scientific issue – where scientific consensus *is* important – is whether the method actually produces figures no larger than those that would be obtained with perfect knowledge of the reference population and its structure. The NRC committee, according to Lander (1993), thought 'that the chosen upper bound sufficed to eliminate serious scientific objections', and there seems to be little, if any, dispute over the propo-

[6] The material considered by the court in excluding the DNA evidence included an affidavit from Elizabeth Thompson branding the ceiling principle "data-driven, interest-ridden, pseudo-statistical, ad hoc methodology, to which no statistician or scientist should be a party." Thompson (1994, note 21). (Dr. Thompson informs me that her criticism was directed not just at the ceiling principle, but also at various other attempts by both sides to adjust the uses of the ceiling principle to their perceived goals.) In a more recent case, Eric Lander provided an affidavit stating that: As a scientist, the [*Hollis*] court's analysis is troubling to me. The court focuses on the fact that scientists disagree about the *best* method of estimating population frequencies, but seems to miss the point that scientists can disagree about the *best* estimate of a frequency, while agreeing that a formula (or even several alternative formulas) provide an upper bound on the frequency. This is the case for the ceiling principle, which makes adequate allowance for variation among populations and results in estimates that would not overestimate the case against the defendant. State v. Dyer, No. 93-1-00489-0 (Superior Ct. King County) (Affidavit of Dr. Eric S. Lander, Sept. 20, 1993).

sition that for a structured population with each subpopulation in equilibrium, the ceiling methods produce generous estimates of binotype frequencies.[7] In fact, the very perception that the methods are enormously generous evokes antagonism on the part of the scientists and statisticians who see the ceiling computations as inappropriate.

Several courts have recognized this distinction between the scientific question of whether the method gives conservative results and the policy question of whether these results are too conservative. The *Porter* court explained that:

> [I]t is not necessary for the prosecution to prove . . . that there is a scientific consensus as to the precise probability of a coincidental match. So long as there is a consensus that the chances of such a match are no greater than some very small fraction, then the evidence is probative and should be admitted on an appropriately conservative basis.

More recently, in *State v. Alt*, 504 N.W.2d at 51 n. 21, the Minnesota court noted that 'the recent debate does not affect the admissibility in this case of DNA evidence under the NRC approach' because the report 'does not purport to yield a statistical frequency which is more scientifically defensible than either the DNA proponents or opponents would advocate'. The court thus concluded that 'Insofar as the recent debate reflects disagreement with the policy judgments of the NRC Committee, or reiterates the view of DNA proponents that population subgroups are not a significant problem, the debate does not deprive the report of sufficient scientific acceptance . . .'.

Demanding numbers

Another part of the NRC report that courts have attended to is the statement (p. 74) that '[t]o say that two patterns match, without providing any scientifically valid estimate (or, at least, an upper bound) of the frequency with which such matches might occur by chance, is meaningless'. *State v. Cauthron*, 846 P.2d 502 (Wash. 1993), is illustrative. At trial, Cellmark's experts testified not only that defendant's DNA matched that in semen samples taken from rape victims, but also that 'the DNA could not have come from anyone else on earth'. On appeal, citing other opinions and quoting the NRC report, the Washington Supreme Court reversed the conviction on the ground that no 'probability statistics' accompanied this opinion. (It relied as well on the NRC committee's recommendation (p. 92) that '[r]egardless of the calculated frequency, an expert should – given the relatively small number of loci used and the available population data – avoid assertions in court that a particular genotype is unique in the population'.)[8]

As a legal matter, a completely unexplained statement of a 'match' should be inadmissible because it is too cryptic to be weighed fairly by the jury, and an overblown characterization like the one in *Cauthron* should be excluded as, on balance, unfairly prejudicial. Conclusions like these follow from the policies behind the pertinent rules of evidence. They are not scientific judgments. Yet, the NRC report seems to have prompted some courts to think that the admissibility of nonnumerical characterizations must be resolved by reference to scientific norms. The dominant *Daggett* plurality, for instance, disparaged expert testimony that used phrases like 'highly likely' but did not include numbers because the state 'cited no authorities and presented no testimony . . . that the use of such terms is generally accepted by the scientific community in evaluating the significance of a match' 622 N.E.2d at 275 n.4. 'The point is', the plurality insisted, 'not that this court should require a numerical frequency, but that the scientific community clearly does'. Likewise, in *State v. Bible*, 858 P.2d 1152 (Ariz. 1993), the Arizona Supreme Court, misconstruing *Cauthron*, suggested that the NRC report's 'meaningless' remark demonstrates that the law's general acceptance requirement precludes presenting evidence of a match without an estimate of the binotype frequency.

This view is plainly mistaken. The general acceptance standard addresses the validity and reliability of the methodology that produces evidence of identity. The fact of a match is scientifically valid evidence of identity as long as it can be shown from theory and data that the binotype is not ubiquitous in the relevant population. But how to present to a jury valid scientific evidence of a match is a legal rather than a scientific issue falling far outside the domain of the general acceptance test and the fields of statistics and popula-

[7] Slimowitz and Cohen (1993) question the premise that all the subpopulations are in equilibrium and discuss the conditions under which the ceiling binotype frequency is an upper bound.

[8] Other opinions citing the NRC report's demand for quantitative assessments include *Daggett*, *Watson*, and still others cited by the plurality of justices who prevailed in *Daggett*, 622 N.E.2d at 274 n.2.

tion genetics. Thus, it would not be 'meaningless' to inform the jury that two samples match and that this match makes it more probable, in an amount that is not precisely known, that the DNA in the samples comes from the same person. Nor, when all estimates of the frequency are in the millionths or billionths, would it be meaningless to inform the jury that there is a match that is known to be extremely rare in the general population.[9] Courts may reach differing results on the legal propriety of qualitative as opposed to quantitative assessments, but they only fool themselves when they act as if scientific opinion automatically dictates the correct answer.

Acknowledgments

This work was supported by a stipend from the Arizona State University College of Law Research Fund.

References

Berry, D., 1991. Inferences using DNA profiling in forensic identification and paternity cases. Statistical Science 6: 175–205.

Chakraborty, R., *et al.*, 1992. Effects of population subdivision and allele frequency differences on interpretation of DNA typing data for human identification, pp. 205–222 in Proceedings of the Third International Symposium on Human Identification. Promega, Madison, Wisconsin.

Devlin, B., N. Risch & K. Roeder, 1994. Comments on the statistical aspects of the NRC's report on DNA typing. Journal of Forensic Science 39: 28–40.

Devlin, B., N. Risch & K. Roeder, 1993. Statistical evaluation of DNA fingerprinting: a critique of the NRC's report. Science 259: 748–749, 837.

Devlin, B., N. Risch & K. Roeder, 1992. Forensic inference from DNA fingerprints. Journal of the American Statistical Association 87: 337–350.

Evett, I.W., J. Scranage & R. Pinchin, 1993. An illustration of efficient statistical methods for RFLP analysis in forensic science. American Journal of Human Genetics 52: 498–505.

Panel on Statistical Assessments as Evidence in the Courts, 1989. The Evolving Role of Statistical Assessments as Evidence in the Courts, edited by S.E. Fienberg. Springer-Verlag, N.Y.

Kaye, D.H., 1993. DNA evidence: probability, population genetics, and the courts. Harvard Journal of Law & Technology 7: 101–172.

Lewontin, R.C. & D.L. Hartl, 1991. Population genetics in forensic DNA typing. Science 254: 1745–1750.

Lander, E.S., 1993. Letter: DNA fingerprinting: the NRC report. Science 260: 1221.

Lempert, R., 1993. DNA, science and the law: two cheers for the ceiling principle. Jurimetrics Journal 34: 41–57.

Morton, N.E., 1992. Genetic structure of forensic populations. Proceedings of the National Academy of Sciences USA 89: 2556–2560.

McCormick, C., 1992. McCormick on Evidence, volume 1, 4th edition, edited by J. Strong. West Publishing Company, Minneapolis, MN.

Roeder, K., 1994. DNA fingerprinting: a review of the controversy. Statistical Science 9: 222–247.

Slimowitz, J.R. & J.E. Cohen, 1993. Violations of the ceiling principle: exact conditions and statistical evidence. American Journal of Human Genetics 53: 314–323.

Thompson, W.C., 1993. Evaluating the admissibility of new genetic identification tests: lessons from the "DNA war". Journal of Criminal Law and Criminology 84: 22–104.

Weir, B.S., 1992. Population genetics in the forensic DNA debate. Proceedings of the National Academy of Sciences USA 89: 11654–11659.

Weir, B.S., 1993a. Letter: forensic population genetics and the NRC. American Journal of Human Genetics 52: 437.

Weir, B., 1993b. Letter: DNA fingerprinting report. Science 260: 473.

Editor's comments

Scientists caught up in the debate over the use of DNA for human identification may not be fully aware of the legal consequences of their pronouncements. The author provides a very useful discussion of these consequences in showing how courts have interpreted the 1992 NRC report.

[9] This is not to say that binotype frequencies are the best quantitative expression of the probative value of the evidence. To the contrary, they have shortcomings that the presentation of a suitably computed likelihood ratio or posterior probability might overcome. (Berry, 1991; Devlin, Risch & Roeder, 1992; Evett, Scranage & Pinchin, 1993; Kaye, 1993.)

B. S. Weir (ed.), Human Identification: The Use of DNA Markers, 107–117, 1995.

Applications of the Dirichlet distribution to forensic match probabilities

Kenneth Lange
Department of Biostatistics, School of Public Health, University of Michigan, Ann Arbor, MI 48109-2029, USA

Received 7 February 1994 Accepted 26 July 1994

Key words: genetic equilibrium, racial admixture, Monte Carlo, upper bounds

Abstract

The Dirichlet distribution provides a convenient conjugate prior for Bayesian analyses involving multinomial proportions. In particular, allele frequency estimation can be carried out with a Dirichlet prior. If data from several distinct populations are available, then the parameters characterizing the Dirichlet prior can be estimated by maximum likelihood and then used for allele frequency estimation in each of the separate populations. This empirical Bayes procedure tends to moderate extreme multinomial estimates based on sample proportions. The Dirichlet distribution can also be employed to model the contributions from different ancestral populations in computing forensic match probabilities. If the ancestral populations are in genetic equilibrium, then the product rule for computing match probabilities is valid conditional on the ancestral contributions to a typical person of the reference population. This fact facilitates computation of match probabilities and tight upper bounds to match probabilities.

Introduction

The introduction of DNA profile evidence in criminal cases has sparked one of the most spirited and, at times, acrimonious debates in recent scientific history (Chakraborty & Kidd, 1991; Devlin, Risch & Roeder, 1992; Jeffreys, Wilson & Thien, 1985; Lander, 1989; Lewin, 1989; Lewontin & Hartl, 1991). At the heart of this debate are disagreements about techniques for computing match probabilities. Defense experts contend that match probabilities as currently computed are meaningless because of the failure of the product rule in contemporary American populations. Application of the product rule requires that the reference population for the evidentiary DNA be in genetic equilibrium. Because of the racial heterogeneity in the United States, no one can honestly claim that the population as a whole is in genetic equilibrium. However, this fact should not signal the end of the debate. There is a conditional form of the product rule that holds even in the presence of genetic heterogeneity (Lange, 1993). This conditional form provides a wedge for valid calculation of match probabilities. Of course, to make this proposal workable, we must introduce some approximations about the nature of the genetic contributions from the various subpopulations. This is where the Dirichlet distribution proves useful.

The Dirichlet distribution provides a flexible way of parameterizing the contributions of the subpopulations to a proposed reference population. By the reference population for a particular crime, we simply mean the collection of people who could have conceivably contributed the evidentiary DNA. This vague hypothetical construct may vary from locale to locale and from case to case. The particular ancestry of the alleged contributor of the evidentiary DNA, be he perpetrator or victim, is largely irrelevant to the choice of the reference population. Of course, the postulated racial composition of the reference population should be broad enough to reflect the ancestry of the alleged contributor.

The mathematical attractiveness of the Dirichlet distribution stems from the fact that it makes necessary expectations trivial to compute. The following sections document the mathematical manipulations involved in computing match probabilities via the Dirichlet distribution. Readers primarily interested in general conclusions can skip these details and turn directly to the discussion at the end of the paper. Some limitations of the Dirichlet model are mentioned there.

The Dirichlet distribution is also relevant to the related problem of allele frequency estimation. If one adopts a Bayesian framework for estimation, then a Dirichlet prior for the multinomial distribution of allele counts leads to a Dirichlet posterior. Classical statisticians will object that the choice of any particular prior from the Dirichlet family is bound to be arbitrary. When data from several populations are available, one can choose the prior empirically from the data. This empirical Bayes procedure incorporates some of the best features of classical and Bayesian statistics. The resulting Bayesian estimates of allele frequencies tend to moderate the extremes seen in classical estimates based on sample proportions. We develop this perspective in a preliminary digression that may be of independent interest to many readers.

Empirical Bayes estimation of allele frequencies

Consider a locus with k codominant alleles. To estimate the frequencies $p_1, \ldots, p_k$ of these alleles in some population, suppose one takes a random sample from the population and observes n_i genes of type i. Then $n_i/n_.$ is the maximum likelihood estimate of p_i, where the abbreviation $n_. = \sum_{i=1}^k n_i$ relies on the usual summation convention. This frequentist estimate based on the multinomial distribution can be contrasted to a Bayesian estimate using a Dirichlet prior for the allele frequencies (Good, 1965). The Dirichlet distribution (Kingman, 1993) with parameters $\gamma_1, \ldots, \gamma_k > 0$ has density

$$\frac{\Gamma(\gamma_.)}{\prod_{i=1}^k \Gamma(\gamma_i)} \prod_{i=1}^k p_i^{\gamma_i - 1} \tag{1}$$

on the simplex

$$\Delta_k = \left\{ (p_1, \ldots, p_k) : p_1 > 0, \ldots, p_k > 0, \sum_{i=1}^k p_i = 1 \right\}$$

endowed with the uniform measure. One of the virtues of the Dirichlet distribution is the elegant moment formula

$$E\left(\prod_{i=1}^k p^{t_i}\right) = \frac{\Gamma(\gamma_.)}{\prod_{i=1}^k \Gamma(\gamma_i)} \int_{\Delta_k} \prod_{i=1}^k p_i^{t_i + \gamma_i - 1} dp$$

$$= \frac{\Gamma(\gamma_.)}{\Gamma(t_. + \gamma_.)} \prod_{i=1}^k \frac{\Gamma(t_i + \gamma_i)}{\Gamma(\gamma_i)}. \tag{2}$$

The Dirichlet prior is a conjugate prior for the multinomial distribution (Lee, 1989). In the current context, this means that if the allele frequency vector $p = (p_1, \ldots, p_k)$ has a Dirichlet prior with parameters $\gamma_1, \ldots, \gamma_k$, then based on the sample, p has a Dirichlet posterior with revised parameters $n_1 + \gamma_1, \ldots, n_k + \gamma_k$. This fact follows from an application of the moment formula (2) in the conditional density computation

$$\frac{\dfrac{\Gamma(\gamma_.)}{\prod_{i=1}^k \Gamma(\gamma_i)} \binom{n}{n_1 \ldots n_k} \prod_{i=1}^k p_i^{n_i + \gamma_i - 1}}{\dfrac{\Gamma(\gamma_.)}{\prod_{i=1}^k \Gamma(\gamma_i)} \binom{n}{n_1 \ldots n_k} \int_{\Delta_k} \prod_{i=1}^k q_i^{n_i + \gamma_i - 1} dq}$$

$$= \frac{\Gamma(n_. + \gamma_.)}{\prod_{i=1}^k \Gamma(n_i + \gamma_i)} \prod_{i=1}^k p_i^{n_i + \gamma_i - 1}.$$

A second application of (2) now implies that $(n_i + \gamma_i)/(n_. + \gamma_.)$ is the posterior mean of p_i. The posterior mean is a strongly consistent, asymptotically unbiased estimator of the true p_i.

The primary disadvantage of taking a Bayesian stance on allele frequency estimation is that there is no obvious way of selecting a reasonable prior. However, if data from several distinct populations are available, then one can select an appropriate prior empirically. Consider the marginal distribution of the allele counts $(N_1, \ldots, N_k)$ in a sample of genes from a single population. Integrating out the prior on the allele frequency vector $p = (p_1, \ldots, p_k)$ yields the predictive distribution (Mosimann, 1962)

$$\Pr(N_1 = n_1, \ldots, N_k = n_k)$$

$$= \binom{n}{n_1 \ldots n_k} \frac{\Gamma(\gamma_.)}{\Gamma(n_. + \gamma_.)} \prod_{i=1}^k \frac{\Gamma(n_i + \gamma_i)}{\Gamma(\gamma_i)}. \tag{3}$$

This distribution is known as the Dirichlet-multinomial distribution. Its parameters are the γ's rather than the p's

With independent data from several distinct populations, one can estimate the parameter vector $\gamma = (\gamma_1, \ldots, \gamma_k)$ of the Dirichlet-multinomial distribution by maximum likelihood. Newton's method offers by far the fastest means of finding the maximum likelihood estimate. To implement Newton's method, one needs the loglikelihood $L(\gamma)$, the score vector $dL(\gamma)$, and the observed information matrix $-d^2L(\gamma)$ for each population. Elementary calculus based on the likelihood (3) shows that the score has entries.

$$\frac{\partial}{\partial \gamma_i}L(\gamma) = D(\gamma_.) - D(n_. + \gamma_.) + D(n_i + \gamma_i) - D(\gamma_i), \quad (4)$$

where $D(s) = d/ds \ln \Gamma(s)$ is the digamma function (Hille, 1959). The observed information has entries

$$-\frac{\partial^2}{\partial \gamma_i \partial \gamma_j}L(\gamma) = -T(\gamma_.) + T(n_. + \gamma_.) - \chi_{\{i=j\}} \times [T(n_i + \gamma_i) - T(\gamma_i)], \quad (5)$$

where $\chi_{\{i=j\}}$ is the indicator function of the event $\{i = j\}$, and where $T(s)$ is the trigamma function $d^2/ds^2 \ln \Gamma(s)$ (Hille, 1959). The digamma and trigamma functions appearing in the expressions (4) and (5) should not be viewed as a major barrier to computation since good software for evaluating these transcendental functions exists (Bernardo, 1976; Schneider, 1978).

Extending the above notation for the loglikelihood, score, and observed information from a single population to the entire random sample from several populations, Newton's method updates the current parameter iterate γ^m by

$$\gamma^{m+1} = \gamma^m - d^2L(\gamma^m)^{-1}dL(\gamma^m). \quad (6)$$

For Newton's method to move in an uphill direction, the observed information matrix $-d^2L(\gamma^m)$ should be positive definite. This may not always be the case. The obvious remedy is to replace the observed information $-d^2L(\gamma^m)$ by an approximating matrix that is positive definite.

Equation (5) for a single population evidently can be summarized in matrix form by

$$-d^2L(\gamma) = D - c\mathbf{1}\mathbf{1}^t, \quad (7)$$

where D is a diagonal matrix with ith diagonal entry $d_i = T(\gamma_i) - T(n_i + \gamma_i)$, c is the constant $T(\gamma_.) - T(n_. + \gamma_.)$, and $\mathbf{1}$ is a column vector of all 1's. Because the trigamma function is decreasing (Hille, 1959), $d_i > 0$ when $n_i > 0$; the constant $c > 0$ always. Since the representation (7) is preserved under finite sums, it holds, in fact, for the entire sample.

The idea now is to approximate $-d^2L(\gamma)$ by the right side of (7) with a decreased value of the constant c if necessary. If the right side of (7) is to be positive definite, then

$$\mathbf{1}^tD^{-1}(D - c\mathbf{1}\mathbf{1}^t)D^{-1}\mathbf{1} = \mathbf{1}^tD^{-1}\mathbf{1}(1 - c\mathbf{1}^tD^{-1}\mathbf{1})$$

must be positive. This implies

$$1 - c\mathbf{1}^tD^{-1}\mathbf{1} = 1 - c\sum_{i=1}^{k}\frac{1}{d_i} > 0. \quad (8)$$

Conversely, inequality (8) is sufficient for the right side of (7) to be positive definite. This fact can be most easily demonstrated by noting the Sherman-Morrison formula (Miller, 1987)

$$(D - c\mathbf{1}\mathbf{1}^t)^{-1} = D^{-1} + \frac{c}{1 - c\mathbf{1}^tD^{-1}\mathbf{1}}D^{-1}\mathbf{1}\mathbf{1}^tD^{-1} \quad (9)$$

Formula (9) proves that $(D - c\mathbf{1}\mathbf{1}^t)^{-1}$ exists and is positive definite under assumption (8). Since the inverse of a positive definite matrix is positive definite, it follows that $D - c\mathbf{1}\mathbf{1}^t$ is positive definite.

These results suggest that c be replaced by $\min\left\{c, (1-\epsilon)/\left(\sum_{i=1}^{k} d_i^{-1}\right)\right\}$, where ϵ is a small positive constant. With this substitution and with occasional backtracking to avoid overshooting the maximum of $L(\gamma)$ along the current Newton direction, Newton's method can proceed safely. Near the maximum likelihood point, $-d^2L(\gamma)$ will be positive definite, and no adjustment of it is necessary. Throughout the iterations the Sherman-Morrison formula can be used to invert $-d^2L(\gamma)$ or its substitute.

Example of the empirical Bayes procedure

Edwards *et al.* (1992) gathered population data in Houston, Texas on the eight alleles of the HUMTH01 locus on chromosome 11. This is a tandem repeat locus whose allele names refer to numbers of repeat units. From the four separate subpopulations of Caucasians, blacks, Chicanos, and Asians, the eight γ's are estimated by maximum likelihood to be .11, 4.64, 7.33, 2.97, 5.32, 5.26, .27, and .10. Using these estimated Dirichlet parameters, Table 1 compares the maximum likelihood estimates (top row) and posterior mean estimates (bottom row) of the allele frequencies within each subpopulation. It is noteworthy that all posterior means are within one standard error of the maximum likelihood estimates. (These standard errors are given in Table 2 of Edwards *et al.*, 1992.) Nonetheless, the empirical Bayes procedure does tend to moderate the

Table 1. Classical and Bayesian allele frequency estimates.

Estimator	Allele	Caucasian	Black	Chicano	Asian
Classical	5	.0054	.0000	.0000	.0000
Bayesian	5	.0053	.0003	.0003	.0006
Classical	6	.2258	.1351	.2083	.1039
Bayesian	6	.2227	.1380	.2064	.1147
Classical	7	.1586	.3703	.3333	.2597
Bayesian	7	.1667	.3645	.3301	.2630
Classical	8	.1102	.2108	.0677	.0519
Bayesian	8	.1105	.2045	.0707	.0609
Classical	9	.1425	.1459	.1432	.4416
Bayesian	9	.1465	.1498	.1471	.4073
Classical	10	.3522	.1378	.2474	.0909
Bayesian	10	.3424	.1421	.2445	.1070
Classical	11	.0054	.0000	.0000	.0455
Bayesian	11	.0057	.0007	.0007	.0404
Classical	12	.0000	.0000	.0000	.0065
Bayesian	12	.0002	.0002	.0002	.0061
Sample size *n*		372	370	384	154

extremes in estimated allele frequencies seen in the different subpopulations. In particular, all posterior mean estimates are positive. The maximum likelihood estimates suggest that those alleles failing to appear in a sample are nonexistent in the corresponding subpopulation. The posterior mean estimates suggest more reasonably that such alleles are simply rare in the subpopulation.

Match probabilities for a population at equilibrium

For a single population in Hardy-Weinberg equilibrium, forensic match probabilities can be computed from either a frequentist or a Bayesian perspective. If the allele frequencies are known without error, then the match probabilites for homozygous *i/i* genotypes and heterozygous genotypes *i/j* are p_i^2 and $2p_ip_j$, respectively. In practice, the frequencies p_i can only be estimated. Assuming codominant alleles and the maximum likelihood estimates $\hat{p}_i = n_i/n_.$, the frequentist genotype estimates have the following sampling properties (Chakraborty, Srinivasan & Daiger, 1993):

$$E(\hat{p}_i^2) = p_i^2 + \frac{p_i(1-p_i)}{n_.}$$

$$Var(\hat{p}_i^2) = \frac{4p_i^3(1-p_i)}{n_.} + O\left(\frac{1}{n_.^2}\right)$$

$$E(2\hat{p}_i\hat{p}_j) = 2p_ip_j - \frac{2p_ip_j}{n_.}$$

$$Var(2\hat{p}_i\hat{p}_j) = \frac{4p_ip_j}{n_.}(p_i + p_j - 4p_ip_j) + O\left(\frac{1}{n_.^2}\right).$$

From the Bayesian perspective, the genotype probabilities are random variables whose distributions depend on the posterior distribution of the allele frequencies. Based on the moment formula (2), it is straightforward to compute that

$$E(p_i^2|N_1 = n_1, \ldots, N_k = n_k) = \frac{(n_i+\gamma_i)^{\bar{2}}}{(n_. + \gamma_.)^{\bar{2}}}$$

$$Var(p_i^2|N_1 = n_1, \ldots, N_k = n_k) = \frac{(n_i+\gamma_i)^{\bar{4}}}{(n_. + \gamma_.)^{\bar{4}}} - \left[\frac{(n_i+\gamma_i)^{\bar{2}}}{(n_. + \gamma_.)^{\bar{2}}}\right]^2$$

$$E(2p_ip_j|N_1 = n_1, \ldots, N_k = n_k) = 2\frac{(n_i+\gamma_i)(n_j+\gamma_j)}{(n_. + \gamma_.)^{\bar{2}}}$$

$$Var(2p_ip_j|N_1 = n_1, \ldots, N_k = n_k) = \frac{4(n_i+\gamma_i)^{\bar{2}}(n_j+\gamma_j)^{\bar{2}}}{(n_. + \gamma_.)^{\bar{4}}}$$

$$-\left[\frac{2(n_i+\gamma_i)(n_j+\gamma_j)}{(n_.+\gamma_.)^{\bar{2}}}\right]^2,$$

where $x^{\bar{r}} = x(x+1)\cdots(x+r-1)$ denotes a rising power. It is interesting that the above mean expressions entail

$$E(p_i^2|N_1 = n_1, \ldots, N_k = n_k) > \tilde{p}_i^2$$
$$E(2p_ip_j|N_1 = n_1, \ldots, N_k = n_k) < 2\tilde{p}_i\tilde{p}_j,$$

where $\tilde{p}_i$ and $\tilde{p}_j$ are the posterior means of p_i and p_j.

As a numerical example, consider the 5/5 and 5/6 genotypes for Caucasians at the HUMTH01 locus of Table 1. Using the maximum likelihood estimates of p_5 and p_6, these genotypes have predicted frequencies of .0000289 and .00243, respectively. These values can be compared to the Bayesian values $E(p_5^2) = .0000412$ and $E(2p_5p_6) = .00235$ and their standard deviations $\sqrt{Var(p_5^2)} = .0000608$ and $\sqrt{Var(2p_5p_6)} = .00162$. Let us emphasize here that the reference population for the two matches is Caucasian and that conditional expectations and variances are abbreviated for the sake of convenience as ordinary expectations and variances.

Chakraborty, Srinivasan and Daiger (1993) point out that it is probably more relevant to consider the sampling distribution of $\ln p_i^2$ and $\ln 2p_ip_j$, since the log transformation focuses attention on the order of magnitude of a match probability and makes the central limit theorem applicable when a match probability is computed over several independent loci. In the Bayesian context, one can compute the moments of the random vector $(\ln p_1, \ldots, \ln p_k)$ through its multivariate moment generating function

$$E\left(e^{\sum_{i=1}^k t_i \ln p_i}|N_1 = n_1, \ldots, N_k = n_k\right)$$
$$= E\left(\prod_{i=1}^k p_i^{t_i}|N_1 = n_1, \ldots, N_k = n_k\right),$$

which is given explicitly by equation (2) with $n_i + \gamma_i$ replacing γ_i. Entirely straightforward, but slightly tedious calculations show that

$$E(\ln p_i^2|N_1 = n_1, \ldots, N_k = n_k)$$
$$= 2[D(n_i+\gamma_i) - D(n_.+\gamma_.)]$$
$$Var(\ln p_i^2|N_1 = n_1, \ldots, N_k = n_k)$$
$$= 4[T(n_i+\gamma_i) - T(n_.+\gamma_.)]$$
$$E(\ln 2p_ip_j|N_1 = n_1, \ldots, N_k = n_k)$$
$$= \ln 2 + D(n_i+\gamma_i) + D(n_j+\gamma_j) - 2D(n_.+\gamma_.)$$
$$Var(\ln 2p_ip_j|N_1 = n_1, \ldots, N_k = n_k)$$
$$= T(n_i+\gamma_i) + T(n_j+\gamma_j) - 4T(n_.+\gamma_.).$$

Consider again the 5/5 and 5/6 genotypes at the HUMTH01 locus among Caucasians. Using the maximum likelihood estimates of p_5 and p_6, these genotypes have predicted log frequencies of −10.45 and −6.02 to the base e, respectively. These classical values can be compared to the Bayesian values $E(\ln p_5^2) = -10.99$ and $E(\ln 2p_5p_6) = -6.31$ and their standard deviations $\sqrt{Var(\ln p_5^2)} = 1.55$ and $\sqrt{Var(\ln 2p_5p_6)} = .78$.

Now contemplate m codominant loci in Hardy-Weinberg and linkage equilibrium. Let G_l be the probability of an observed genotype at locus l. Thus, each G_l corresponds to either an expression p_i^2 for a homozygote or to an expression $2p_ip_j$ for a heterozygote. Given a random sample of genotypes at locus l, the moments of the random variables G_l and $\ln G_l$ can be calculated as indicated above based on a Dirichlet posterior distribution of allele frequencies. Since the random variables G_l are independent under the assumption of linkage equilibrium, a multilocus match probability $\prod_{l=1}^m G_l$ has posterior mean and variance

$$E\left(\prod_{l=1}^m G_l\right) = \prod_{l=1}^m E(G_l)$$
$$Var\left(\prod_{l=1}^m G_l\right) = \prod_{l=1}^m E(G_l^2) - \prod_{l=1}^m E(G_l)^2$$
$$= \prod_{l=1}^m [Var(G_l) + E(G_l)^2]$$
$$- \prod_{l=1}^m E(G_l)^2.$$

Similarly, $\ln \prod_{l=1}^m G_l$ has posterior mean and variance

$$E\left(\sum_{l=1}^m \ln G_l\right) = \sum_{l=1}^m E(\ln G_l)$$
$$Var\left(\sum_{l=1}^m \ln G_l\right) = \sum_{l=1}^m Var(\ln G_l).$$

Of course, all of these means and variances are conditional on the sampled data.

Match probabilities in admixed populations

Once the assumptions of Hardy-Weinberg and linkage equilibrium fail, calculation of match probabilities become problematic. In particular, the product rule for combining match probabilities across separate loci no longer holds. One device for rescuing the product rule is to condition on the ancestry of a typical person from the reference population (Lange, 1993; Mickey *et al.*, 1983). This ancestry should consist of contributions from a finite number of specified ancestral populations that individually are assumed to be at equilibrium even when the reference population is not. A convenient way of parameterizing these contributions is to postulate that a proportion x_i of the maternal genes and y_i of the paternal genes of the typical person originate from ancestral population *i*. The mother and father of the typical person are assumed to be unrelated.

Assuming that there are *n* ancestral populations, it is again convenient to assign Dirichlet distributions to the random vectors $x = (x_1, \ldots, x_n)$ and $y = (y_1, \ldots, y_n)$. If the common Dirichlet parameter associated with the components x_i and y_i is α_i, then ancestral population *i* contributes a proportion $\alpha_i/\alpha_.$ of the genes observed in the reference population. Of course, the exact ancestral contributions will vary from person to person in the reference population. If the total $\alpha_.$ of the α_i is close to 0, then most people will exhibit maternal and paternal vectors *x* and *y* having a single component close to 1 and the remaining components close to 0. This is consistent with little mixing of the races in the reference population. Large values of $\alpha_.$ suggest a thorough mixing of the races.

Two extremes relating *x* and *y* are apt to be important in practice. On one hand, the maternal contributions *x* might be independent of the paternal contributions *y*. On the other hand, total endogamy might prevail, in which event *x* and *y* coincide. In between these two extremes is partial endogamy, where *x* and *y* are independent with probability β and coincide with probability $1 - \beta$.

To compute a multilocus match probability, we adopt the allele frequency notation p_{ijk} with three subscripts *i, j,* and *k*, indicating population, locus, and allele, respectively. The posterior Dirichlet parameter that corresponds to p_{ijk} we denote by θ_{ijk}, condensing into a single symbol the sum of the prior Dirichlet parameter and the number of genes sampled for this combination of population, locus, and allele. Because sampling to estimate allele frequencies in ancestral populations has little to do with determining the racial composition of a hypothetical reference population, it is sensible to assume that the racial admixture proportions are independent of the posterior allele frequencies. It is also reasonable to assume that posterior allele frequencies are independent from one ancestral population to the next and in view of linkage equilibrium from one locus to the next within an ancestral population.

Now consider a multilocus genotype defined by genotype k_j/l_j at locus *j*, where *j* ranges over some prescribed set of *m* loci situated on *m* different chromosomes. The conditional probability of observing this multilocus genotype is

$$\begin{aligned} & w \prod_j \Bigg[\left(\sum_u x_u p_{ujk_j} \right) \left(\sum_v y_v p_{vjl_j} \right) \\ & + \chi_{\{k_j \neq l_j\}} \left(\sum_u x_u p_{ujl_j} \right) \left(\sum_v y_v p_{vjk_j} \right) \Bigg] \\ & + (1 - w) \prod_j \Bigg[(1 + \chi_{\{k_j \neq l_j\}}) \left(\sum_u x_u p_{ujk_j} \right) \\ & \times \left(\sum_v x_v p_{vjl_j} \right) \Bigg], \end{aligned} \tag{10}$$

where the random variable *w* is independent of all other random variables in sight and indicates whether the ancestral proportions *x* and *y* are independent ($w = 1$) or agree ($w = 0$). In similar fashion, the function $\chi_{\{k_j \neq l_j\}}$ indicates whether the observed genotype at locus *j* is heterozygous. The expectation of (10) is by definition the multilocus match probability.

The best methods of evaluating the match probability all hinge on first computing the conditional expectation of (10) with respect to the ancestral proportions *x* and *y*. Owing to the various independence assumptions and to the fact that $E(w) = \beta$, this conditional expectation reduces to

$$\begin{aligned} & \beta \prod_j \Bigg[\sum_u \sum_v x_u y_v E(p_{ujk_j} p_{vjl_j}) \\ & + \chi_{\{k_j \neq l_j\}} \sum_u \sum_v x_u y_v E(p_{ujl_j} p_{vjk_j}) \Bigg] \\ & + (1 - \beta) 2^h \prod_j \Bigg[\sum_u \sum_v x_u x_v E(p_{ujk_j} p_{vjl_j}) \Bigg] \end{aligned} \tag{11}$$

if *h* heterozygous genotypes are observed among the *m* loci. The expression (11) evidently reflects the fact that conditional on ancestry, match probabilities obey the product rule. Motivation and explanation for this conditional product rule is given in Lange (1993) and will not be repeated here.

The ordinary expectations appearing in (11) fortunately yield to the moment formula (2). Indeed,

$$E(p_{ujk_j}p_{vjl_j}) = \begin{cases} \dfrac{\theta_{ujk_j}}{\theta_{uj.}}\dfrac{\theta_{vjl_j}}{\theta_{vj.}} & u \neq v \\ \dfrac{\theta_{ujk_j}\theta_{ujl_j}}{\theta_{uj.}^{\bar{2}}} & u = v\ k_j \neq l_j. \\ \dfrac{\theta_{ujk_j}^{\bar{2}}}{\theta_{uj.}^{\bar{2}}} & u = v\ k_j = l_j \end{cases}$$

As before, the dot subscript indicates summation over an omitted index.

If we now naively take the expectation of (11) and use the distributive rule, we are faced with evaluating n^{2m} terms of type

$$E\left(\prod_j x_{u_j}x_{v_j}\right)\prod_j E(p_{u_jjk_j}p_{v_jjl_j}) \qquad (12)$$

and $2^h n^{2m}$ terms of type

$$E\left(\prod_j x_{u_j}\right)E\left(\prod_j y_{v_j}\right)\prod_j E(q_{u_jj}r_{v_jj}), \qquad (13)$$

where $q_{u_jj} = p_{u_jjk_j}$ and $r_{v_jjl_j} = p_{v_jjl_j}$, or $q_{u_jj} = p_{u_jjl_j}$ and $r_{v_jj} = p_{v_jjk_j}$. To evaluate $E\left(\prod_j x_{u_j}\right)$, suppose the variable m_k counts the number of $u_j = k$. Then

$$E\left(\prod_j x_{u_j}\right) = \frac{\prod_k \alpha_k^{\overline{m_k}}}{\alpha_.^{\overline{m}}}.$$

Similarly, if m_k counts the number of $u_j = k$ plus the number of $v_j = k$, then

$$E\left(\prod_j x_{u_j}x_{v_j}\right) = \frac{\prod_k \alpha_k^{\overline{m_k}}}{\alpha_.^{\overline{2m}}}.$$

Thus, the principal barrier to computation is not evaluation of the various expectations, but rather the sheer number of terms that must be summed.

If we are satisfied with an upper bound on the match probability, we can consider the x_u and y_v appearing in (11) to be parameters rather than random variables and then find the maximum of (11) with respect to x and y. This is the point of view taken in Lange (1993). The resulting bound is valid regardless of the joint distribution assigned to x and y. A better bound is available under our current Dirichlet assumptions about x and y. This improved bound follows from the simple observation that the function $s \to (a+s)/(b+s)$ is increasing provided $0 < a \leq b$. Given this fact, the expectation $E(p_{u_jjk_j}p_{v_jjl_j})$ satisfies the inequality

$$\begin{aligned} E(p_{u_jjk_j}p_{v_jjl_j}) &\leq c_{u_jjk_j}d_{v_jjl_j} \\ &\leq d_{u_jjk_j}d_{v_jjl_j}, \end{aligned} \qquad (14)$$

where $c_{u_jjk_j} = \theta_{u_jjk_j}/\theta_{u_jj.}$ and $d_{v_jjl_j} = (\theta_{v_jjl_j}+1)/(\theta_{v_jj.}+1)$. Both upper bounds in (14) will be close to $c_{u_jjk_j}c_{u_jjl_j}$ if the random gene sample from population v_j at locus j is reasonably large.

When the first upper bound in (14) is substituted in (12) and (13), it follows that the match probability is bounded above by

$$\begin{aligned} &\beta E\Bigg[\prod_j \Bigg\{\left(\sum_u x_u c_{ujk_j}\right)\left(\sum_v y_v d_{vjl_j}\right) \\ &+\chi_{\{k_j \neq l_j\}}\left(\sum_u x_u c_{ujl_j}\right)\left(\sum_v y_v d_{vjk_j}\right)\Bigg\}\Bigg] \\ &+(1-\beta)2^h E\Bigg[\prod_j\left(\sum_u x_u c_{ujk_j}\right) \\ &\times\left(\sum_v x_v d_{vjl_j}\right)\Bigg]. \end{aligned} \qquad (15)$$

In the absence of endogamy, this bound is vastly simpler to evaluate than the original match probability. Indeed, the expectation (15) now splits into 2^h terms of type

$$\begin{aligned} &E\left[\prod_j\left(\sum_u x_u e_{uj}\right)\left(\sum_v y_v f_{vj}\right)\right] \\ &= E\left[\prod_j\left(\sum_u x_u e_{uj}\right)\right]E\left[\prod_j\left(\sum_v y_v f_{vj}\right)\right], \end{aligned}$$

where $e_{uj} = c_{ujk_j}$ and $f_{vj} = d_{vjl_j}$, or $e_{uj} = c_{ujl_j}$ and $f_{vj} = d_{vjk_j}$. Evaluation of

$$E\left[\prod_j\left(\sum_u x_u e_{uj}\right)\right]$$

$$= \sum_{u_1} \cdots \sum_{u_m} E\left(\prod_j x_{u_j}\right) \prod_j e_{u_j j}$$

involves summing only n^m terms rather than n^{2m} terms. It is easy to imagine evaluating match probabilities involving $n = 10$ or 20 populations at $m = 5$ or 6 loci.

As an alternative to exact computation of an upper bound, one can estimate the match probability by Monte Carlo simulation based on expression (11). To simulate the ancestral proportions x and y, it is helpful to note that if $Z_1, \ldots, Z_n$ are independent random variables with Z_i having gamma density $z_i^{\alpha_i - 1} e^{-z_i} / \Gamma(\alpha_i)$, then the random proportions $W_1, \ldots, W_n$ defined by

$$W_i = \frac{Z_i}{\sum_{j=1}^{n} Z_j}$$

follow a Dirichlet distribution with parameters $\alpha_1, \ldots, \alpha_n$ (Kingman, 1993). Gamma distributed random variables can be simply and economically simulated by an acceptance-rejection method for $\alpha_i \leq 1$ (Ahrens & Dieter, 1974) or by a ratio method for $\alpha_i > 1$ (Cheng & Feast, 1979).

Sample match probability calculations

As an example of the above calculations, consider the four loci HUMHPRTB, HUMTH01, HUMRENA, HUMFABP featured in Edwards *et al.* (1992). These tandem repeat loci occur on chromosomes X, 11, 1, and 4, respectively. For the purposes of this example, we will pretend that locus HUMHPRTB is autosomal and seek the match probability of the typical multilocus genotype 6/12, 7/9, 8/8, and 10/11. As already described for the HUMTH01 locus, it is possible to estimate the parameters of the Dirichlet-multinomial distribution for each locus from the data of Table 2 of Edwards *et al.* (1992). Once this is done, the posterior allele frequency parameters θ_{ijk} are immediately available. For the ancestral parameters, we select the approximate American proportions

$$\frac{\alpha_{\text{Caucasian}}}{\alpha_.} = .7$$

$$\frac{\alpha_{\text{black}}}{\alpha_.} = .15$$

$$\frac{\alpha_{\text{Chicano}}}{\alpha_.} = .1$$

$$\frac{\alpha_{\text{Asian}}}{\alpha_.} = .05.$$

The total $\alpha_.$ of the ancestral parameters is more difficult to determine. As mentioned earlier, large values of $\alpha_.$ are consistent with widespread racial admixture and small values of $\alpha_.$ with little admixture. The parameter β controls the degree of endogamy between the parents of a typical person of the reference population. Because of doubt about the exact values of $\alpha_.$ and β, it is crucial to test the sensitivity of the match probability to a range of values of these two parameters.

Table 2 presents the results of our calculations. The Monte Carlo estimates are based on 10,000 samples each of the ancestral proportions x and y. As noted above, the Dirichlet upper bound is relevant only when endogamy is absent ($\beta = 1$). By way of comparison, the upper bound to the match probability previously suggested in Lange (1993) is 6.3×10^{-6}, about double the Dirichlet upper bound. If one naively ignores the racial stratification of the population, then one arrives at a match probability of 3.5×10^{-6} or 1.9×10^{-6} (Chakraborty & Kidd, 1991). The first of these depends on allele frequencies estimates derived by pooling the sampled genes from the four separate subpopulations. The second depends on allele frequencies that are weighted averages of the maximum likelihood allele frequencies calculated for the separate subpopulations, with weight $\alpha_i / \alpha_.$ given to subpopulation i.

It is noteworthy how close the naive match probabilities are to the match probabilities based on the Dirichlet model. The parameters $\alpha_.$ and β appear to have little effect on computed values under the Dirichlet model.

Discussion

The empirical Bayes approach to allele frequency estimation has much to offer. Because populations are never totally isolated, allele information from one population is bound to be relevant to other populations. Empirical Bayes procedures permit propagation of partial knowledge from the whole to its parts. At the same time, Bayesian estimates conform to the dictates of data as more data are gathered. The practical effect of these adaptive features is a moderation of the extreme allele estimates produced by the sample proportions in each population. In particular, small allele frequency estimates tend to increase under the Bayesian procedures. Since match probabilities are sensitive to small allele frequencies, Bayesian multilocus match probabilities

Table 2. Calculations for a multilocus match probability.

Ancestral Total $\alpha_.$	Monte Carlo $\beta = 1$	Monte Carlo $\beta = .5$	Monte Carlo $\beta = 0$	Dirichlet Bound
.1	2.7×10^{-6}	2.5×10^{-6}	2.4×10^{-6}	2.8×10^{-6}
1	2.9×10^{-6}	2.8×10^{-6}	2.7×10^{-6}	3.0×10^{-6}
10	3.1×10^{-6}	3.1×10^{-6}	3.1×10^{-6}	3.3×10^{-6}
100	3.2×10^{-6}	3.2×10^{-6}	3.2×10^{-6}	3.3×10^{-6}

are less influenced by rare genotypes than classical match probabilities.

In our example dealing with just four populations, the empirical Bayes estimates show these desirable properties. Doubtless, sampling a larger number of populations would enhance the empirical choice of a prior. The empirical Bayes procedure is apt to be most advantageous when the number of populations is large and the sample size per population is relatively small. Economic constraints on data collection suggest that this situation is likely to occur as more populations are sampled.

Stochastic variation enters at two levels in computing match probabilities. First, match probabilities are usually computed under the assumption that allele frequencies are known with infinite precision. When limited samples are available to estimate allele frequencies, it is helpful to include this uncertainty in match probability estimation (Chakraborty, Srinivasan & Daiger, 1993). We have shown how to accomplish this in the Bayesian context for a population at genetic equilibrium. Our explicit expressions for posterior means and variances of genotype probabilities and their logarithms are straightforward to evaluate.

The second source of stochastic variation arises from our uncertainty about the degree of racial admixture in contemporary populations. Even within groups that society lumps into a single ethnic category, there can be considerable genetic heterogeneity. Hispanics are a good case in point. Because genetic equilibrium is required for application of the product rule, it is necessary to revert to a weaker form of the product rule. As argued in Lange (1993), the product rule is valid conditional on the ancestry of a typical person in the reference population. In other words, if we imagine tracing back the pedigree of the typical person to his ancestors in defined populations at equilibrium, then the presence of equilibrium in the ancestral populations implies independence of his genotypes at loci occurring on different chromosomes. In practice, we replace this hypothetical pedigree by the proportion x_i of his maternal genes and y_i of his paternal genes contributed by the *i*th ancestral population.

If we make the further assumption that these proportions follow Dirichlet distributions, then we can write explicit expressions for multilocus match probabilities. Even with a moderate number of loci, these expressions involve large numbers of terms. Nonetheless, it is possible to evaluate them accurately by simulation and to provide simple upper bounds amenable to exact evaluation. The example in Table 2 illustrates the close agreement between simulated values and upper bounds that can be achieved in practice. Although this particular example suggests that naive use of the product rule is acceptable, it is premature to make this generalization. As more allele frequency data accumulate from different ancestral populations, further comparisons of the Dirichlet match probabilities and those computed by the product rule should be undertaken.

Although the Dirichlet model does provide a more sophisticated basis for calculation, it still entails approximations and intentional simplifications. For instance, the model omits laboratory errors, the possibility of confused DNA samples, and fraud within the criminal justice system. The later two issues are hard to quantify, but juries need to bear them in mind. Incorporation of data from the currently used VNTR loci necessarily involves an analysis of laboratory measurement errors (Devlin, Risch & Roeder, 1992). It is debatable whether the additional information content of these highly polymorphic loci is worth their price in phenotypic ambiguity. One can argue that substituting more loci with less information content per locus is preferable (Lange, 1991). This substitution would help, for example, in distinguishing the culprit or victim from his close relatives such as siblings, an important issue not directly addressed by the Dirichlet model. Evett (1992) suggests computational remedies for the sibling problem.

One can also criticize the Dirichlet model as insufficiently flexible in parameterizing the contributions of the various ancestral populations to a typical person of the reference population. While this may be true, application of the Dirichlet model surely is better than incorrectly assuming the validity of the product rule when genetic equilibrium fails. Note that numerically bounding a match probability by adjusting ancestral gene proportions entails no distributional assumptions at all (Lange, 1993). Unfortunately, there are no obvious alternatives to the Dirichlet model, either mechanistic or phenomenological, that permit exact calculation of match probabilities in the absence of genetic equilibrium. The fact that the Dirichlet model involves substantial computation should not be a deterrent to its use. These calculations can easily be done on a personal computer. Whether a judge or jury can understand the nature of the calculations is another matter. However, this objection is irrelevant if a scientific consensus develops affirming the usefulness of the model and the feasibility of its attendant calculations. Finally, the Dirichlet model fails to be fully Bayesian. A match probability is not, after all, a posterior probability that the suspect or victim contributed the evidentiary DNA.

Despite these reservations, the Dirichlet model advances our approximate understanding of how to compute match probabilities. It is not ideal, but no scientific theory or technique ever is. Doubtless the Dirichlet model can be refined and improved. At some point, however, the legal and scientific professions need to reach a consensus about how to compute match probabilities; otherwise, DNA profiling can serve only to exonerate the innocent and never to convict the guilty. Our legal system constantly contends with approximations to the truth. This attitude, embodied in the phrase 'beyond a reasonable doubt', is as much a part of our scientific heritage as it is of our legal heritage. A rigid insistence on infallible procedures is antithetical to both professions.

Acknowledgments

I thank Ethan Lange for computing assistance and Thomas Belin and David Gjertson for reading a preliminary version of this manuscript and suggesting clarifications. This work was supported in part by the University of California, Los Angeles, and USPHS Grant CA 16042.

References

Ahrens, J.H. & U. Dieter, 1974. Computer methods for sampling from gamma, beta, Poisson and binomial distributions. Computing 12:223–246.

Bernardo, J.M., 1976. Algorithm AS 103: psi (digamma) function. Appl. Stat. 25:315–317.

Chakraborty, R. & K.K. Kidd, 1991. The utility of DNA typing in forensic work. Science 254:1735–1739.

Chakraborty, R., M.R. Srinivasan & S.P. Daiger, 1993. Evaluation of standard error and confidence interval of estimated multilocus genotype probabilities, and their applications in DNA forensics. Am. J. Hum. Genet. 52:60–70.

Cheng, R.C.H. & G.M. Feast, 1979. Some simple gamma variate generators. Appl. Stat. 28:290–295.

Devlin, B., N. Risch & K. Roeder, 1992. Forensic inference from DNA fingerprints. J. Am. Stat. Assoc. 87:337–350.

Edwards, A., H.A. Hammond, L. Jin, C.T. Caskey & R. Chakraborty, 1992. Genetic variation at five trimeric and tetrameric tandem repeat loci in four human population groups. Genomics 12:241–253.

Evett, I.W., 1992. Evaluating DNA profiles in a case where the defence is 'It was my brother'. J. Forensic Sci. Soc. 32:5–14.

Good, I.J., 1965. The Estimation of Probabilities: An Essay on Modern Bayesian Methods. MIT Press, Cambridge, MA.

Hille, E., 1959. Analytic Function Theory Vol. 1. Blaisdell Ginn, New York.

Jeffreys, A.J., V. Wilson & S.L. Thein, 1985. Individual-specific 'fingerprints' of human DNA. Nature 316:76–79.

Kingman, J.F.C., 1993. Poisson Processes. Oxford University Press, Oxford.

Lander, E., 1989. DNA fingerprinting on trial. Nature 339:501–505.

Lange, K., 1991. Comment on 'Inferences using DNA profiling in forensic identification and paternity cases' by D.A. Berry. Stat. Science 6:190–192.

Lange, K., 1993. Match probabilities in racially admixed populations. Am. J. Hum. Genet. 52:305–311.

Lee, P.M., 1989. Bayesian Statistics: An Introduction. Edward Arnold, London.

Lewin, R., 1989. DNA typing on the witness stand. Science 244:1033–1035.

Lewontin, R.C. & D.L. Hartl, 1991. Population genetics in forensic DNA typing. Science 254:1745–1750.

Mickey, M.R., J. Tiwari, J. Bond, D. Gjertson & P.I. Tersaki, 1983. Paternity probability calculations for mixed races. pp. 325–347 in Inclusion Probabilities in Parentage Testing, edited by R.H. Walker, Amer. Assoc. Blood Banks, Arlington, VA.

Miller, K.S., 1987. Some Eclectic Matrix Theory. Robert E. Krieger Publishing, Malabar, FL.

Mosimann, J.E., 1962. On the compound multinomial distribution, the multivariate β-distribution, and correlations among proportions. Biometrika 49:65–82.

Schneider, B.E., 1978. Algorithm AS 121: trigamma function. Appl. Stat. 27:97–99.

Editor's comments

The author continues the formal Bayesian analysis introduced by Gjertson & Morris in this volume. He invokes Dirichlet distributions, and so brings rigor to the discussion of the effects of population structure on match probabilities. The increased computational burden this approach entails should not be regarded as a hindrance.

B. S. Weir (ed.), Human Identification: The Use of DNA Markers, 119–124, 1995.

The honest scientist's guide to DNA evidence

Received 18 May 1994 Accepted 22 July 1994

Dear Bruce,

Thank you for your invitation to participate in the DNA symposium. As you know DNA has never been a prime research focus of mine, and I have been so preoccupied with my own work on ITPT (intertemporal personal transportation) that I thought I must decline. Happily, however, the two projects came together, for I recently had an amazing breakthrough during which by coincidence I stumbled across a book entitled *A Century of DNA Testing* and holocopied (a fancy form of Xeroxing) the following few pages for you.

Even by 1996, before the second NRC Report, which fostered the switch to Bayesian methods, appeared and several years before the general abandonment of RFLP procedures, the statistical issues that had so roiled the early days of DNA testing had largely been resolved. This occurred not because the statistical and genetic disputes we describe above had all been resolved, but rather because the need to confirm identifications from burgeoning data bases stimulated several different innovations that increased the informativeness of the RFLP approach to DNA testing, thus reducing the chances of coincidental matches to minuscule proportions even when the suspect population included relatives or had the same narrow ethnic heritage. In one Minnesota case, for example, the state crime lab simply tested alleles at nine loci. But Minnesota at the time was laboring under a short-sighted (and soon overturned) court ruling which precluded attaching statistical probabilities to DNA matches, and thus, without resorting to statistics, the lab's scientists had to be able to testify that an identification was virtually certain.

Nevertheless, problems stemming from what Professor Junipurr, some fifty years ago, in his study of forensic laboratories called the 'infallibility complex' continued to appear. Some of the most important of these problems and ways that were suggested at the time for meeting them are captured in the following extracts from the original edition (Jack Point, ed., 1996) of the still essential handbook, *The Honest Scientist's Guide to DNA Evidence*. We can in this history do no better than reproduce the relevant portions, leaving the original spelling and grammar, including archaic gendered pronouns, untouched. The somewhat preachy nature of the original handbook was, we believe, influenced by the religious fundamentalism resurgent at that period. Also, though it may not be evident to readers today, many of the Handbook's suggestions were regarded at the time as quite Utopian.

Section 7.1: (Subjectivity)

The honest scientist recognizes that she herself is a test instrument, and a fallible one at that. Subjectivity inescapably enters into any human endeavor, and should not be denied. DNA testing is rife with subjective elements, no place more so than at the crucial stage of deciding whether a match exists (1). On the one hand, non-matching extraneous bands may sometimes be properly disregarded and patterns that do not quite meet objective matching criteria may be appropriately regarded as incriminatory matches. On the other hand, band patterns that do meet objective matching criteria may be treated as exonerative depending on how they deviate from perfect matches. The DNA expert should not hide behind the cloak of science to deny the role of human judgement. White coats should not be worn into the courtroom either literally or figuratively.

At the same time, the honest scientist tries to be as objective as possible in her judgements. She realizes that this is inconsistent with a strong *a priori* belief that the donor of a suspect sample is guilty. Thus she avoids any information suggesting the involvement or uninvolvement of the accused until after she has prepared her report and, in the ideal case, until after she has testified. Laboratories should cooperate to make this easy. Crime-related information should be stripped from all information sent to the analyst unless it is essential to the test and its interpretation (e.g. information that several people are suspected of participating in a rape or that the suspected rapist is the victim's brother).

Scrupulous laboratories should also realize that being continually identified with a side may also bias judgements. Thus, they should seek out business from defense attorneys who seek to exonerate clients and encourage their scientists to take the rare opportunities that come along to testify on behalf of defendants.

An additional and highly recommended way of coping with subjectivity is to transform DNA tests from true-false into multiple-choice tests by routinely providing scientists with three test samples. One or two of these, varying on some random schedule, should be the suspect's and the other(s) should come from one or two non-suspects. This marginally increases the cost of DNA tests, but we are dealing with decisions that can take human lives. Moreover, multiple testing carries the added benefit of providing laboratory-specific proficiency information.

Section 9.3: (Proficiency Testing—continued)

The honest scientist works for the Virtuous laboratory. The Virtuous laboratory maintains a rigorous system of proficiency testing, both to identify and correct sources of error and as an incentive for continual high quality staff work. Four types of samples may be analyzed as part of a proficiency testing program: known samples, unknown samples of pure quality, unknown samples of case quality, and apparent case samples. While these different types of samples all have a role to play in measuring and maintaining laboratory proficiency, only the last, test samples regarded by both the laboratory and analyst as true casework samples, provides a true picture of the quality of laboratory procedures, for only apparent case samples are certain to be treated by laboratories as if they were casework samples. Conversely, if a laboratory knows that any apparent case sample it analyzes may be a test, the incentive is to treat each sample submitted for analysis as if it were a test sample, and the quality of all the laboratory's work should increase.

It is, even today, difficult to generate convincing casework test samples, for such samples must appear to have come from specific police departments or prosecutors' offices, and any communications with the police or prosecutor's office during the course of analysis must be consistent with the otherwise apparent genuineness of the sample. For this reason, laboratories, police and prosecutors should agree to join the proposed APTLAB (Accredited Proficiency Tested Laboratory Analysis Bureau) organization. As proposed APTLAB would serve as an intermediary between parties seeking DNA tests and laboratories that do them. APTLAB would receive all DNA test samples from police departments or other agencies and after removing unnecessary case-related information that might accompany samples (see Section 7.1) would forward the material to the laboratory that the agency chose to conduct the test. Approximately one out of every eight samples forwarded to a laboratory, on some random schedule, would be a casework test sample. Although the problem of false positive matches has been the target of most concern, APTLAB, according to the proposal, would forward as many matching test samples as non-matching ones. Missing matches also reveals weaknesses or biases in laboratory procedures, and the error of mistakenly freeing an actual rapist or murderer may be as socially harmful as mistakenly convicting an innocent person. Only after test results were communicated to APTLAB would the testing laboratory receive the name of the submitting agency. Thereafter all communications would be between the laboratory and the agency. Statistics on error rates, aggregated across laboratories using similar procedures and also by laboratories, would be made freely available.

At the moment it does not appear that APTLAB will get off the ground, for the suggestion has met with fierce resistance from forensic scientists and laboratory directors. Three arguments are made against it. Least is made of the first argument, but we suspect it is the strongest of the stated motivations behind the heated objections. This is that APTLAB will interfere with the relations that many laboratories and forensic scientists have with the agencies and individuals that submit samples to them. While this is true, the honest scientist should regard this as a virtue of the APTLAB proposal. Relations between laboratories, scientists and police agencies today may include conversation about the heinousness of the crime, which can cause a scientist to overly-identify with the police function, and other evidence that incriminates the accused. Ordinarily this information is at best irrelevant to the scientific validity of the test, and it can bias judgements and so detract from it.

A second objection one often hears is that interposing an agency between the law enforcement body and the testing laboratory simply adds a source of error. This objection cannot be dismissed out of hand, but since all test samples will be sent by courier under seal to APTLAB and the seal will not be broken in the transmission process, the possibility of introducing additional error will be minimal.

The third objection concerns costs. This objection is substantial, for running a new agency, even a genuine non-profit agency costs money as does the proposed $12\frac{1}{2}\%$ increase in the number of DNA tests conducted. These costs will, of course, be reflected in the price of DNA tests and may decrease somewhat the propensity of policy agencies to order DNA tests. However, the cases where testing will be forgone will be most likely those in which evidence of guilt apart from DNA is overwhelming and DNA evidence is window dressing not needed for a conviction. The cost issue also appears different when one thinks in terms of total social costs. It can cost between $20,000 and $50,000 to keep a rapist in prison for a year and many times that to keep a person on death row and eventually execute him. This is totally apart from the costs to the wrongfully convicted man. Conversely, mistaken acquittals can lead to the rape or murder of future victims. Thus, if the APTLAB procedures reduce errors as expected, they will be cost justified. If after some experience with APTLAB, its procedures do not appear cost justified, the consortium can be terminated, and we can have some confidence that there is no problem here.

There are also objections to APTLAB that have not figured in the debate because no one dares make them publicly. The first is the fear which some scientists privately voice, that uncomfortably high laboratory error rates will be revealed, and the second is that APTLAB will displace several other organizations that in a less rigorous but nonetheless serious fashion have begun to develop proficiency testing programs. Perhaps for these reasons more than the three that have been articulated, objections to the APTLAB proposal have been so fierce in the forensic science community that absent legal mandate the best that can be hoped for in the way of proficiency testing may be procedures, like those that have been implemented in a number of conscientious laboratories, consistent with the recommendations for proficiency testing in the first NRC report. If so, it may be that the honest scientist will find that there is no place that quite meets her standards.

Section 13.2: (Statistics and Error—continued)

The honest scientist recognizes that no matter what the odds that a random man's DNA would match evidence DNA, the probative value of a reported match between the defendant's DNA and evidence DNA is always affected by the likelihood that a match will be mistakenly reported (2). This probability is greater than the likelihood that a test error might somehow mistakenly yield a match, for it includes the chance that either inadvertently or intentionally a second sample of evidence DNA has been substituted for or contaminated the DNA of the defendant in a way that will not be obvious in the testing (3). In testifying, the honest scientist should never attach more probative weight to the evidence that this probability, which we will call the false positive rate, reflects. In particular, the extremely low probabilities often associated with random matches are likely to mislead the trier of fact and should not be mentioned or, if mentioned, should be placed very carefully in context so that they are clearly subordinate to the false positive rate.

Such careful, limited testimony poses, however, a difficulty. Unscrupulous defense attorneys may seek to interpret false positive statistics so as to induce the jury to make what Thompson and Schumann (4) have called the 'defense attorney's fallacy.' The defense attorney's fallacy falsely presupposes that there is no reason to think the defendant is guilty apart from the identification evidence. Without stating this presupposition, defense attorneys argue that statistical identification evidence simply locates the defendant as one member of a large group who might have committed the crime, and so does little to finger the suspect.

In the example that follows, drawn from the transcript of the case called *State v. Evett,* the DNA expert is careful not to overweight the probability of a reported DNA match. The cross-examiner then tries to take advantage of this honest scientist by attempting to induce the defense attorney's fallacy. On redirect examination, this effort to mislead the jury is successfully countered. In a portion of the transcript we omit, the scientist testified that a false positive rate of one in a hundred, which she later uses, is a conservative estimate. She also explains that while ordinarily the police agencies she works with divide evidence DNA and send it to several laboratories for independent testing (thus justifying a far lower false positive estimate), division was impossible in this case because of the limited amount of available evidence DNA. To conserve space we have also eliminated other aspects of the direct, cross and redirect examinations that do not bear on the issue discussed in this section.

Direct examination

Prosecutor [P]:
When you conducted the analysis you described, what exactly were your findings?

Scientist [S]:
I found that, within the limits I have already mentioned, each of the seven alleles identified in the DNA extracted from the defendant's blood matched an allele found in the DNA extracted from the semen found in the victim's vagina.

P: Did you find any alleles that did not match?

S: No, every allele we identified in the DNA extracted from the defendant's blood corresponded in length to an allele extracted from the semen.

* * *

P: Can you give the jury some idea of how unusual it is to find the seven allele match you have reported?

S: If DNA tests were perfect, the chance that one would find DNA matching that in the evidence sample by testing one Caucasian man drawn randomly from the country's population is about one in fifty million. However, as with all human endeavors DNA tests are not perfect. To the best we can estimate, in about one test of every hundred we run a DNA match is erroneously reported due to some mix up or mistake in the field or in the laboratory. So the best way of thinking about this evidence is that there is about one chance in 100 that somewhere a mistake was made, and the defendant's DNA did not match the DNA extracted from the semen. But if we did not make a mistake, then there is only about one chance in fifty million that some random person, which is to say some white person not related to the defendant, could have left the evidence DNA. Ultimately, however, the former figure is more important than the latter, so I will summarize my testimony by saying that the defendant's DNA appeared to match the evidence DNA but there is about one chance in 100 that this conclusion is mistaken. If the defendant's DNA did match, it is very unlikely he is the victim of coincidence.

P: Thank you, no further questions.

Cross examination

Defense Counsel [DC]:
Now Dr., you said that about one time in 100 when you test for DNA you report a match as you did in this case, but that match turns out to be false—isn't that so?

S: Well, not exactly. The one in a hundred figure is a conservatively estimated error rate based on what are known as 'double blind' proficiency tests.

DC: But one in a hundred is your best estimate of the false positive rate, isn't that so?

S: Best conservative estimate.

DC: Now there are about two million Caucasian males living within 40 miles of where the crime occurred, isn't that so?

S: If you say so; I don't live here.

DC: Well you can take my word on it. Now if any of these men had been chosen at random and tested there is about a one in a hundred chance that you would be here telling me there was a match, isn't that so?

S: Yes, assuming they had been suspected as the defendant was.

DC: Dr. please just answer 'yes' or 'no.' Now it follows doesn't it that if all two million Caucasian men had been tested, there would be about 20,000 positive tests reported?

S: Well, it's more complicated that that.

DC: Dr. 1/100 times two million is 20,000 is it not?

S: Yes.

DC: So basically what your test establishes is that there is about one chance in 20,000 that my client is the culprit.

S: No, that's . . .

DC: Dr. I didn't make any mathematical mistakes, did I?

S: No.

DC: Thank you, no further questions.

Redirect examination

P: Dr. do you feel that the defense counsel was accurately stating the import of the match you reported, when she suggested that it meant that there was one chance in 20,000 that the defendant was the culprit?

S: No, I don't.

P: Can you explain why?

S: Yes, the defense counsel was trying to induce the jury to buy into what is known in the literature as the defense attorney's fallacy. It misleads because it assumes that there is no other evidence in the case.

P: Can you say more?

S: Yes, the defense counsel's calculations would not be misleading had the defendant's name simply been drawn from the phone book and had he then been required to undergo a DNA test. If the police drew 100 such names from the phone book and processed each individual as they would if he were the prime suspect, handling his DNA and the evidence DNA as they would that of a suspect, proficiency test results suggest that even if all the people tested were innocent, there is a good chance that we would mistakenly conclude in the case of one person that there was a DNA match. If we tested more people at random the chance of finding additional false positive results would increase, with the expectation being that one such result would occur in every hundred people tested, provided the DNA and tests were handled exactly as they are when actual suspects are tested.

But this defendant's name was not drawn at random from a phone book. Defense counsel's fundamental error is to ignore this fact; the defendant was not selected for the DNA test at random. Instead there was a reason the defendant's DNA was tested. As I understand the state's evidence in this case . . .

DC: Objection.Your Honor, the witness's understanding of the evidence is irrelevant and interferes with the province of the jury.

Judge [J]

Were you going to interpret the evidence in this case Dr. or just list what has been introduced?

S: I was just going to list it, Your Honor.

J: In that case you may proceed. Please do not give your opinion of the evidence or mention any evidence the state has not introduced, for if you do I shall have to sustain defense counsel's objection.

S: As I was saying the defendant was not just a random person selected for testing. In this case the defendant was picked up because the last three digits on his license plate matched the information provided by a woman who witnessed the abduction, the victim testified that though the defendant had worn a ski mask, she recognized his eyes and his voice, when he was arrested the defendant had scratches on his face, which he refused to explain to the police, and a hair was found in the defendant's car consistent with hair that might have come from the victim. In these circumstances, even with a false positive rate of one in one hundred, the evidence of a DNA match I presented is powerful confirmatory evidence. Had the defendant been innocent despite the other evidence against him 99 times out of 100 the DNA test would have demonstrated his innocence. If his alibi were true, the DNA test would fail to confirm it only one time in a hundred.

Another way to think about this is to think about the number of innocent Caucasian males, among the two million who live in the area, who might by coincidence be associated with as much incriminatory evidence as the defendant in this case. How many others, for example, would share the same three license plate digits, have unexplained scratches, sound like the defendant and have hair that might have come from the victim in their car or have an equivalent amount of other incriminatory evidence pointing to them? What is this number? One other? Five others? Ten others? While it is for the jury to say and not I, my hunch is that the number is so small that a test with a false positive rate of one in one hundred would be very unlikely to mistakenly identify anyone.

P: Thank you, Dr., no further questions.

Considering when it was written, in some respects *The Honest Scientist's Guide to DNA Evidence* was indeed Utopian, but as we document in the pages that follow, in other respects the *Guide* sparked debates that led to needed reforms.

Well Bruce, that's it; at least that's all I could copy and bring back with me given existing mass constraints and the danger of temporal disruption. Of course, before I submit this to your volume, I must rewrite this material so it looks like these are my own ideas. And please keep what I have told you regarding the true source of my comments confidential until the equations for intertemporal transportation appear in *Nature*, and I have taken the first steps towards patenting my machine.

Sincerely,
Richard Lempert

(1) Thompson & Ford, The Meaning of a Match: Sources of Ambiguity in the Interpretation of DNA Prints. In *Forensic DNA Tech.* Ch. 7 (M. Farley & J. Harrington, eds. 1991).

(2) Lempert, Some Caveats Concerning DNA as Criminal Identification Evidence: With Thanks to the Reverend Bayes, 13 *Cardozo Law Review* 303 (1991).

(3) Establishing the likelihood of false positive errors and their implications for the probative value of DNA matches is no simple matter even putting aside the number of proficiency tests that should be conducted to achieve reliable error rate estimates. For example, the first external proficiency tests reported in the literature were conducted by the California Association of Crime Laboratory Directors which sent 50 mock forensic samples to several of the laboratories that pioneered in the forensic analysis of DNA identification evidence. In one set of tests, for example, a laboratory made a false positive error which a number of early commentators, particularly in law reviews, took as indicating that the best empirically grounded false positive error rate estimate was 2%. However, the laboratory was charged with comparing each sample with every other, which suggests that 1225 comparisons were made, leaving the impression of a far higher accuracy rate, a point also made in the literature. Neither portrait is correct. First, the laboratory in question was able to compare only 45 samples, and the presence of matching samples, including some that were matched by three other samples, further limited the number of false positives that could occur. Moreover, the likely source of the false positive was the inadvertent transfer of DNA in a sample in one test tube to an adjacent DNA containing test tube. This type of error could not affect most comparisons in the proficiency test because the test tubes were not all adjacent to each other. In actual forensic testing, however, suspect and evidence samples might routinely occupy adjacent test tubes in the test preparation process, suggesting that the 2% error rate estimate could be valid after all. However, the laboratory involved quite sensibly changed its procedures following the proficiency test results (and it claimed it had ordinarily done so in forensic work before the test) so little can be made of this figure as an estimate of the laboratory's future false positive rate. The story of this test and the error rate estimates it generated emphasizes the importance of double blind proficiency tests in which laboratories cannot distinguish proficiency test samples from ordinary forensic samples. For a discussion of further complications regarding the implications of error probabilities, see Thompson's discussion in this volume.

(4) Thompson & Schumann, Interpretation of Statistical Evidence in Criminal Trials, 11 *Law & Human Behavior* 167 (1987).

Editor's comments

The author's 'piece of whimsy' (his phrase) adds a light touch to this volume, even though it contains a serious discussion of the issue of false positives.

B. S. Weir (ed.), Human Identification: The Use of DNA Markers, 125–138, 1995.

A comparison of tests for independence in the FBI RFLP data bases

P.J. Maiste & B.S. Weir*
Program in Statistical Genetics, Department of Statistics, North Carolina State University, Raleigh, NC 27695-8203, USA
**Author for correspondence*

Received 1 July 1994 Accepted 1 December 1994

Key words: Hardy-Weinberg, forensic databasses, statistical tests, FBI data

Abstract

Several tests of independence of allelic frequencies within and between loci have been compared, and it has been found that Fisher's exact test is the best test to use. When this test is applied to RFLP databases established by the FBI, paying no attention to the single-band problem, there is generally evidence for independence at one locus but not at two loci. When the test is restricted to double-banded entries in the databases; there is overall evidence for independence.

Introduction

Independent allelic frequencies can be multiplied together to provide estimates of genotypic frequencies. This paper discusses a range of tests for independence of allelic frequencies at one or two loci, and serves as an introduction to the paper by Zaykin, Zhivotovsky and Weir in this volume that concentrates on exact tests at arbitrary numbers of loci. The tests described are applied to RFLP data collected by the FBI, so that the paper also serves to amplify the discussion given by Weir (1992).

The task of testing hypotheses about independence for loci with many alleles is not as straight-forward or well-established as might be supposed. Although the basic population-genetic law concerning independent alleles, the Hardy-Weinberg law (HWE), was enunciated in 1908, the statistical difficulties in testing for consistency of observations with this law continue to exercise statisticians. This paper extracts the material most directly relevant to forensic databases from a wider study by Maiste (1993), which examined several test procedures. If a statistical test is conducted to present an answer that data either support or do not support a hypothesis, one of two types of error may be committed. Either support will be declared when the hypothesis is false, or non-support will be declared when the hypothesis is true. Since the truth or falsity of the hypothesis is unknown, it is important to use tests with low expected error rates. The work of Maiste (1993) compares tests on the basis of such rates. In particular, attention is paid to size, or significance level, and power.

The significance, or α, level of a test is the probability that it will lead to false rejection of a true hypothesis. Generally this level is specified and leads to a rule by which a given data set will cause rejection. The significance level is also called the size of a test, and there may be problems in achieving a desired size simply because of the nature of the test. If test statistic has only a small number of possible values it may not be possible to have the size be exactly 0.05, for example. The probability of the value of a test statistic, if the data on which it is based are chosen from a population obeying the hypothesis, is usually called the p-value, so that rejection corresponds to $p \leq \alpha$.

The power of a test is the probability that a false hypothesis will be rejected. Although it is desirable to compare test procedures on the basis of power, the task is difficult because power determination requires a specification of exactly how the population fails to satisfy the hypothesis. At the least, some single parameter of departure needs to be identified, and a range of values of this parameter considered.

The hypotheses

Testing for independence of the two allelic frequencies at one locus is the same as testing for Hardy-Weinberg equilibrium. If locus **A** has alleles A_i with frequencies p_i, the Hardy-Weinberg relation states that genotypes A_iA_j have frequencies $\Pr(A_iA_j)$ given by

$$\Pr(A_iA_j) = \begin{cases} p_i^2 & i = j \\ 2p_ip_j & i \neq j \end{cases}$$

and there is a rich literature addressing appropriate ways to test whether a sample of genotypes from a population has frequencies consistent with this relation. Reviews have been given by Emigh and Kempthorne (1975), Elston and Forthofer (1977), Emigh (1978, 1980), Haber (1981) and Hernández and Weir (1989). Almost all the discussion in these papers has been for loci with two alleles, with some mention of the case of three or four alleles per locus. However, the loci being used for human identification, especially the RFLP systems, are characterized by having many more than two alleles. Indeed, this is the reason they have been used. These multi-allele loci have prompted a new examination of tests of independence.

The first careful study of the issue was given by Guo and Thompson (1992), who showed the feasibility of conducting exact tests with significance levels generated by permutation. Weir (1992) discussed likelihood-ratio tests. Maiste (1993) provided a numerical comparison of a range of different tests, and full details of those comparisons will be published elsewhere.

At two loci, the issue of interest in the use of genetic markers for human identification is whether two-locus genotype frequencies can be represented as the products of four allelic frequencies — two at each locus. If loci **A**, **B** have alleles A_i with frequencies p_i and B_k with frequencies q_k respectively, then the hypothesis is

$$\Pr(A_iA_jB_kB_l) = \begin{cases} p_i^2q_k^2 & i = j, k = l \\ 2p_ip_jq_k^2 & i \neq j, k = l \\ 2p_i^2q_kq_l & i = j, k \neq l \\ 4p_ip_jq_kq_l & i \neq j, k \neq l \end{cases}$$

Test statistics

Sampling theory

Locus **A** has alleles A_i, $i = 1, 2, \ldots a$ with population frequencies p_i and sample frequencies written as $\tilde{p}_i$. In a sample of size n individuals, the count for genotype A_iA_j is written x_{ij}. Although it is convenient to suppose that $i \leq j$, a sum such as $\sum_{j>i} x_{ij}$ is taken to be over all heterozygotes for allele A_i, and to count every such heterozygote only once. The number of heterozygotes in a sample is written as H

$$H = \sum_i \sum_{j>i} x_{ij}$$

When sampling is at random, the probability of a set of genotypic counts is given by the multinomial distribution. If P_{ij} is the population frequency of genotype A_iA_j, the probability of the set of counts $\vec{x} = \{x_{11}, x_{12}, \ldots, x_{aa}\}$ is

$$\Pr(\vec{x}) = n! \prod_{i=1}^{a} \prod_{j \geq i} \frac{P_{ij}^{x_{ij}}}{x_{ij}!}$$

Under the null hypothesis H_0 of Hardy-Weinberg frequencies, this becomes

$$\Pr(\vec{x}|H_0) = \frac{n! 2^H \prod_{i=1}^{a} (p_i)^{2n\tilde{p}_i}}{\prod_{i=1}^{a} \prod_{j \geq i} x_{ij}!} \tag{1}$$

For two loci, the genotype $A_iA_jB_kB_l$ has count x_{ijkl} and population frequency P_{ijkl}. The numbers of alleles at the two loci are written as a_1 and a_2, respectively. The number of heterozygotes in the sample is the sum of the numbers of heterozygotes at each locus

$$H = \sum_{i=1}^{a_1-1} \sum_{j>i} \sum_{k=1}^{a_2} x_{ijkk} + \sum_{i=1}^{a_1} \sum_{k=1}^{a_2-1} \sum_{l>k} x_{iikl} + 2 \sum_{i=1}^{a_1-1} \sum_{j>i} \sum_{k=1}^{a_2-1} \sum_{l>k} x_{ijkl}$$

Under random sampling, the multinomial probability of the sample array is

$$\Pr(\vec{x}) = n! \prod_{i=1}^{a_1} \prod_{j \geq i} \prod_{k=1}^{a_2} \prod_{l \geq k} \frac{P_{ijkl}^{x_{ijkl}}}{x_{ijkl}!}$$

and, under the null hypothesis of independence, this becomes

$$\Pr(\vec{x}|H_0) = \frac{n!2^H \prod_{i=1}^{a_1}(p_i)^{2n\tilde{p}_i} \prod_{k=1}^{a_2}(q_k)^{2n\tilde{q}_k}}{\prod_{i=1}^{a_1}\prod_{j\geq i}\prod_{k=1}^{a_2}\prod_{l\geq k} x_{ijkl}!} \quad (2)$$

Pearson's goodness of fit statistic: X^2

At the beginning of this century, Pearson (1900) proposed a statistic to measure goodness of fit to a model. It has since become one of the most widely used, discussed, and debated statistics in the literature (Cressie & Read, 1989). In general, it can be defined as

$$X^2 = \sum_{i=1}^{K} \frac{(X_i - E_i)^2}{E_i}$$

where K is the number of cells or categories in the model, X_i is the observed count for cell i, and E_i is the expected count for cell i under some null hypothesis. In the notation of this paper, where cells are genotypes,

$$X^2 = \sum_{i=1}^{a} \frac{(x_{ii} - n\tilde{p}_i{}^2)^2}{n\tilde{p}_i{}^2} + \sum_{i=1}^{a-1}\sum_{j>i} \frac{(x_{ij} - 2n\tilde{p}_i\tilde{p}_j)^2}{2n\tilde{p}_i\tilde{p}_j}$$

for one locus, and

$$\begin{aligned} X^2 = & \sum_{i=1}^{a_1}\sum_{k=1}^{a_2} \frac{(x_{iikk} - n\tilde{p}_i{}^2\tilde{q}_k{}^2)^2}{n\tilde{p}_i{}^2\tilde{q}_k{}^2} \\ & + \sum_{i=1}^{a_1-1}\sum_{j>i}\sum_{k=1}^{a_2} \frac{(x_{ijkk} - 2n\tilde{p}_i\tilde{p}_j\tilde{q}_k{}^2)^2}{2n\tilde{p}_i\tilde{p}_j\tilde{q}_k{}^2} \\ & + \sum_{i=1}^{a_1}\sum_{k=1}^{a_2-1}\sum_{i>k} \frac{(x_{iikl} - 2n\tilde{p}_i{}^2\tilde{q}_k\tilde{q}_l)^2}{2n\tilde{p}_i{}^2\tilde{q}_k\tilde{q}_l} \\ & + \sum_{i=1}^{a_1-1}\sum_{j>i}\sum_{k=1}^{a_2-1}\sum_{l>k} \frac{(x_{ijkl} - 4n\tilde{p}_i\tilde{p}_j\tilde{q}_k\tilde{q}_l)^2}{4n\tilde{p}_i\tilde{p}_j\tilde{q}_k\tilde{q}_l} \end{aligned}$$

for two loci.

Asymptotically, X^2 was proven to have a chi-square distribution with $K - s - 1$ degrees of freedom by Fisher (1924), contrary to Pearson's (1990) claim that it had $K - 1$ degrees of freedom. The s here is the number of independently estimated parameters. For HWE testing, $s = a - 1$ and therefore there are $a(a + 1)/2 - a = a(a - 1)/2$ degrees of freedom. When testing for independence at two loci, $s = a_1 + a_2 - 2$ and so the degrees of freedom are $a_1(a_1 + 1)a_2(a_2 + 1)/4 - a_1 - a_2 + 1$.

When does this asymptotic result hold? The long standing and very general rule of thumb, discussed by Cochran (1954), is that expected counts in most cells should be at least five. If this does not hold, the suggestion is that cells should be combined until the rule can be satisfied. However, when combining genotypic classes, departures from HWE may be masked by such action and, in general, this idea leads to a loss in the power of the test statistic (Cochran, 1954). These are among the reasons that it is necessary to study exact tests as well as approximate tests.

Log-likelihood ratio statistic: G^2

Another well known and well used test throughout statisics is the likelihood ratio test. In general, this test compares the values of likelihood functions under the null and under the alternative or full model. Typically, the logarithm of the ratio of these likelihoods is taken to produce the log-likelihood ratio statistic.

For the one-locus multinomial case, and for the Hardy-Weinberg hypothesis, from Equation 1 the likelihood under the null hypothesis is

$$\begin{aligned} \mathrm{L}_0 & \equiv \mathrm{L}(\vec{p}|\vec{x}, \mathrm{H}_0) \\ & \propto \left(\prod_{i=1}^{a}(\tilde{p}_i{}^2)^{x_{ii}}\right)\left(\prod_{i=1}^{a-1}\prod_{j>i}(2\tilde{p}_i\tilde{p}_j)^{x_{ij}}\right) \\ & = 2^H \prod_{i=1}^{a}(\tilde{p}_i)^{2n\tilde{p}_i} \end{aligned}$$

and the likelihood not constrained by the null hypothesis is

$$\begin{aligned} \mathrm{L}_1 & \equiv \mathrm{L}(\vec{P}|\vec{x}) \\ & \propto \prod_{i=1}^{a}\prod_{j\geq i}(\tilde{P}_{ij})^{x_{ij}} \end{aligned}$$

Therefore

$$\begin{aligned} \log\left(\frac{\mathrm{L}_0}{\mathrm{L}_1}\right) & = H\log 2 + 2n\sum_{i=1}^{a}\tilde{p}_i\log\tilde{p}_i \\ & \quad - \sum_{i=1}^{a}\sum_{j\geq i} x_{ij}\log\tilde{P}_{ij} \end{aligned}$$

where $\tilde{P}_{ij} = x_{ij}/n$.

An analogous argument for two loci, from Equation 2, leads to

$$\log\left(\frac{L_0}{L_1}\right) = H\log 2 + 2n\sum_{i=1}^{a_1}\tilde{p}_i\log\tilde{p}_i + 2n\sum_{k=1}^{a_2}\tilde{q}_k\log\tilde{q}_k - \sum_{i=1}^{a_1}\sum_{j\geq i}\sum_{k=1}^{a_2}\sum_{l\geq k} x_{ijkl}\log\tilde{P}_{ijkl}$$

The log-likelihood ratio statistic is denoted by

$$G^2 = -2\log\left(\frac{L_0}{L_1}\right)$$

Notice that G^2 will be large when the data give evidence against the null and small when the null is supported.

Similar to X^2, it is well known that G^2 has a limiting chi-square distribution with the same degrees of freedom as for X^2 (e.g. Casella & Berger, 1990). Once again, however, this approximation is suspect under large, sparse multinomial sampling. In fact, previous results seem to show that G^2 is often less well approximated by chi-square than is X^2. For example, Larntz (1978), Koehler and Larntz (1980), and Koehler (1986) have found that X^2 is more robust to smaller sample sizes and larger tables than is G^2, where robustness is with respect to the chi-square approximation. They also claim that chi-square is not usually a good approximation for G^2 when $n/K < 5$, and G^2 can often be either very liberal or very conservative when this is true. So, as for Pearson's X^2, G^2 will not necessarily follow a chi-square distribution for the genotypic data discussed here.

Power-divergence statistic: $I^{2/3}$

Cressie and Read (1984) defined what they termed power-divergence statistics. This is a family of statistics for use in goodness of fit testing, and is indexed by a single parameter λ. This family is quite interesting; through this one parameter, many previously studied goodness of fit statistics are tied together in a continuum. The definition, as laid out in Cressie and Read (1984), is

$$2nI^{\lambda}\left(\frac{1}{n}\vec{x} : \Pi_0\right) = \frac{2}{\lambda(\lambda+1)}\sum_{i=1}^{K} X_i\left\{\left(\frac{X_i}{E_i}\right)^{\lambda} - 1\right\}, \lambda \text{ real}$$

where for $\lambda = 0$ and -1, $2nI^{\lambda}$ is defined by continuity, as in

$$2nI^{0}\left(\frac{1}{n}\vec{x} : \Pi_0\right) = \lim_{\lambda\to 0} 2nI^{\lambda} = 2\sum_{i=1}^{K} X_i\log\frac{X_i}{E_i}$$

In their notation, X_i is the observed count for cell i, E_i is the expected count under some hypothesis, K is the total number of cells, and Π_0 represents the null parameter space. From here on, this statistic will be donated simply by I^{λ} instead of $2nI^{\lambda}\left(\frac{1}{n}\vec{x} : \Pi_0\right)$.

Notice that I^0 reduces to the log-likelihood ratio statistic and I^1 is Pearson's X^2. In addition, other goodness of fit statistics that have been proposed fall into this family. For example, $I^{-\frac{1}{2}}$ is the Freeman-Tukey statistic (Fienberg, 1979; Moore, 1986), I^{-1} is the modified log-likelihood ratio statistic (Kullback, 1959, 1985), and I^{-2} is the Neyman-modified statistic (Newman, 1949).

What are the properties of the power-divergence family? Read (1982) showed that all members of this family are asymptotically equivalent. That is, they all converge to the same chi-square distribution under the null hypothesis. This was proven to be true even for complex hypotheses with nuisance parameters. Thus, intuition would lead one to believe that if asymptotics for X^2 or G^2 are not working in a certain situation, then other members of the family would also have poor asymptotic approximations.

Cressie and Read (1984) give recommendations for choosing a value of λ. These recommendations have apparently been based on models with small samples, small numbers of cells, reasonably high expected counts, and rate of convergence to the chi-square distribution. Although no single λ value performed best in all situations, they deem $\lambda = 2/3$ as a strong competitor and an 'excellent compromise' to the usual X^2 ($\lambda = 1$) and G^2 ($\lambda = 0$) statistics. This recommendation leads to consideration of the statistic

$$I^{2/3} = \frac{9}{5}\left\{\sum_{i=1}^{a} x_{ii}\left[\left(\frac{x_{ii}}{n\tilde{p}_i{}^2}\right)^{\frac{2}{3}} - 1\right] + \sum_{i=1}^{a-1}\sum_{j>i} x_{ij}\left[\left(\frac{x_{ij}}{2n\tilde{p}_i\tilde{p}_j}\right)^{\frac{2}{3}} - 1\right]\right\}$$

for one locus, and

$$I^{2/3} = \frac{9}{5}\Bigg\{\sum_{i=1}^{a_1}\sum_{k=1}^{a_2} x_{iikk}\left[\left(\frac{x_{iikk}}{n\tilde{p}_i{}^2\tilde{q}_k{}^2}\right)^{\frac{2}{3}} - 1\right]$$

$$+\sum_{i=1}^{a_1-1}\sum_{j>i}\sum_{k=1}^{a_2} x_{ijkk}\left[\left(\frac{x_{ijkk}}{2n\tilde{p}_i\tilde{p}_j\tilde{q}_k{}^2}\right)^{\frac{2}{3}} - 1\right]$$

$$+\sum_{i=1}^{a_1}\sum_{k=1}^{a_2-1}\sum_{l>k} x_{iikl}\left[\left(\frac{x_{iikl}}{2n\tilde{p}_i{}^2\tilde{q}_k\tilde{q}_l}\right)^{\frac{2}{3}} - 1\right]$$

$$+\sum_{i=1}^{a_1-1}\sum_{j>i}\sum_{k=1}^{a_2-1}\sum_{l>k} x_{ijkl}\left[\left(\frac{x_{ijkl}}{4n\tilde{p}_i\tilde{p}_j\tilde{q}_k\tilde{q}_l}\right)^{\frac{2}{3}} - 1\right]\Bigg\}$$

for two loci.

Fisher's exact test statistic: F

Fisher (1935) paved the way for the use of *probability tests.* Such tests are based on the multinomial probability of occurrence of all possible samples given the observed allele counts. No asymptotics are used; hence the term 'exact.' Fisher's exact test is carried out as a conditional probability test. Until recent years, it could be used only for very small contingency tables with small sample sizes because of the computational difficulty of enumerating all possible samples.

Guo and Thompson (1992) give a good discussion on the use of Fisher's exact test for HWE in the presence of an arbitrary number of alleles. The probability of any sample of size n under the null, conditional on observed allele counts (and correcting a small error in Guo and Thompson, 1992) is

$$\Pr(\vec{x} \mid \text{allele counts}) = \frac{2^H n! \prod_{i=1}^{a}(2n\tilde{p}_i)!}{(2n)! \prod_{i=1}^{a}\prod_{j\geq i} x_{ij}!}$$

$$\propto \frac{2^H}{\prod_{i=1}^{a}\prod_{j\geq i} x_{ij}!} \qquad (3)$$

Notice that this proportionality holds since $\prod_{i=1}^{a}(2n\tilde{p}_i)!$ is a constant under the conditional argument ($2n\tilde{p}_i$ are the observed allele counts). It may not be immediately obvious why conditioning is done on allele counts in the case of genetic data. However, noticing that $2n\tilde{p}_i$, $i = 1, \ldots, a$, are sufficient statistics for p_i makes this situation analogous to conditioning on the margins of a 2×2 contingency table. As can be seen from Equation 3, the attraction of conditioning is that the allele frequencies have been conditioned out and are no longer a part of this probability. Of course, this is as desired since the frequencies themselves are of no interest; there is interest only in there being certain relationships among them as expressed by the null hypothesis. Therefore, there is no further dependence on these nuisance parameters.

Now, after taking logarithms, the test statistic for one locus is defined as

$$F = H\log 2 - \sum_{i=1}^{a}\sum_{j\geq i}\log x_{ij}!$$

and

$$F = H\log 2 - \sum_{i=1}^{a_1}\sum_{j\geq i}\sum_{k=1}^{a_2}\sum_{l\geq k}\log x_{ijkl}!$$

for two loci.

When using Fisher's exact test to calculate a p-value, the probability of seeing the observed data or a more extreme dataset under the null, all possible conditional samples are ranked based on their conditional probability of occurrence under the null. That is, a sample's departure from the null is measured by its probability of having occurred, not on a typical goodness of fit measure such as X^2, G^2, or $I^{2/3}$.

Exact multinomial test statistic: M

The exact multinomial test, discussed by Cressie and Read (1989), is a probability test and is very similar in principle to Fisher's exact test. All possible datasets are ranked according to exact multinomial probabilities. However, contrary to Fisher's test, an unconditional analysis is used. This probability, under the null, comes directly from Equations 1 and 2 for one and two loci, respectively.

Therefore, for one locus, the test statistic is defined as

$$M = H\log 2 + 2n\sum_{i=1}^{a}\tilde{p}_i\log\tilde{p}_i - \sum_{i=1}^{a}\sum_{j\geq i}\log x_{ij}!$$

and for two loci as

$$M = H\log 2 + 2n\sum_{i=1}^{a_1}\tilde{p}_i\log\tilde{p}_i$$

$$+2n\sum_{k=1}^{a_2}\tilde{q}_k\log\tilde{q}_k - \sum_{i=1}^{a_1}\sum_{j\geq i}\sum_{k=1}^{a_2}\sum_{l\geq k}\log x_{ijkl}!$$

Notice that because this statistic is used in an unconditional analysis, the probabilities p_i in the multinomial formula have been replaced by their observed values (their maximum likelihood estimates).

Conditional versus unconditional tests

There has been much debate on whether it is appropriate to perform conditional tests on contingency tables. The same arguments carry over to multinomial testing with nuisance parameters. Berkson (1978), Upton (1982), Suissa and Shuster (1985), D'Agostino, Chase and Belanger (1988) and Storer and Kim (1990) have shown quite conclusively through extensive Monte Carlo studies that Fisher's exact test is very conservative. Consequently, it has much lower power in many cases than an asymptotic test statistic such as Pearson's X^2.

The number of possible datasets after conditioning becomes quite large even for small numbers of alleles and small sample sizes. For example, for $a = 4$ and $n = 50$, there can be as many as one million configurations of the genotypic counts conditional on allele counts. This calculation was made with the help of an algorithm by Louis and Dempster (1987). A second argument for the use of unconditional tests was given by Suissa and Shuster (1985). They claim that the p-values from unconditional tests are easier to interpret. This remark seems reasonable and has not been debated.

On the other side of the argument, authors such as Yates (1984), Little (1989), Camilli (1990) and Greenland (1991) defend Fisher's exact test, relying on arguments of sufficiency, ancillarity, and plain common sense. In addition, arguments such as those by Little (1989) have been used to undermine the Monte Carlo studies showing conservatism of Fisher's exact test.

Both sides seem to sidestep other important issues in the realm of hypothesis testing. First, the word 'conditional' when speaking of conditional tests means that only datasets which satisfy the observed values of the sufficient statistics (i.e. margins of a 2×2 contingency table, or allele frequencies) are considered in calculating p-values. This does not say anything about which of these datasets are to be considered as evidence against H_0 with respect to the observed data. Despite this, the prevailing strategy has always been to base departure from H_0 on the conditional probability of observing a particular dataset. This leads to Fisher's exact test. However, it is not necessary that less likely datasets signify evidence against H_0 (Radlow & Alf, 1975; Gibbons & Pratt, 1975; Agresti & Wackerly, 1977; Cressie & Read, 1989). An alternative, then, is to base departure from H_0 on statistics which were designed to measure goodness of fit, namely X^2, G^2 and $I^{2/3}$. In fact, for a locus with three alleles, Louis and Dempster (1987) showed that there can be a significant difference between the probability method and the X^2 method of ranking datasets (see also the paper by Zaykin, Zhivotovsky and Weir in this volume).

The method used here of implementing unconditional tests was used by Storer and Kim (1990). They called it the *appropriate unconditional test*. The main idea is to obtain estimates for the nuisance parameters (allele frequencies), typically by maximum likelihood. Then a goodness of fit statistic can be computed as if these estimates were the true values, and a p-value can be calculated. They found that the obtained significance level of this test rarely exceeds the nominal level for 2×2 tables. Obviously, as n increases and the estimates better reflect the true population, this approximate method becomes closer to the exact unconditional method mentioned earlier.

Conditional tests are performed by conditioning on a sufficient statistic for the nuisance parameter. As mentioned previously, the allele counts, $2n\tilde{p}_i$, are sufficient for allele frequencies under H_0. Thus, a conditional test enumerates all possible genotypic count vectors $\vec{x}_j$ that have the same allele counts as the observed data $\vec{x}$.

Exact versus asymptotic tests

Conducting a test using the asymptotic distribution of the statistic is straightforward. It is just a matter of calculating the statistic and looking up the p-value in a table or with a computer package.

Performing an exact test is not as easy. The exact, finite sample size distribution of the statistic under the null is needed, and this is typically impossible to write down. Thus, a computer program is required to estimate the tail of the distribution of the statistic, or directly calculate the p-value for the test.

The word 'exact' is actually misleading. None of the tests considered here are truly exact. Approximations are used at many levels. For example, as alluded to in the previous section, the true distribution of exact unconditional tests depends on the nuisance parameter $\vec{p}$. Since these allele frequencies are of course unknown, the true distribution can only be approximated by substituting maximum likelihood estimates for $\vec{p}$.

Table 1. Summary of tests of independence.

Test	Approximate unconditional	Exact conditional	Asymptotic
Log-likelihood ratio	${G^2}_{AU}$	${G^2}_{EC}$	${G^2}_{AS}$
Multinomial	M_{AU}	—	—
Fisher's exact	—	F_{EC}	—
Pearson	${X^2}_{AU}$	${X^2}_{EC}$	${X^2}_{AS}$
Power-divergence	${I^{2/3}}_{AU}$	${I^{2/3}}_{EC}$	${I^{2/3}}_{AS}$

Other levels of approximation are more computer oriented. First, since all possible conditional or unconditional datasets cannot possibly be generated for the large multinomials under consideration, only a sample of these can be taken. Similarly, only a small proportion of all possible samples from the population can be analyzed. These approximations, however, can be as good as one likes, with the only restriction being the amount of computer time involved, as discussed by Guo and Thompson (1992) and Zaykin, Zhivotovsky and Weir (this volume).

It can be seen that exact tests are computationally much more demanding than asymptotic tests. However, good computer methods can make the extra time worthwhile if the exact test is the better test. For applications such as to the FBI databases treated here, it is much more important to be correct than fast.

Testing summary

The various testing strategies are summarized in Table 1. There are five statistics, X^2, G^2, $I^{2/3}$, F, M, and three classes of tests, approximate unconditional, exact conditional, and asymptotic. From here on, the abbreviations AU, EC, and AS will be used to refer to these three classes, respectively. These will also be used as subscripts on X^2, G^2, $I^{2/3}$, F and M to refer to a specific class of one of these statistics. Table 1 summarizes this notation. Note that, by definition, F is evaluated only as an EC test and M is evaluated only as an AU test. Although these two tests are both based on the same principle, that is multinomial probabilities, they are given different notation here in the interest of historical accuracy with respect to Fisher's exact test.

Comparison of one-locus tests

A full investigation of the comparative properties of the tests in Table 1, when used to test for independence of allelic frequencies at one or two loci, was given by Maiste (1993) and will be published elsewhere. It is convenient here to summarize the main conclusions of Maiste's study.

In choosing which version of a test to use, power does not seem to be a consideration. As long as the respective versions, AU, EC, and AS, have the appropriate size, they have virtually the same power when considering a particular statistic. Therefore, the importance lies in choosing the test statistic. It seems that the probability tests, F and M, are superior to G^2, X^2, and $I^{2/3}$, in small samples. In larger samples, X^2 is comparable to these two.

For the specific case of large multinomials with many nuisance parameters in H_0, the EC tests are generally superior to AU tests. Although the power of an EC test is the same as the power for the corresponding AU test, the EC test is guaranteed to have the correct α-level. For small samples, AU tests may or may not attain the correct level, and this will depend on factors such as the unknown true allele frequencies. For larger samples, these two classes of tests give virtually identical results in both power and attained size. Therefore, EC tests are never beaten by corresponding AU tests.

There is one very imposing reason to wish to use asymptotic tests over exact tests, namely computer time. Using the asymptotic approximations of G^2, X^2, or $I^{2/3}$ to conduct a Hardy-Weinberg test is virtually instantaneous and does not require a specially written computer program. The opposite is true of exact tests. So, if it is the case that the asymptotic version of a test compares equally with the exact version, there is certainly good reason to use the chi-square approximation. If the opposite is true, however, it is worth the effort to obtain the capability to perform exact tests for important problems.

There are negligible differences in power between the EC and AS versions of a test statistic. Thus, the deciding factor is again the ability to attain the correct significance level. Also, EC tests have the advantage since AS tests are not guaranteed to have the correct size. $G^2{}_{AS}$ is always very unreliable, $I^{2/3}{}_{AS}$ is unreliable for small n, and $X^2{}_{AS}$ is unreliable for small n and unequal $\vec{p}$.

However, for large samples, $X^2{}_{AS}$ and $I^{2/3}{}_{AS}$ are dependable in their size and equivalent in terms of power. Therefore, because of the advantage of computer speed, AS tests become a smart choice as n increases. The next objective is deciding which AS test is best, and does it beat the probability test, F_{EC}.

Among the goodness of fit tests, G^2, X^2, $I^{2/3}$, and the probability tests F and M, X^2 and F are the best all around. When sample sizes are small, that is $n \leq 250$ when $a = 20$ and $n \leq 100$ when $a = 10$, F is certainly the best choice. However, for n larger than this, X^2, and in particular $X^2{}_{AS}$, matches F in power and does well in terms of size. Therefore, if the sample size is larger than the above cutoffs, $X^2{}_{AS}$ seems to be a perfectly legitimate statistic for this goodness of fit testing situation. A problem point is that $X^2{}_{AS}$ does seem quite sensitive to unequal allele frequencies, and care must be taken in cases where the frequencies may be even more extreme than those used in the simulations of Maiste (1993).

Therefore, F_{EC} is recommended for small samples, as defined above, and $X^2{}_{AS}$ is recommended for larger samples. Because allele frequencies are quite often very extreme in forensic databases, and to be conservative, it is recommended that F_{EC} be used when $n \leq 500$ and a is near 20. Otherwise, $X^2{}_{AS}$ seems preferable.

The fact that Fisher's exact test compared very favorably to the other exact conditional statistics rebuts the arguments of authors (Radlow & Alf, 1975; Gibbons & Pratt, 1975; Cressie & Read, 1989) who claim that probability methods of ranking the departures of datasets from H_0 are inferior to distance measures such as X^2.

Comparison of two-locus tests
Because of the large number of multinomial categories compared to the sample size when considering two loci, receiving good results from any goodness of fit test would seem unlikely. This has been confirmed in terms of the size the tests achieve, except for the exact conditional tests which are guaranteed to have nearly the correct size. Also, the chi-square approximation for the asymptotic statistics is generally very bad even for quite large sample sizes. Therefore, for two locus independence testing, it is recommended that only EC tests be used.

As to which of the EC tests should be used, the power comparisons show that Fisher's exact test, F_{EC}, is the best for all parameter combinations studied here and is therefore recommended. Each of the other tests seems to be competitive in different situations, but never overtake F_{EC}. The log-likelihood statistic G^2_{EC} was found to perform very well for large numbers of alleles per locus $(a_1 = a_2 = 20)$.

Interestingly, the absolute powers for the two locus tests were found to be quite good, even for moderate sample sizes.

The FBI databases

Description of data
The FBI has collected blood samples from many individuals of different ethnic backgrounds, Caucasian, Black and Hispanic, and different geographic regions (Budowle *et al.*, 1991). Here, samples from different geographic regions have been combined. DNA profiles over six VNTR loci, D1S7, D2S44, D4S139, D10S28, D14S13, and D17S79, were analyzed for each individual. See Budowle *et al.* (1991) for details on these loci and the molecular methods used. Table 2 contains a summary of these databases. Further references to a particular database will use the locus name followed by the letter C, B, or H to designate the Caucasian, Black or Hispanic databases respectively.

One locus p-values
P-values have been calculated for each database using each of the eleven tests in Table 1, and the results are shown in Table 3.

What can be seen from these results? One would hope, of course, that the estimated *p*-values from one dataset would differ very little from one test to another. Unfortunately, this is not the case as can be seen from some of the large values in the 'Range' column of Table 6.3. Thus, it is necessary to rely on the statistic known to perform well in this hypothesis testing situation.

On the other hand, with $\alpha = .05$ specified, for the most part there is agreement as to the decision of rejection or acceptance of the hypothesis of independence. Notably, there are three situations for which this is not true: D1S7B, D10S28H and D17S79C. Other than these three, there is generally very good evidence in

Table 2. Summary statistics for the FBI databases.

Locus	Database	Sample size	No. of alleles	No. possible genotypes	No. observed genotypes
D1S7	Caucasian	593	26	351	217
	Black	359	26	351	190
	Hispanic	520	24	300	197
D2S44	Caucasian	790	21	231	166
	Black	474	24	300	180
	Hispanic	514	19	190	144
D4S139	Caucasian	593	17	153	112
	Black	446	18	171	131
	Hispanic	521	16	136	105
D10S28	Caucasian	428	23	276	180
	Black	287	24	300	172
	Hispanic	439	21	231	165
D14S13	Caucasian	750	24	300	169
	Black	523	25	325	201
	Hispanic	493	23	276	198
D17S79	Caucasian	775	13	91	47
	Black	549	15	120	78
	Hispanic	521	9	45	39

Table 3. One locus p-value estimates for the FBI databases (ignoring single-band problem).

	Test statistic											
	G^2			M, F		X^2			$I^{2/3}$			
Locus	AU	EC	AS	AU	EC	AU	EC	AS	AU	EC	AS	Range
D1S7C	.388	.474	.494	.245	.353	.425	.430	.457	.397	.437	.755	.510
D1S7B	.383	.513	.612	.125	.206	.077	.076	.051	.140	.169	.572	.561
D1S7H	.001	.001	.000	.001	.001	.001	.000	.000	.000	.000	.002	.002
D2S44C	.622	.668	.641	.440	.565	.423	.436	.470	.495	.530	.729	.306
D2S44B	.018	.023	.052	.006	.003	.001	.000	.000	.000	.000	.006	.052
D2S44H	.715	.740	.499	.592	.748	.746	.750	.768	.737	.749	.808	.309
D4S139C	.530	.576	.568	.308	.381	.524	.525	.596	.499	.514	.699	.391
D4S139B	.610	.642	.370	.531	.683	.653	.664	.678	.646	.653	.708	.338
D4S139H	.391	.426	.295	.245	.281	.351	.362	.375	.367	.380	.488	.243
D10S28C	.307	.371	.288	.110	.139	.527	.541	.573	.441	.472	.716	.606
D10S28B	.692	.808	.493	.383	.739	.381	.381	.404	.614	.669	.830	.449
D10S28H	.185	.214	.055	.103	.143	.030	.026	.021	.064	.067	.128	.164
D14S13C	.074	.097	.377	.065	.055	.078	.070	.018	.052	.055	.296	.359
D14S13B	.633	.719	.662	.416	.636	.108	.098	.077	.333	.360	.677	.642
D14S13H	.323	.389	.072	.053	.072	.200	.203	.205	.215	.229	.332	.336
D17S79C	.005	.005	.118	.018	.014	.163	.163	.065	.055	.056	.129	.158
D17S79B	.000	.000	.000	.000	.000	.000	.000	.000	.000	.000	.000	.000
D17S79H	.000	.000	.000	.000	.000	.000	.000	.000	.000	.000	.000	.000

Table 4. One locus p-value estimates for the FBI databases (accounting for single-band problem).

Locus	Caucasian M_{AU}	Caucasian F_{EC}	Black M_{AU}	Black F_{EC}	Hispanic M_{AU}	Hispanic F_{EC}
D1S7	.519	.542	.678	.908	.009	.003
D2S44	.800	.776	.532	.577	.752	.766
D4S139	.766	.628	.628	.634	.465	.282
D10S28	.665	.816	.663	.932	.365	.379
D14S13	.671	.380	.772	.885	.420	.328
D17S79	.974	.835	.289	.075	.009	.002

either direction. Those databases for which independence is not rejected have very high p-values across the board. Those which provide rejections have all of the p-values very near zero.

The three cases D1S7B, D10S28H and D17S79C are of most interest here. What decision should be made when there is evidence pointing in both directions? In the case of D1S7B, the p-value calculated using $X^2{}_{AS}$ is .051 and using $G^2{}_{AS}$ it is .612. The obvious answer is to use the tests recommended. D1S7B and D10S28H fall under the recommendation of F_{EC}, and $X^2{}_{AS}$ is recommended for D17S79C. The first two would certainly cause rejection in this case, while the third remains a borderline case if the agreed on p-value is .065. This is of course not the only solution to this problem. It may be suggested, for the sake of conservatism, to reject the null hypothesis if any of the statistics lead to a rejecion. This, however, disregards the results of Maiste (1993) and requires in each situation that a p-value be calculated for each possible test. This seems to be an extreme action when there is a good idea as to which tests are doing the best job.

There are some other interesting patterns that can be seen among the p-values in Table 3. The EC and AU versions of G^2, X^2, and $I^{2/3}$ always give nearly identical p-values. This is not true of the probability tests. In a couple of datasets, the p-value from M_{AU} is much smaller than that of F_{EC}. This goes back to Maiste (1993) where it was seen that in many cases, M_{AU} often rejects much too often under the null hypothesis.

In addition, while $X^2{}_{AS}$ always returns a p-value similar to $X^2{}_{AU}$ and $X^2{}_{EC}$, this does not hold for $G^2{}_{AS}$ and $I^{2/3}{}_{AS}$. The p-values from the $I^{2/3}{}_{AS}$ test are often much higher than from its AU and EC counterparts. Maiste (1993) found that $I^{2/3}{}_{AS}$ was often seen to be very conservative and thus reject too few times. On the other hand, $G^2{}_{AS}$ is also often very different from $G^2{}_{AU}$ and $G^2{}_{EC}$, but there is no set pattern. In a few databases, it is much lower, and in others it is much higher. Again, Maiste found that $G^2{}_{AS}$ was very unpredictable and unreliable in the Monte Carlo simulations. These p-values seem to support that statement.

The single-band problem

Although the discussion of the results in Table 3 is of great statistical interest, it is almost irrelevant for issues concerning human identification. The rejection of the Hardy-Weinberg law for any genetic marker with no known biological function runs counter to population genetic expectations. A population geneticist therefore seeks an explanation, and is quickly led to the phenomenon of single-bands for the RFLP systems discussed here.

As discussed at length in the recent literature (e.g. Weir, 1992; Chakraborty & Li, this volume), there are many instances where the apparent presence of a single allele ('single-band pattern') at a locus does not signify a homozygote. Alleles of similar size may fail to be distinguished, or one allele may be outside the detection system being used. Without further laboratory analysis, it is not possible to determine the cause of such single-allele patterns, and for this reason the FBI does not invoke the Hardy-Weinberg law in such cases. It is invoked only for unambiguous heterozygotes ('double-band patterns'), suggesting a need to test for independence for such patterns.

The two probability tests F_{EC} and M_{AU} were applied to the double-banded patterns for each locus in each database. When shuffling was used to establish p-values, only those shuffled arrays with all double-banded patterns were used. The results are displayed in Table 4, and are more consistent with biological expec-

Table 5. Two locus p-value estimates for the FBI Caucasian database (ignoring single-band problem).

	Test statistic										
	G^2			M, F		X^2			$I^{2/3}$		
Loci†	AU	EC	AS	AU	EC	AU	EC	AS	AU	EC	AS
1,2	.101	.013	1.000	.107	.013	.528	.579	1.000	.334	.450	1.000
1,3	.134	.063	1.000	.155	.051	.088	.100	.000	.069	.075	1.000
1,4	.031	.000	1.000	.024	.000	.793	.867	1.000	.423	.635	1.000
1,5	.118	.006	1.000	.120	.005	.516	.595	1.000	.451	.592	1.000
1,6	.170	.095	1.000	.119	.004	.837	.880	1.000	.722	.818	1.000
2,3	.158	.073	1.000	.186	.059	.294	.330	.075	.213	.247	1.000
2,4	.099	.034	1.000	.079	.011	.666	.722	1.000	.408	.564	1.000
2,5	.119	.008	1.000	.143	.000	.033	.021	.000	.009	.008	1.000
2,6	.245	.223	1.000	.165	.016	.271	.280	.004	.275	.288	1.000
3,4	.068	.006	1.000	.066	.001	.542	.614	1.000	.294	.397	1.000
3,5	.185	.109	1.000	.178	.030	.329	.356	1.000	.209	.256	1.000
3,6	.042	.001	1.000	.057	.000	.612	.666	1.000	.435	.510	1.000
4,5	.035	.000	1.000	.054	.000	.083	.107	.000	.033	.067	1.000
4,6	.060	.001	1.000	.052	.000	.843	.901	1.000	.698	.845	1.000
5,6	.280	.318	1.000	.159	.018	.885	.908	1.000	.780	.850	1.000

†1 = D1S7, 2 = D2S44, 3 = D4S139, 4 = D10S28, 5 = D14S13, 6 = D17S79

tation. It is only D1S7 and D17S79, in the Hispanic database, for which there is apparent dependence. This may be a reflection of the heterogeneity of that database, as it is based on people drawn from Houston, TX and Miami, FL who may be expected to have different ethnic backgrounds (B. Budowle, personal communication).

Two locus p-values

Two locus p-values for each pair of VNTR loci have also been calculated for each of the 11 statistics. These results are presented in Tables 5, 6 and 7.

The p-values seen in these tables are very interesting, as there is utter discrepancy between the different tests. The asymptotic tests are completely unreliable. They nearly always return a p-value of 1.0 due to the vast incorrectness of the chi-square approximation. The exception is $X^2{}_{AS}$ which usually returns p-values in the other extreme, that is, near or exactly 0. This can be explained by the huge variance of $X^2{}_{AS}$. Of course, the p-values from these statistics should not be used.

Very generally, the AU and EC versions of the different statistics result in similar p-values, at least in terms of the ultimate conclusion to reject or not reject. However, between the statistics, there is no such similarity. X^2 and $I^{2/3}$ often return quite high p-values, leading to certain acceptance of the null. This is especially true for the Caucasian and Hispanic databases. On the other hand, the F and G^2 p-values are often quite low, with the EC versions being very near 0 in all but a few cases. In fact, $G^2{}_{EC}$ and F_{EC} lead to the same rejection decision in all but four of the 45 two-locus datasets.

The closeness between $G^2{}_{EC}$ and F_{EC} concurs with the results of Maiste (1993), where it was seen that for large a_1 and a_2, these two tests had nearly equivalent powers. The fact that $X^2{}_{EC}$ was seen to reject even less than $I^{2/3}{}_{EC}$ also backs up the results of Maiste since $X^2{}_{EC}$ was seen to reject even less than $I^{2/3}{}_{AC}$.

The recommendation from Maiste (1993) was that F_{EC} is the best test to use in this situation. Using this leads to rejection for almost all of the locus-pairs in all the datasets. Only a few of the Hispanic cases resulted in fairly large p-values from this test.

Once again, it is necessary to stress that Tables 5, 6 and 7 have no relevance for the applications of RFLPs to human identification since they ignore the single-band problem. In Table 8, p-values are shown when the probability tests F_{EC} and $_{AU}$ are applied to double-banded patterns only. These values show

Table 6. Two locus p-value estimates for the FBI Black database (ignoring single-band problem).

	Test statistic										
	G^2			M,F		X^2			$I^{2/3}$		
Loci†	AU	EC	AS	AU	EC	AU	EC	AS	AU	EC	AS
1,2	.004	.000	1.000	.005	.000	.227	.234	.000	.081	.137	1.000
1,3	.034	.003	1.000	.030	.001	.176	.156	.000	.084	.118	1.000
1,4	.005	.000	1.000	.004	.000	.040	.022	.000	.006	.007	1.000
1,5	.011	.000	1.000	.014	.000	.054	.047	.000	.008	.010	1.000
1,6	.002	.000	1.000	.009	.000	.001	.001	.000	.000	.000	1.000
2,3	.030	.000	1.000	.027	.000	.186	.177	.000	.057	.079	1.000
2,4	.006	.000	1.000	.007	.000	.212	.228	.000	.046	.084	1.000
2,5	.006	.000	1.000	.008	.000	.025	.014	.000	.005	.005	1.000
2,6	.024	.000	1.000	.022	.000	.098	.095	.000	.025	.028	1.000
3,4	.027	.019	1.000	.034	.026	.414	.442	.988	.248	.440	1.000
3,5	.091	.044	1.000	.076	.015	.003	.000	.000	.004	.003	1.000
3,6	.040	.000	1.000	.046	.000	.024	.015	.000	.004	.004	1.000
4,5	.012	.000	1.000	.012	.000	.197	.177	.000	.052	.109	1.000
4,6	.030	.000	1.000	.031	.000	.019	.015	.000	.005	.004	1.000
5,6	.023	.000	1.000	.032	.000	.026	.023	.000	.004	.005	1.000

†1 = D1S7, 2 = D2S44, 3 = D4S139, 4 = D10S28, 5 = D14S13, 6 = D17S79

Table 7. Two locus p-value estimates for the FBI Hispanic database (ignoring single-band problem).

	Test statistic										
	G^2			M,F		X^2			$I^{2/3}$		
Loci†	AU	EC	AS	AU	EC	AU	EC	AS	AU	EC	AS
1,2	.189	.419	1.000	.172	.298	.419	.446	.918	.292	.415	1.000
1,3	.220	.427	1.000	.241	.528	.792	.825	1.000	.574	.724	1.000
1,4	.092	.088	1.000	.101	.129	.236	.215	.000	.116	.185	1.000
1,5	.115	.055	1.000	.122	.046	.206	.197	ce .000	.059	.098	1.000
1,6	.168	.129	1.000	.177	.087	.261	.270	.000	.206	.251	1.000
2,3	.120	.036	1.000	.142	.023	.591	.607	1.000	.347	.442	1.000
2,4	.064	.020	1.000	.071	.008	.125	.117	.000	.071	.100	1.000
2,5	.046	.000	1.000	.056	.000	.147	.142	.000	.030	.034	1.000
2,6	.085	.019	1.000	.075	.006	.218	.226	.000	.085	.099	1.000
3,4	.048	.000	1.000	.068	.001	.041	.030	.000	.021	.025	1.000
3,5	.076	.003	1.000	.097	.003	.315	.314	.000	.133	.153	1.000
3,6	.266	.292	1.000	.219	.109	.008	.007	.000	.009	.008	1.000
4,5	.064	.004	1.000	.055	.001	.020	.015	.000	.007	.007	1.000
4,6	.046	.006	1.000	.063	.001	.057	.048	.000	.009	.012	1.000
5,6	.032	.003	1.000	.051	.000	.091	.071	.000	.010	.008	1.000

†1 = D1S7, 2 = D2S44, 3 = D4S139, 4 = D10S28, 5 = D14S13, 6 = D17S79

Table 8. Two locus p-value estimates for the FBI Hispanic database (accounting for single-band problem).

	Caucasian		Black		Hispanic	
Loci†	M_{AU}	F_{EC}	M_{AU}	F_{EC}	M_{AU}	F_{EC}
1,2	.607	.366	.266	.380	.546	.763
1,3	.829	.164	.366	.670	.722	.292
1,4	.335	.659	.136	1.000	.403	.357
1,5	.862	.204	.268	.167	.617	.463
1,6	.957	.900	.623	.141	.845	.345
2,3	.917	.511	.452	nspace .736	.760	.115
2,4	.510	.783	.195	1.000	.506	.795
2,5	.950	.671	.398	.073	.675	.547
2,6	.988	.754	.913	.980	.862	.253
3,4	.697	.768	.204	.154	.626	.053
3,5	.974	.717	.485	.934	.792	.098
3,6	.946	.051	.813	.084	.945	.583
4,5	.701	.100	.188	.222	.575	.934
4,6	.867	.607	.573	.956	.771	.119
5,6	.995	.974	.827	.023	.884	.271

†1 = D1S7, 2 = D2S44, 3 = D4S139, 4 = D10S28, 5 = D14S13, 6 = D17S79

the expected consistency with independence of allele frequencies within *and* between pairs of RFLP loci in the FBI databases.

Conclusions

There are a variety of methods for testing for independence of allelic frequencies at one or two loci, but the best method seems to be that of Fisher's exact test, which is a test conditional on allelic frequencies and requires extensive computing to establish p-values. Details of the computing algorithms are given by Zaykin, Zhivotovsky and Weir (this volume).

The applications of a battery of tests to the FBI data make it very clear that not all tests behave similarly in practice, even if they have the same asymptotic distributions. The discrepancies between different tests are particularly pronounced for two-locus tests.

When tests are applied to RFLP databases established by the FBI, it can be seen that most of the one-locus datasets do not cause rejection of the null hypothesis of independent frequencies, while many of the two-locus datasets do lead to rejection. However, when the tests are confined to individuals known to be heterozygous, there is nearly complete support for independence of alleles within or between loci. The same conclusions have been reached previously (e.g. Weir 1992, 1993; Chakraborty & Li, this volume and references therein).

A possible cause of any apparent association of allelic frequencies is known as the *Wahlund effect*. This occurs when there are various reproductively independent subpopulations that make up a larger population. For example, it could be argued that the present day Caucasian population in the United States is a combination of various ethnic groups which, especially in the past, tended to stay separate in terms of marriage. Geography is another factor which can separate subpopulations. Similar arguments hold for other populations of individuals. The extent to which such substructuring would be detected by tests for association is discussed by Zaykin, Zhivotovsky and Weir (this volume). A way for accommodating for substructuring, by means of F-statistics, is given by Weir (1994).

Acknowledgements

This work was supported in part by NIH grant GM32518. FBI data were kindly provided by Dr B. Budowle.

References

Agresti, A. & D. Wackerly, 1977. Some exact conditional tests of independence for RxC cross-classification tables. Psychometrika 42: 111–125.

Berkson, J., 1978. In dispraise of the exact test. J. Stat. Planning and Inf. 2: 27–42.

Budowle, B., K.L. Monson, K.S. Anoe, F.S. Baechtel, D.L. Bergman, E. Buel, P.A. Campbell, M.E. Clement, H.W. Coey, L.A. Davis, A. Dixon, P. Fish, A.M. Guisti, T.L. Grant, T.M. Gronert, D.M. Hoover, L. Jankowski, A.J. Kilgore, W. Kimoto, W.H. Landrum, H. Leone, C.R. Longwell, D.C. MacLaren, L.E. Medlin, S.D. Narveson, M.L. Pierson, J.M. Pollock, R.J. Raquel, J.M. Reznicek, G.S. Rogers, J.E. Smerick & R.M. Thompson, 1991. A preliminary report on binned general population data on six VNTR loci in Caucasians, Blacks and Hispanics from the United States. Crime Lab. Digest 18: 9–26.

Camilli, G., 1990. The test of homogeneity for 2×2 contingency tables: a review of and some personal opinions on the controversy. Psych. Bull. 108: 135–145.

Casella, G. & R.L. Berger, 1990. Statistical Inference. Wadsworth and Brooks/Cole. Pacific Grove, CA.

Cochran, W.G., 1954. Some methods for strengthening the common χ^2 tests. Biometrics 10: 417–451.

Cressie, N. & T.R.C. Read, 1984. Multinomial goodness of fit tests. J. Roy. Stat. Soc. B 46: 440–464.

Cressie, N. & T.R.C. Read, 1989. Pearson's X^2 and the loglikelihood ratio statistic G^2: a comparative review. Int. Stat. Rev. 57: 19–43.

D'Agostino, R.B., W. Chase & A. Belanger, 1988. The appropriateness of some common procedures for testing the equality of two independent binomial populations. Am. Statist. 42: 198–202.

Elston, R.C. & R. Forthofer, 1977. Testing for Hardy Weinberg equilibrium in small samples. Biometrics 33: 536–542.

Emigh, T.H., 1978. The power of tests for random mating in genetics. Biometrics 34: 730.

Emigh, T.H., 1980. A comparison of tests for Hardy-Weinberg equilibrium. Biometrics 36: 627–642.

Emigh, T.H. & O. Kempthorne, 1975. A note on goodness-of-fit of a population to Hardy-Weinberg structure. Am. J. Hum. Genet. 27: 778–783.

Fienberg, S.E., 1979. The use of chi-square statistics for categorical data problems. J. Roy. Stat. Soc. B 41: 54–64.

Fisher, R.A., 1924. The conditions under which χ^2 measures the discrepancy between observation and hypothesis. J. Roy. Stat. Soc. 87: 442–450.

Fisher, R.A., 1935. The logic of inductive inference. J. Roy. Stat. Soc. 98: 39–54.

Gibbons, J.D. & J.W. Pratt, 1975. P-values: interpretation and methodology. Am. Statist. 29: 20–25.

Greenland, S., 1991. On the logical justification of conditional tests for two-by-two contingency tables. Am. Statist. 45: 248–251.

Guo, S.W. & E.A. Thompson, 1992. Performing the exact test of Hardy-Weinberg proportion for multiple alleles. Biometrics 48: 361–372.

Haber, M., 1981. Exact significance levels of goodness-of-fit tests for the Hardy-Weinberg equilibrium. Hum. Hered. 31: 161–166.

Hernández, J.L. & B.S. Weir, 1989. A disequilibrium coefficient approach to Hardy-Weinberg testing. Biometrics 45: 53–70.

Koehler, K.J., 1986. Goodness of fit tests for log linear models in sparse contingency tables. J. Am. Stat. Assoc. 81: 483–493.

Koehler, K. & K. Larntz, 1980. An empirical investigation of goodness-of-fit statistics for sparse multinomials. J. Am. Stat. Assoc. 75: 336–344.

Kullback, S., 1959. Information Theory and Statistics. Wiley, New York.

Kullback, S., 1985. Minimum discrimination information (MDI) estimation. In Encyclopedia of Statistical Sciences 5: 527–529, Ed. S. Kotz and N.L. Johnson. Wiley, New York.

Larntz, K., 1978. Small sample comparisons of exact levels for chi-squared goodness-of-fit statistics. J. Am. Stat. Assoc. 73: 253–263.

Little, R.J.A., 1989. Testing the equality of two independent binomial proportions. Am. Statist. 43: 283–288.

Louis, E.J. & E.R. Dempster, 1987. An exact test for Hardy-Weinberg equilibrium and multiple alleles. Biometrics 43: 805–811.

Maiste, P.J., 1993. Comparison of Statistical Tests for Independence at Genetic Loci with Many Alleles. Ph.D. Thesis, North Carolina State University, Raleigh, NC.

Moore, D.S., 1986. Tests of chi-squared type. In Goodness-of-Fit Techniques, Ed. R.B. D'Agostino and M.A. Stephens, pp.63–95, Marcel Dekker, New York.

Neyman, J., 1949. Contribution to the theory of the χ^2 test. Proc. 1st Berk. Symp. Math. Stat. Prob. pp. 239–273.

Pearson, K., 1900. On the criterion that a given system of deviations from the probable in the case of a correlated system of variables is such that it can be reasonable supposed to have arisen from random sampling. Phil Mag. Series (5) 50: 157–172.

Radlow, R. & E.F. Alf, 1975. An alternate multinomial assessment of the accuracy of the χ^2 test of goodness of fit. J. Am. Stat. Assoc. 70: 811–813.

Read, T.R.C., 1982. On choosing a goodness-of-fit test. Ph.D. Thesis, Flinders University, South Australia.

Storer, B.E. & C. Kim, 1990. Exact properties of some exact test statistics for comparing two binomial proportions. J. Am. Stat. Assoc. 85: 146–155.

Suissa, S. & J.J. Shuster, 1985. Exact unconditional sample sizes for the 2×2 binomial trial. J. Roy. Stat. Soc. A 148: 317–327.

Upton, G.J.G., 1982. A comparison of alternative tests for the 2×2 comparative trial. J. Roy. Stat. Soc. A 145: 86–185.

Weir, B.S., 1990. Genetic Data Analysis. Sinauer, Sunderland, MA.

Weir, B.S. 1992., Independence of VNTR alleles defined as fixed bins. Genetics 130: 873–887.

Weir, B.S., 1993. Independence tests for VNTR alleles defined as quantile bins. Am. J. Hum. Genet. 53: 1107–1113.

Weir, B.S., 1994. The effects of inbreeding on forensic calculations. Ann. Rev. Genet. 28: 597–621.

Yates, F., 1984. Tests of significance for 2×2 contingency tables (with discussion). J. Roy. Stat. Soc. A 147: 426–463.

B. S. Weir (ed.), Human Identification: The Use of DNA Markers, 139–144, 1995.

Alternative approaches to population structure

Newton E. Morton
Human Genetics Centre, University of Southampton, Southampton, UK

Received 5 January 1994 Accepted 1 August 1994

Key words: DNA identification, forensic DNA, kinship, likelihood ratio, population structure

Abstract

There are three approaches to DNA identification: tectonic, halieutic and icarian, of which the tectonic is sensible, the halieutic impractical, and the icarian idiotic. The rationale and consequences of these approaches are detailed.

Introduction

DNA typing for human identification typically involves material from two sources, S and C, which in the simplest case represent single individuals and satisfy tests of that hypothesis. Often S is a suspect and C is an evidentiary sample that may represent a culprit. Sometimes S is a missing individual and C is an unidentified victim. There are several statistical decisions that may be made (Table 1). Only if exclusion, coincidence, and kinship are rejected can the inference of identity be accepted (Collins *et al.*, 1994). The appropriate method at each step is a specific likelihood ratio of the form

$$LR = \frac{P(C|S, \text{closer relationship})}{P(C|S, \text{looser relationship})} \qquad (1)$$

For half a century it has been generally realised that likelihood ratios are the optimal basis for statistical decisions (Table 2). Although appealing to Bayesians, no Bayesian inference is implied. Within this framework or in conflict with it three approaches have been made, called tectonic, halieutic, and icarian for reasons that will be apparent. I assume that each locus is examined separately, since multilocus cocktails of linked or unlinked loci introduce approximations that should be avoided.

Table 1. Statistical decisions on S and C.

Decision	Implication
exclusion	different genotypes
coincidence	same genotype, random individuals
kinship	same genotype, relatives
identity	same genotype, same individual

The tectonic approach

This expression, derived from the Greek *tecton* (a skilful builder or wright), was applied by Haldane (1964) to the work of Sewall Wright who developed much of the current theory about population structure. His first fundamental equation expressed genotype frequencies under random differentiation in a given population in terms of gene frequencies and a scalar called inbreeding:

$$P(Q_rQ_s) = \begin{cases} q_r^2 + q_r(1-q_r)F & \text{if } r = s \\ 2q_rq_s(1-F) & \text{if } r \neq s \end{cases} \qquad (2)$$

Table 2. Chronology of likelihood ratios.

1928	Neyman/Pearson	Statistical decision
1947	Wald	Sequential analysis
1959	Kullback	Information theory

Table 3. Tectonic formulations of population structure.

Parameter	Symbols Wright	Cockerham
inbreeding	F_{IT}	F
random kinship	F_{ST}	θ
conditional kinship	F_{IS}	f

Table 4. Chronology of population structure.

1908	Hardy-Weinberg	Genotype frequencies under panmixia
1921	Wright	Genotype frequencies under inbreeding
1940	Cotterman	Genotype frequencies of regular relatives
1943	Wright	Genotype frequencies in hierarchic populations
1948	Malecot	Inbreeding generalized to kinship matrix and isolation by distance
1968	Yasuda	Genotype frequencies of mates

His second fundamental equation expressed kinship in a hierarchical structure as

$$1 - F_{IT} = (1 - F_{IS})(1 - F_{ST}) \tag{3}$$

where I denotes a local population belonging to a subpopulation S within a major population T (Table 3). Choice of population determines both gene frequencies and F. In practice we try to avoid populations corresponding either to a small sample (hence unreliable gene frequencies and uncertain F) or to a large sample with a high value of F_{IT}.

The followers of Wright extended population structure theory to any relationship between S and C (Table 4). Malecot (1948) generalized inbreeding to a kinship matrix, giving for each pair of individuals or subpopulations i, j the probability φ_{ij} that a gene drawn at random from one be identical by descent with a random allele from the other. Inbreeding is the special case of kinship when i, j are mates. Isolation by distance d between subpopulations is described approximately by

$$\varphi(d) = (1 - L)ae^{-bd} + L \tag{4}$$

and so

$$F_{IT} = ae^{-bd}$$

$$F_{ST} = -L/(1 - L)$$

where L is kinship as bd becomes large and therefore for the least related subpopulations. An expert witness should present LRs under various hypotheses, but the court must decide which is the most relevant on the basis of other evidence, including estimates of these parameters from genealogies, migration, isonymy, and kinship bioassay in human populations (Morton, 1992). For nearly all forensic situations the value of F_{ST} is much less than .01. F_{IS} is usually negligible but is much more variable and in the most extreme isolates (seldom encountered in court) approaches the value of .15 observed for expressed loci in major races (Morton & Keats, 1976). Nonexpressed loci are much less variable between major races (Morton, Collins & Balazs, 1993), either because they are not exposed to different selection in different environments or because mutation regenerates pre-existing alleles.

The tectonic approach is well founded in population genetics, and so it satisfies all mathematical and biological constraints (Table 5). Since kinship does not induce linkage disequilibrium, the LRs are multiplicative over loci unless replicates or relatives are inadvertently included in the data base on which gene frequencies are estimated. A method to test and allow for dependence is available (Morton, 1992, Eq. 15). Unbiased estimates of mean matching probabilities have been given, making calculations robust against sampling errors and choice of population at a modest cost that can easily be determined by LR theory (Collins *et al.*, 1994). The current shift from manual assay of RFLPs to automated determination of PCR fragment lengths will augment the large body of evidence on F_{ST} and F_{IS} in human populations.

Although the tectonic approach is uniquely suited to forensic inference, there is no consensus about its presentation in court. Should an expert witness select a particular hypothesis about the culprit? If called by the prosecution, should he be a salesman for the Hardy-Weinberg assumption and withhold kinship calculations that he would present if called by the defense?

Table 5. Three approaches to population structure.

	Tectonic Wright/ Malecot	Halieutic Lewontin/ Hartl	Icarian Lander/ NRC
Kinship explicit	yes	no	no
Gene frequencies	estimated	mongrelized	fabricated
Frequencies depend on population	yes	yes	no
Allowance for sampling bias	yes	no	no
Misuse of confidence interval	no	no	yes
Population	forensic	singular	imaginary
Used in population genetics	yes	no	no
Supported by migration, genealogy, etc.	yes	no	no
Results are probabilities	yes	yes	no
Likelihood ratios applicable	yes	yes	no

Should he make only those calculations requested by the court? It is often argued that courts are confused by multiple estimates, each corresponding to a different population and relationship. Should the expert witness capitulate to that concern or attempt to educate the court? The risk of confusion can be minimized by expert testimony, but the expert cannot claim to know the identity of the culprit. Therefore one model of expert testimony requires that multiple estimates be volunteered by the witness. The alternative model requires that such testimony be elicited through cross-examination by adequately prepared defense counsel, to which the prosecution could reply by presenting evidence excluding close relatives, against the kinship favoured by the defense and/or in favour of a large likelihood ratio even against that hypothesis. Such tactical considerations are of no scientific relevance. Any serious program of quality control over forensic calculations must incorporate several relationships and populations and be supported by empirical estimates of kinship in man.

The halieutic approach

In deriving this term from the Greek *halieutes* (a fisher) Haldane had in mind R.A. Fisher, who made great contributions to mathematical genetics but was not much interested in structured populations. Fishing also implies searching without any clearly stated objective, as in 'fishing expedition.' Both senses are invoked here.

The halieutic approach was spawned by Lewontin (1972) in an attempt to apportion human diversity among populations and ethnic groups. He applied the Shannon information measure without allowance for sampling error to a compendium of blood groups published twenty years earlier, heavily weighted with small samples of isolated populations and with typing errors (Mourant, 1954). His conclusion that 'human races and populations are remarkably similar to each other' attracted little attention at a time when many investigators were engaged in more detailed studies (e.g. Morton, 1973; Kirk & Thorne, 1976).

The same data were trawled again twenty years further on, ignoring voluminous new information (Mourant, Kopec & Domaniewska-Sobczak, 1976; Tills, Kopec & Tills, 1983; Roychoudhury & Nei, 1988). Although the selected data were unchanged, the conclusion was reversed: diversity among populations within ethnic groups is so great that the probability of a random match 'cannot be estimated,' and 'each particular individual may require a different reference group composed of appropriate ethnic or geographic subpopulations' (Lewontin & Hartl, 1991). This conclusion has been criticized from several points of view. Eq. 1 shows that the subpopulation of the suspect is formally irrelevant unless the culprit is a relative. Usually there is little information about the culprit's subpopulation, and even the assignment of the suspect to a narrowly defined group may be poorly supported (Caskey, 1991; Wooley, 1991). DNA evidence should not be dismissed because the suspect claims to be the last of the Mohicans. These problems, easily handled by the tectonic approach, are troublesome if the population is finely partitioned to justify the Hardy-Weinberg assumption that F = 0. The alleged 'typical' gene frequencies cited by Lewontin and Hartl are strongly biased (Chakraborty & Kidd, 1991; Morton, Collins & Balazs, 1993), and less than 2% of the diversity selected by them is due to the national kinship to which they attribute it.

The icarian approach

Icarian implies hubris and inadequacy to bring about an ambitious project, from the Greek myth of Icarus who flew so high on man-made wings that the sun melted them and sent him to destruction. It has therefore been applied to failed utopias and unsound proposals.

There was much to criticise in early presentation of DNA evidence. The possibility of laboratory error must always be considered, but with technical advances and the introduction of quality control the debate centred on population structure. This led to the maiden flight of Icarus (Lander, 1991): 'regardless of the defendant's ethnic background, each allele frequency used (should) be the maximum observed in various ethnic samples.' The latter were not specified, except that they be 'a dozen or so well-separated ethnic population samples,' not necessarily including the population of the suspect and making no claim of relevance to the culprit. No explanation was offered as to why a court in Kansas should accept gene frequencies in Cambodia as evidence. At a single locus the culprit might be assumed to be a Lapp for one allele and a Hottentot for the other. However extreme a sample might be, another sample could always be found with an apparently higher gene frequency, either through drift, mistyping, or sampling error.

In the event, this flight was too modest for Icarus. He persuaded the National Research Council Committee on Forensic Science to soar higher: a committee should choose 'random samples of 100 persons from 15–20 populations, each representing a group relatively homogeneous genetically,' taking 'as the ceiling the highest frequency in any of these populations or 5%, whichever is larger.' During the time when these samples are being collected, an equally arbitrary but more stringent rule would apply. Likelihood ratios, miscalled Bayesian, were not accepted and could not be used because the numbers calculated are not probabilities. Since there was no minority report, we must presume that none of the committee realized that they were neglecting established genetic principles, misleading the courts, and disgracing the National Academy. However, one member who performs DNA identification has not followed their recommendation (Ginther, Isseltarver & King, 1992).

The published report created a storm of opposition. The 'ceiling principle' is neither a ceiling (Slimowitz & Cohen, 1993) nor a principle (Devlin, Risch & Roeder, 1993; Morton, 1993; Weir, 1993). On what grounds would Senegalese be chosen and Nigerians rejected, how can a population represent a group, how can any population except the smallest be sampled randomly, how can 'relatively homogeneous genetically' be applied to any human population (Lewontin, 1972), and what justifies this absurd rule? Lacking any logical basis, it necessarily leads to conflict about which samples and why, with no possibility of scientific resolution. If it were accepted, advances in molecular biology could not be introduced in court until forensic scientists donned their pith helmets and bled pygmies, or whatever other group the autocratic committee might choose. Icarus can fly no higher.

Where are we?

The National Academy charged the National Research Council to inform, but the forensic committee attempted to legislate. It is not the responsibility of scientists to protect a suspect from the evidence by arguments without scientific validity. Among a set of credible likelihood ratios the court may wish to choose the one most favourable to the suspect, but the expert witness should not usurp the prerogative of the court. A second committee with statistical and population genetic expertise has now been formed and we may hope for improvement. However, institutions on reflection tend to repeat their blunders. It will take an informed effort by the second committee to dispel the confusion sown by their predecessors. A solution must satisfy seven principles:

1. There is no connection between a matching probability in one population and gene frequencies in an unrelated population.
2. An upper bound to a probability is not a probability.
3. Of the indefinitely large number of ways in which such an upper bound may be estimated, few conform to generally accepted theory.
4. For every genotype-specific matching probability there is a genotype-specific likelihood ratio, mean

Table 6. Desiderata in forensic loci.

1. Easy typing
2. Reliable typing: alleles well separated
3. No null alleles
4. Highly polymorphic
5. Not expressed
6. Not highly mutable
7. Not linked to other forensic loci
8. Little racial variation

matching probability, and a corresponding likelihood ratio under credible hypotheses about the population of the culprit and his relationship to the suspect. Consideration of these alternatives by the court protects adequately against excessive reliance on evidence of identity.

5. An acceptable bound must not violate statistical or genetic principles or known values of gene frequency or kinship.
6. Gene frequencies should be estimated in large samples to minimize sampling error.
7. In the absence of evidence to the contrary, the suspect and culprit should be assumed to be randomly drawn from a forensic population. Contrary evidence may be accommodated by the affinal or other model of population genetics and by an appropriate estimate of kinship, without altering gene frequencies in the reference population.

Testimony that violates any of these principles should not be admissible in court.

As we await the second NRC attempt to reflect scientific consensus, there are two grounds for hope. First, the loci used in forensic testing are becoming ideal (Table 6) and can in principle be assayed automatically with an error less than 1 bp. Exclusion will soon be categorical. Secondly, icarians have so far been limited to the United States. Elsewhere, advances made when America led the world in statistics and population genetics are the basis of a tectonic approach (Evett, 1992). American courts must also be reconciled to science.

References

Aldhous, P., 1993. Geneticists attack NRC report as cientifically flawed. Science 259: 755–756.

Caskey, C.T., 1991. Comments on DNA based forensic analysis. Am. J. Hum. Genet. 49: 893–894.

Chakraborty, R. & K.K. Kidd, 1991. The utility of DNA typing in forensic work. Science 254: 1735–1739.

Collins, A. & N.E. Morton, 1994. Likelihood ratios for DNA identification. Proc. Nat. Acad. Sci. USA 91: 6007–6011.

Committee on DNA Technology in Forensic Science, National Research Council, 1992. DNA Technology in Forensic Science. Nat. Academy Press, Washington.

Cockerham, C.C., 1969. Variance of gene frequencies. Evolution 23: 72–84.

Cotterman, C.W., 1940. A calculus for statico-genetics. Thesis, Ohio State University, Ohio. Reprinted in P. Ballonoff, (ed.), Genetics and Social Structure (1974). Dowden, Hutchinson, and Ross, Inc., Stroudsburg, Pa.

Devlin, B., N. Risch, & K. Roeder, 1993. Statistical evaluation of DNA fingerprinting: a critique of the NRC's report. Science 259: 748–749, 837.

Evett, I.W., 1992. Evaluating DNA profiles in a case where the defence is "it was my brother." J. Forens. Sci. Soc. 32: 5–14.

Ginther, C., L. Isseltarver & M.C. King, 1992. Identifying individuals by sequencing mitochondrial DNA from teeth. Nature Genet. 2: 135–138.

Haldane, J.B.S., 1964. A defense of beanbag genetics. Perspect. Biol. Med., 7: 343–359.

Hardy, G.H., 1908. Mendelian proportions in a mixed population. Science 28: 49–50.

Kirk, R.L. & Thorne, A.G. 1976. The Origin of the Australians. Humanities Press, New Jersey.

Kullback, S., 1959. Information Theory and Statistics. Wiley, New York.

Lander, E.S., 1991. Research on DNA typing catching up with courtroom application. Am. J. Hum. Genet. 48: 819–823.

Lewontin, R.C., 1972. The apportionment of human diversity. Evol. Biol. 6: 381–398.

Lewontin, R.C. & D.L. Hartl, 1991. Population genetics in forensic DNA typing. Science 254: 1745–1750.

Malecot, G., 1948. Les mathématiques de l'hérédité. Paris, Masson.

Morton, N.E., 1973. Genetic Structure of Populations. University Press, Honolulu.

Morton, N.E. & Keats, 1976. Human microdifferentiation in the Western Pacific. In R.L. Kirk & A.G. Thorne (eds.) The origin of the Australians. Human Biol. Series 6, Australian Institute of Aboriginal Studies, Canberra. Humanities Press, NJ.

Morton, N.E., 1992. Genetic structure of forensic populations. Proc. Nat. Acad. Sci. USA 89: 2556–2560.

Morton, N.E., A. Collins, & I. Balazs, 1993. Kinship bioassay on hypervariable loci in Blacks and Caucasians. Proc. Nat. Acad. Sci., USA 90: 1892–1896.

Mourant, A.E., 1954. The Distribution of the Human Blood Groups, Blackwell, Oxford.

Mourant, A.E., A.C. Kopec & K. Domaniewska-Sobczak, 1976. The Distribution of the Human Blood Groups and Other Polymorphisms. Oxford University Press.

Neyman, J. & E.S. Pearson, 1928. On the use and interpretation of certain test criteria for purposes of statistical inference. Biometrica 20A: 175, 263.

Roychoudhury, A.K. & M. Nei, 1988. Human Polymorphic Genes. World Distribution. Oxford University Press.

Slimowitz, J.R. & J.E. Cohen, 1993. Violations of the ceiling principle: exact conditions and statistical evidence. Am. J. Hum. Genet. 53: 314–323.

Tills, D., A. Kopec & R.E. Tills, 1983. The Distribution of the Human Blood Groups and Other Polymorphisms. Suppl. 1. Oxford University Press.

United States v. Yee (1991). 134 F.R.D. 161 (N.D. Ohio).

Wald, A., 1947. Sequential Analysis. Wiley, New York.
Weinberg, W., 1908. Uber den Nachweis der Vererbung beim Menschen. Jahresh Verein f Vaterl Naturk Wurttem 64: 368–382. Eng. trans. in S.H. Boyer (ed.), Papers on Human Genetics (1963). Prentice-Hall, Englewood Cliffs, New Jersey.
Weir, B.S., 1993. Forensic population genetics and the National Research Council (NRC). Am. J. Hum. Genet. 52: 437–440.
Wooley, J.R., 1991. A response to Lander: The courtroom perspective. Am. J. Hum. Genet. 49: 892–893.
Wright, S., 1921. Systems of mating. Genetics 6: 111–178.
Yasuda, N., 1968. An extension of Wahlund's principle to evaluate mating type frequency. Am. J. Hum. Genet. 20: 1–23.

Editor's comments

The author captures the harsh tone that has often characterized the debate over the use of DNA for human identification. It should be mentioned that D.L. Hartl, E.S. Lander and R.C. Lewontin were invited to respond. The positions of these three authors are contained in their papers, listed in the Bibliography. Readers should note, in particular, Budowle and Lander (1994).

B. S. Weir (ed.), Human Identification: The Use of DNA Markers, 145–152, 1995.

DNA evidence: wrong answers or wrong questions?

Bernard Robertson[1] & G.A. Vignaux[2]
[1]*Department of Business Law, Massey University, P.O. Box 11222, Palmerston North, New Zealand*
[2]*Institute of Statistics and Operations Research, Victoria University of Wellington, P.O. Box 600, Wellington, New Zealand*

Received 22 April 1994 Accepted 15 August 1994

Key words: DNA, forensic science, significance tests, likelihood ratios

Abstract

Much of the controversy over DNA evidence is due to the way in which forensic scientific evidence has classically been presented. The orthodox approach is to consider whether two samples match according to a predetermined criterion. If they do, the fact of match is reported along with an estimate of the frequency of the characteristics. This method fails to address the questions raised in court cases, diverts argument into irrelevancies and stultifies research. Presentation of evidence in the form of likelihood ratios, on the other hand, forces the witness to answer the questions the court is interested in and makes apparent lines of research required to increase our understanding.

Introduction

The introduction of DNA evidence has stirred up fierce controversy over interpretational problems which have always been lurking in the way forensic scientific evidence has been presented. We examine controversies which have arisen in the context of DNA evidence and aim to show that these controversies are actually about how to reason in the face of uncertainty. We adopt a logical approach to probability and conclude that many of the difficulties stem from witnesses asking the wrong questions and sometimes getting the 'right' answers.

The right questions and the wrong questions

Consider a simple case in which a blood stain is found at the scene of a crime. The blood is analysed by techniques which identify alleles which are either present or absent, such as the ABO factors. The blood of any suspect will also be analysed, and if it matches the mark at the scene the scientist will report that the two samples 'could have come from the same person' and that blood of this type occurs in $x\%$ of the population. This may be expressed in the forms of 'odds against a match by chance'.

This technique is standard and will be familiar to many readers. Unfortunately it answers the wrong question.

The question the court wishes the scientist to answer is the post-data question: 'how much does the evidence from the mark at the scene increase the probability that it was the accused who left it?' This is done by comparing the probability of the evidence (E) supposing that the accused left the mark (H1) with the probability of the evidence supposing an alternative hypothesis, usually the defense case (H2). This will enable the court to assess the value of the evidence – in other words, to determine how much it should change its belief in the prosecution case based on other evidence. There can be no other rational purpose for giving the evidence.

The frequency reported in the classical approach to forensic scientific evidence answers a different question, namely the pre-data question: 'what is the probability of obtaining a match by carrying out this procedure?' Forensic scientists have long been able to give evidence in this way because, in certain simple cases, such as the one above, the wrong question appears to produce a helpful and correct answer.

For example, if the frequency of the characteristics of the mark at the scene were 1% then the scientist would classically report that and perhaps add that the odds against a match by chance were 99 to 1.

If we ask the correct question, $P(E|H1)=1$. $P(E|H2) = 0.01$, equivalent to the frequency of the characteristic. Thus the Likelihood Ratio $P(E|H1)/P(E|H2) = 100$ and whatever the court previously assessed the odds in favor of the accused's guilt as being it should now multiply those odds by 100.

The classical method has survived because, in this simple case, the correct Likelihood Ratio can be derived intuitively from the 1% frequency that the forensic scientist would report. Indeed, giving odds against a match has been defended by Magnusson (1993) precisely on the ground that 'it will be ready for use in a Bayesian argument'.

When the wrong questions give the right answers

It turns out that four conditions have to be satisfied so that asking the wrong question will give the right answer. These are:

A) The test must be for a characteristic which is either present or absent, such as an ABO factor; in other words the sample must either match or not match.
B) There must be only one mark, e.g. one bloodstain or one group of glass fragments.
C) The population from which the frequency is derived must contain both the accused and the perpetrator and be homogeneous so that characteristics are distributed equiprobably throughout it.
D) Comparison is made only with one person – the accused.

When the wrong questions give the wrong answers

Only if all the above conditions are satisfied will asking the wrong question give the right answer. In any other case they will give the wrong answer. Let us see why.

(A) Match-nomatch

The first problem is that $P(E|H1)$ may not equal 1. This difficulty has always existed with glass evidence. Blood tests detect alleles which are either present or absent (although even then there may be false positives and negatives). With glass, however, there may be variations in the refractive index over one window; two samples from the same window may not have precisely the same refractive index. D. Lindley pointed out the problem in 1977 but because the solution proposed was 'Bayesian' and hence controversial, most scientists failed to change their practices.

DNA evidence made the problem acute. If two DNA autoradiographs from the same person are analysed, slight differences in observed bandweights may occur under the best of conditions – and forensic scientists frequently work with degraded and contaminated samples.

In order to carry on giving evidence in the classical way a strategem was adopted. This was to define 'a match'. The criterion used depended upon a significance test, that is to say, the percentage of tests from the same person that would fall within a particular bin. Some purely conventional limit, usually 99% or 95% was selected as the 'threshold for a match'. Based on the standard deviation of the measurements this defined the limits of the bin. It would then be reported that the two samples could have had a common origin and the frequency of alleles in that bin would be stated. If the samples did not fall within this margin then they were declared a 'non-match' and the hypothesis of common origin was rejected.

There are a number of problems with this procedure:

i. *The 'fall off the cliff' effect*: It is clearly artificial to say that a difference of 2.9 standard deviations is evidence of common origin and a difference of 3.1 standard deviations is inconclusive. Obviously, the more similar the samples the stronger the evidence that they had a common origin and the less similar the samples the stronger the evidence that they came from different sources (see Evett, 1991).

ii. *The proper match criteria*: Argument becomes diverted into the irrelevant question of the criteria for a match. This question has vigorously exercised courts (e.g., *People v. Castro* (1989) 545 NY Supp 985 (SC, NY)), commentators and even legislators in the USA but it is a question with no answer. Whatever criterion is chosen is arbitrary.

iii. *Favorable evidence is rejected*: Suppose that the probability of obtaining the difference observed if both samples came from the same source were 0.5%. Suppose, also, that DNA bandweights observed in the scene sample were shared by only 1 in 10,000. The evidence would be 200 times more probable if both samples came from the accused than if the scene sample came from someone else. Despite this clearly being useful evidence, the classical method, which concentrates only on the difference between the two samples, would reject the hypothesis of common origin.

iv. *Evidence is overvalued*: Conversely, this procedure can also lead to evidence being overvalued, because once it is decided that two samples 'match' the effect of any difference between them is disregarded.

v. *Conceptual confusion*: Confusion is caused by considering the fact of a match as the evidence whereas the evidence is correctly stated as two propositions:

$$\text{E1} = \text{the sample from the scene has profile A.}$$
$$\text{E2} = \text{the sample from the accused has profile B.}$$

An example of the confusion which use of a 'match' as evidence can cause is given by Balding and Nichols (1994). They assert that:

$$\text{P}[Match] = p.$$

p is equivalent to the frequency in the population. This leads to a line of argument about how to cope with uncertainty about p, which includes the proposition that under the defense hypothesis there are two people of the same profile, making it more common than previously realised.

Analysis of this argument reveals:

i. an unconditional probability P$[Match]$ is stated. As with any probability, its value will depend upon what the conditioning factors are. In this paper I is used to denote background knowledge, other than the items of evidence specified. We include I throughout as a reminder that there may be other relevant factors.

ii. the value p is an estimate of the frequency of a characteristic in a population and the authors are concerned about uncertainty about p. We are not concerned to estimate the frequency of any genetic characteristic but only to assess P(E1,E2|I) conditioned on the information which is available (this will usually be a survey result 'S') and appropriate hypotheses 'H'. The probability we are concerned with is, in full, P(E1, E2|H, S, I) about which there is no uncertainty. The inclusion of 'S' makes clear that this is a conditional probability with only one value and not an 'estimator' of an elusive 'true probability'. We do not use p in our presentation. Hereafter, S is subsumed into I.

Take

$$\text{H1} = \text{the accused left the mark.}$$
$$\text{H2} = \text{someone else left the mark.}$$
$$N + 1 = \text{the size of the population of possible perpetrators, including the accused.}$$

The prior odds $= 1/N$. This population is defined by the prior information considered and consists of people equally likely to be the perpetrator. If there is little or no prior information this may be the local population, in which case a suitably racially weighted database should be used. If, in Japan, a disguised robber were described as six foot tall, then N consists of six foot people present in Japan. The racial composition of this group (i.e. Japanese, Europeans and others) would have to be estimated and a figure produced for the probability of the DNA analysis given 'six-footness', but, until further evidence is considered, all six footers in Japan are equally probable perpetrators.

The posterior odds ratio for the two hypotheses after the two pieces of evidence have been obtained is the prior odds multiplied by the likelihood ratio

$$\frac{\text{P(H1|I)P(E1,E2|H1,I)}}{\text{P(H2|I)P(E1,E2|H2,I)}}.$$

The numerator of the likelihood ratio, P(E1,E2|H1,I) can be written P(E1|E2,H1,I)P(E2|H1,I). The first component, P(E1|E2,H1,I), is the probability that the scene sample would have the given profile supposing that the accused left the mark and knowing the profile revealed by the accused's sample. This will be determined by the difference between the measurements. Call this p_m. Only in the case where discrete alleles are detected and there is no possible error in measurement will this be 1. The second component, P(E2|H1,I), is the probability of the accused's sample supposing the accused left the mark but having no analysis of the scene sample. This must equal the frequency of the accused's characteristics in the sub-population defined by what is known of the accused. Call this f_a.

The denominator of the likelihood ratio, P(E1,E2|H2,I), can be written P(E1|E2,H2,I)P(E2|H2,I). The first component, P(E1|E2,H2,I), is the probability of the scene sample supposing the accused's profile and that someone else left the mark. This is independent of the accused's profile, thus P(E1|E2,H2,I) = P(E1|H2,I) $= f$ where f is the frequency of the characteristic in the population of possible perpetrators. This will be determined by what is known of the perpetrator. The second component, P(E2|H2,I), is the probability that the accused's sample would have this measurement supposing that the someone else left the mark. This is obviously independent of who left the mark and depends on what information about the accused is taken into account. It is f_a.

The likelihood ratio now leads to $p_m . f_a./f . f_a. = p_m/f$. The posterior odds are therefore: p_m/Nf.

The analysis can be repeated by breaking up the E1,E2 pair in the alternative order, P(E2|E1,H1,I) . P(E1|H1,I) etc. The same end result ensues. The probability of obtaining one sample must, at some stage, be considered in the light of the analysis of the other sample. The concept of a match is unnecessary. Furthermore, use of this concept requires artificiality when it is extended to continuously variable characteristics, whereas the treatment above applies to discrete variables, continuous variables and even to characteristics which may be changed, such as hair length.

An alternative presentation is to regard one of the samples as the evidence and the other as part of the hypothesis. A criminal case could be seen as a process which begins with an accused about whom we have information; for example, as soon as he or she pleads *not-guilty* we can see height, race, sex, weight, etc., while we have no information about the offence. Evidence about the offence is considered in the light of the accused's known characteristics. In this case

H1a = the accused,*a person of known characteristics*, left the mark.
H2 = someone else left the mark.
E1 = the sample from the scene has profile A.

Then we have

$$\frac{P(H1a|I)P(E1|H1a,I)}{P(H2|I)P(E1|H2,I)}.$$

P(E1|H1a,I) $= p_m$, P(E1|H2,I) $= f$. The prior odds are $1/N$. The posterior odds are therefore: p_m/Nf.

Confusion arises if the accused's profile is regarded as the evidence. In this model we know all about the offence and consider the accused's characteristics in the light of that knowledge.

H1b = the accused left a mark of profile A.
H2b = someone else left the mark profile A.
E2 = the sample from the accused has profile B.

This gives

$$\frac{P(H1b|I)P(E2|H1b,I)}{P(H2b|I)P(E2|H2b,I)}.$$

P(E2|H1b,I) is the probability that the accused would have the particular profile supposing that the accused left the mark (which has already been analysed). This corresponds to the measurement probability and equals p_m. P(E2|H2b,I) is the probability that the accused would have the particular profile supposing someone else left the analysed mark, f_a.. This will depend on the information about the accused that is taken into account, as part of I.

The likelihood ratio is therefore p_m/fa.

Whatever information that is taken into account about the accused must also be included in the I in the prior odds. P(H1b|I) is therefore the probability that the accused left a given mark supposing all our background information. This will be $fa/N + 1$. P(H2b|I) is $Nf/N + 1$. The prior odds ratio is no longer $1/N$ but $f_a./Nf$.

The posterior odds are therefore $p_m . f_a./f_a.Nf = p_m/Nf$. The same result is obtained provided we always remember that when the accused's sample is considered any information about the accused will also affect the prior odds ratio.

The Balding and Nichols argument stems from considering an unconditional match probability rather than considering each sample separately, from failure to specify the hypotheses precisely, failing to make clear the conditioning factors of each probability assessment and failure to set out the Likelihood Ratio in full. There are numerous other instances of disputes in forensic science which solve themselves when these steps are correctly taken (see, for example, Koehler, 1992).

vi. *Transposition of the Conditional*: The Match-No Match approach invites the listener to transpose the conditional. This problem was analysed by Thompson and Schumann (1987). The transposition can take two forms:

a) the application of a standard test may cause a high probability for the evidence given the prosecution case to be translated into a high probability for the prosecution case given the evidence. Good examples are provided by behavioural evidence in child sexual abuse cases. In *R v B* [1987] 1 N Z L R 362 a man was accused of sexually assaulting his adopted daughter. A psychologist gave evidence of a number of tests and observations made while interviewing the girl. In discussing each observation the psychologist made some comment such as: "[this] is typical of sexually abused girls" save for the dreams about which she said dreams of this kind are "frequently experienced by sexually abused young people." From this she concluded that the child was telling the truth when she said that she had been abused.

b) Alternatively, the odds against a match by chance become translated into the odds against anyone else having been responsible. Numerous examples can be found of people presented with correct information immediately transposing the conditional in this way, e.g. the statement by the Lord Chief Justice in *R. v. Cannan* (1991) 92 CrAppR 16 (CA) at p 18, that "so far as the DNA evidence was concerned it seems that the chances of anyone else having been responsible for this semen found on the knickers was something like 260 million to one against." Chapter 6 of Grubb and Pearl (1990) consists of an extended discussion in which this error is committed throughout.

Transposing the conditional can be done in a deliberate and grandiose way that obscures the fallacy. A striking example is the procedure for calculating the 'probability of paternity' by applying a 'neutral prior' to the 'paternity index' (which is just a likelihood ratio). This procedure was advocated by Essen-Möller (1938) and is still routinely used despite having been exposed as fallacious in many publications, especially Kaye (1989). In particular, this process can lead to the bizarre result that each of two non-excluded candidates for paternity can be said to have a probability of paternity in excess of 99%. It also leads to confusion about the role of probability in forensic science, as illustrated by State v Skipper 637 A (2nd) 1101 (1994).

The match/no-match procedure clearly leads to an increase in false exclusions. It is often argued that this is acceptable since it errs in favor of the accused, in other words it is a conservative procedure. But work by Evett and others shows that the number of occasions on which falsely inculpatory evidence is obtained is also increased (Evett, Scranage & Pinchin, 1993). Arbitrary adjustments to evidence, motivated by conservativeness, cannot be guaranteed invariably to be conservative.

(B) The mixed-stain problem

Suppose that two or more bloodstains (or a mixed stain) were found at the scene of a crime. If we ask the question "What is the probability that the accused will match one of the stains before we analyse his blood?" we obtain the answer by adding the frequencies of the groups in the stains. This is the form of report recommended by the National Research Council (1992). This cannot possibly be the answer to the correct question "How much should this evidence increase our belief in the accused's guilt?" since it will give the same answer whichever of the blood groups the accused has. Common sense tells us that if one of the groups is very rare and another very common the evidence must be stronger if the accused is of the very rare group. In fact the correct value for the evidence is $1/nf$ where f is the frequency of the characteristics the accused 'matches' and n is the number of stains considered. This solution is fully derived in Evett (1987). For extension of the analysis to DNA testing of mixed stains see Evett *et al.* (1991).

(C) The race of the accused

If we ask the question "What is the probability that the accused will match the mark at the scene before we analyse his blood?" then clearly the answer will depend on what we know of the accused's race. Certain bandweights, for example, may occur together more commonly in particular sub-groups than in the general population so that it is not appropriate in answering this question to use the product rule in the form that assumes independence.

If we ask the correct question we find that the accused's race is irrelevant. Firstly, the accused's race is not relevant to the probability of obtaining the evidence if the accused left the mark since that is assessed simply by comparing the samples. Secondly, the accused's race is not relevant to the probability of obtaining the evidence under the alternative hypothesis if that is that someone else left the mark. We then have the question "what sort of someone else?" and the answer depends on what is known about the *perpetrator*, not the accused. Consider a case in which a man of Maori appearance is seen running away from a place where a murder victim is found. Blood at the scene is analysed and when a man of Maori appearance is arrested his blood is found to resemble that at the scene. This suspect is identified by witnesses as having been the man seen running away. How do we weigh this evidence?

The answer will depend upon the line taken of defense. If the defense is that the accused was not the man seen running away but the victim of mistaken identity, then the perpetrator was presumably the Maori seen running away; the probability of the evidence if the defense hypothesis is correct is the frequency of the characteristics in the local Maori population.

Suppose on the other hand the suspect admits to being the man running away but denies any connection with the murder. If this story is true we have no information as to the perpetrator and so the appropriate population is the population at large.

If, in either case, the suspect were the only immigrant from some remote Pacific island then it would be obvious that there would be no reason to consider the frequency of characteristics in his sub-population.

In a very small number of cases there may be evidence that the perpetrator came from a particular subgroup or even a family. In this case evidence about subgroups is still misleading if it fails to take into account that the suspect pool has been sharply reduced. The overall effect, when properly considered may even be counter-productive to the defense but the emphasis in cross-examination is frequently on producing a number of different figures for the likelihood ratio alone. Only if prosecution counsel is unusually astute will the scientist be given an opportunity in re-examination to point out that changing the alternative hypothesis will change the prior odds ratio.

A logical approach to the questions the court wants answered therefore reveals that where a mark such as a bloodstain is found at the scene of a crime the race of the accused is irrelevant.

(D) Screening and databases

Suppose that we screen a large number of potential suspects or consult a database of DNA profiles. If we ask the question "what is the probability of obtaining a match by carrying out this procedure?" then the larger the database the more likely we are to find a match. If the frequency of a characteristic is 1 in 1000 and we consult a database of 10,000 then we would expect about 10 matches in the database. It is almost certain we would get a match in this case. The probability of one or more matches is $1 - (1 - f)^N$, where f is the frequency of the characteristic and N is the size of the database. If the characteristic is rare and the database small (so that Nf is much smaller than 1) the probability of a match is approximately Nf.

Since the probability of a match increases as the database gets larger, asking the wrong question leads to the conclusion that the larger the database the weaker the evidence. This in turn leads to recommendations about dealing with evidence from database searches, usually aimed at arbitrarily reducing the strength of the evidence. The NRC Report, for example, recommends that the alleles in the database should be ignored for the purpose of assessing the value of a 'match' and only those tests using other probes taken into account.

If we ask the different (but correct) question: "How much does the evidence increase the probability that it was the accused that left the mark?" we see that the size of the database is usually irrelevant. The probability of obtaining the evidence if someone else left the mark is determined by the frequency of the characteristic. But the nature of the database will affect the prior odds to which the likelihood ratio is applied. Whenever a database is searched the alternative hypothesis being investigated is that the perpetrator could have been any member of the population from which the database is chosen. If the database is drawn from the whole of New Zealand (pop 3.5 M) (for example, DNA samples taken from convicted criminals throughout the country) then the prior odds (prior even to eliminating the very elderly, young and infirm) are 3.5 M to 1 against the suspect's guilt. A DNA analysis giving a likelihood ratio of 1 M reduces those odds to 3.5 to 1 against. In other words we expect there to be three or four people in New Zealand who would 'match' those characteristics and more evidence is required to determine which of them is the guilty party. As long as we have strong information about the frequency it does not matter how many 'matches' we find in the database itself. If there were two 'matches' in this case we would still expect there to be one or two more possible 'matches' in the population not included in the database. Each of the three or four people who 'match' on this evidence alone are equally likely to be guilty regardless of whether they are in the database.

On the other hand, the size of the database may be relevant in one situation. This is where we have no prior knowledge of the frequency of the characteristic and are using the database as a sample. Then for any given number of 'matches' the evidence gets stronger as the database gets larger because the estimated frequency gets smaller. Consider a database of the entire population, say the car registration system. If a suspect vehicle is described as a Testarossa and the national registration system reveals only one in New Zealand then obviously we have the right vehicle. On the other hand, a search of the Wellington Council Resident's Parking Permit records might reveal one Testarossa in the 10,000 recorded vehicles. With only this information, one would assume that roughly 1 in 10,000 vehicles in New Zealand are Testarossas with an allowance for background knowledge that Testarossa owners are probably concentrated in the major cities. It cannot be assumed that the 'match' is the guilty vehicle. If two such cars were found the frequency would be estimated as 1 in 5,000 and, again, it is that frequency which determines the strength of the evidence, not the size of the database.

One can fall into some logical traps when screening and using databases. It is easy to transpose the conditional and believe that because under hypothesis H1 the evidence was highly probable, H1 must be correct. It is easy to underestimate the effect of the very low prior that is implied by searching a database. The fear is even expressed that investigators might stop searching at the first match. Where the database is being used to assess the frequency in the population, then clearly failing to search the whole database deprives the investigator of information which should have been used. If the likelihood ratio is based upon external evidence (e.g. a population study of which the database of convicted persons is only a part) then no error will result from stopping at the first match if the correct likelihood ratio and prior odds are used.

What is dangerous is for scientists to take it upon themselves to doctor the evidence to allow for the possibility of these errors on the part of others. These logical traps are simply acute examples of problems general to rational investigation and they are obscured rather than illuminated by the sort of steps recommended by the NRC Report.

Is forensic science a science in its own right?

Measurement of frequencies of occurrence of characteristics is a mechanical process which seems to have stultified consideration of how forensic scientists can be of greater assistance to the court. Once we ask a question such as 'what is the probability that the accused would have this mark on him if he committed the offence?' we realise that it is not enough to know the frequency of occurrence of a type of glass or blood. We also need to know the probability that the offender would have the observed quantity of glass on his clothing at the time of arrest. We must therefore assess the probabilities of 'transfer' and 'persistence', as well as the probability that a person unconnected with such offences would have glass on his clothing (Evett, 1986; Evett & Buckleton, 1990). Further research is needed to obtain the information on which these assessments are based and this form of analysis has stimulated such research in recent years (Pearson, May & Dabbs, 1971; Walsh & Buckleton, 1986; McQuillan & Edgar, 1992).

This analysis forces concentration on the questions of interest to the court. It is this which gives forensic science both unity and distinctiveness. Questions of the value as evidence for particular hypotheses of single non-replicable data sets are not normally considered by chemists, biochemists and statisticians outside the forensic scientific community. The most highly qualified specialists are therefore not necessarily expert in making such assessments. Forensic scientific witnesses, whatever their background, should qualify themselves as forensic scientists, experienced in these inferential problems, rather than as chemists, biochemists or statisticians, in which guise they frequently appear less well qualified than defense witnesses.

Conclusions

a) Forensic scientific evidence should be given by comparing the probabilities of obtaining the evidence under each of two relevant, specific, and positive hypotheses.

b) Forensic science is a discipline concerned with the post-data question of drawing inferences about a particular case from evidence which is actually available. It is these inferential questions which give forensic science unity and which distinguish it from other areas of science. Research and training should be directed towards solving these inferential problems.

c) Orthodox statistical techniques do not address these questions.

d) Forensic scientists should develop a greater sense of professional unity based upon the acceptance of Bayesian inference. They should aim for (and justify) recognition that they have particular expertise in the inferential problems posed by forensic scientific evidence.

Orthodox statistical techniques have survived because in some circumstances they appear to produce the right answers. But it is obviously no defense of a procedure that it sometimes produces the same answer as the correct one. Rather than ask the wrong questions hoping that the right answers will be obtained, surely it is better to ask the right questions in the first place?

References

Balding, D.J. & R.A. Nichols, 1994. DNA Profile Match Probability Calculation: how to allow for population stratification, relatedness, database selection and single bands. Forensic Science International 64: 125–140.

Essen-Möller, E., 1938. Die Biesweiskraft der Ähnlichkeit im Vater Schaftsnachweis; Theoretische Grundlagen. Mitt Anthorop, Ges (Wein) 68: 598.
Evett, I., 1986. A Bayesian Approach to the problem of interpreting glass evidence in forensic science casework. Journal of the Forensic Science Society 26: 3–18.
Evett, I., 1987. On meaningful questions: a two-trace transfer problem. Journal of the Forensic Science Society 27: 375–381.
Evett, I., 1991. Interpretation: a personal odyssey. pp 9–22 in C.G.G. Aitken & D.A. Stoney (eds) The use of statistics in forensic science. Ellis Horwood, Chichester UK.
Evett, I. & J. Buckleton, 1990. The interpretation of glass evidence. A practical approach. Journal of the Forensic Science Society 30: 215–223.
Evett, I.W., C. Buffery, G. Willott & D. Stoney, 1991. A guide to interpreting single locus profiles of DNA mixtures in forensic cases. Journal Forensic Science Society 31: 41–47.
Evett, I.W., J. Scranage & R. Pinchin, 1993. An illustration of the advantages of efficient statistical methods for RFLP analysis in forensic science. Am. J. Human Genetics 52: 498–505.
Grubb, A. & D.S. Pearl, 1990. Blood Testing, AIDS and DNA Profiling; Law and Policy. Jordan and Sons Ltd, Bristol.
Kaye, D., 1989. The probability of an ultimate issue; the strange cases of paternity testing. Iowa Law Review 1: 75–109.
Koehler, J.J., 1992. Probabilities in the Courtroom: An Evaluation of Objections and Policies. pp. 167–183 in D.K. Kagehiro & W.S. Laufer (eds), Handbook of Psychology and Law. Springer-Verlag, New York.
Lindley, D., 1977. A problem in forensic science. Biometrika 64: 207–213.
Magnusson, E., 1993. Incomprehension and Miscomprehension of statistical evidence: an experimental study, Australian Institute of Criminology Conference on Law, Medicine and Criminal Justice, Queensland, July 1993.
McQuillan, J. & K. Edgar, 1992. A survey of the distribution of glass on clothing. Journal of the Forensic Science Society 32: 333–348.
Mills, H., 1991. The Birmingham Six Case – Vital Scientific Evidence Kept From Defence. The (London) Independent, 28 March 1991, p 2.
National Research Council, 1992. DNA Technology in forensic science. National Academy of Sciences.
Pearson, E.F., R.W. May & M.G.D. Dabbs, 1971. Glass and paint fragments found in men's outer clothing – a report of a survey. Journal of Forensic Sciences 13: 283–302.
Thompson, W.C. & E. L. Schumann, 1987. Interpretation of statistical evidence in criminal trials/The prosecutor's fallacy and the defense attorney's fallacy. Law and Human Behaviour 11: 167–187.
Walsh, K.A.J. & J.S. Buckleton, 1986. On the Problem of Assessing the Evidential Value of Glass Fragments Embedded in Footwear. Journal of the Forensic Science Society 26: 55–60.

Editor's comments

The authors reiterate the crucial role of likelihood ratios in presenting forensic evidence, and so continue the crusade mounted by Evett, Buckleton and others.

B. S. Weir (ed.), Human Identification: The Use of DNA Markers, 153–168, 1995.

Subjective interpretation, laboratory error and the value of forensic DNA evidence: three case studies

William C. Thompson
Department of Criminology, Law & Society, University of California, Irvine, CA 92717, USA

Received 5 May 1994 Accepted 10 June 1994

Key words: forensic, DNA, evidence, error, subjective

Abstract

This article discusses two factors that may profoundly affect the value of DNA evidence for proving that two samples have a common source: uncertainty about the interpretation of test results and the possibility of laboratory error. Three case studies are presented to illustrate the importance of the analyst's subjective judgments in interpreting some RFLP-based forensic DNA tests. In each case, the likelihood ratio describing the value of DNA evidence is shown to be dramatically reduced by uncertainty about the scoring of bands and the possibility of laboratory error. The article concludes that statistical estimates of the frequency of matching genotypes can be a misleading index of the value of DNA evidence, and that more adequate indices are needed. It also argues that forensic laboratories should comply with the National Research Council's recommendation that forensic test results be scored in a blind or objective manner.

Introduction

The debate over population genetics has dominated recent commentary on forensic DNA testing. To casual readers of the scientific literature, or of court opinions, it may now appear that the only element of uncertainty about the value of DNA evidence is the rarity of DNA profiles. But this impression is dangerously misleading. There are at least two other sources of uncertainty that may profoundly affect the value of forensic DNA evidence: uncertainty about the interpretation of test results and uncertainty about laboratory error. The purpose of this article is to explain and illustrate these other sources of uncertainty and to comment on how they affect the value of DNA evidence.

Uncertain matches

In cases involving forensic RFLP analysis, there sometimes is uncertainty about the scoring of bands (Lander, 1989; Thompson & Ford, 1991; Thompson, 1993). The bands may be faint, or may be difficult to distinguish from artifacts. Yet accurate scoring is important. In some forensic cases, the scoring of a single ambiguous band can determine the outcome – one interpretation produces damning incrimination of the suspect, another interpretation completely exonerates the suspect.

There is a danger that such judgments will be influenced by examiner bias. Commentators have noted the tendency of forensic analysts to see what they expect to see when scoring ambiguous autorads (Lander, 1989; Thompson & Ford, 1991, p. 140–41). In the widely-discussed *Castro* case, for example, a forensic laboratory recorded three bands in a lane where other experts saw only one, and recorded two bands in a lane where other experts saw four (Lander, 1989). Lander suggested that the analyst 'may have been influenced by making direct comparisons between lanes containing different DNA samples [that were expected to match], rather than by considering each lane in its own right'. (p. 502).

Lander considered cross-lane comparison to be inappropriate for forensic DNA analysis due to the human tendency to see what is expected. This tendency, sometimes called an expectancy effect, has been documented in a variety of contexts (Nisbett & Ross, 1980) and has led scientists in many fields to insist that procedures for interpretation of potentially ambiguous data be either blind or objective. But forensic scientists appear to be committed to procedures that allow analysts considerable discretion in scoring autorads. For example, an official statement of the FBI's Working Group on Statistical Standards for DNA Analysis declares: 'Whether or not there is a match between patterns produced by DNA samples extracted from two or more sources is primarily a qualitative judgment by a knowledgeable investigator based on a careful review of all information pertinent to the tests undertaken'. (Technical Working Group on DNA Analysis Methods, 1990).

The report of the National Research Council (1992, p. 53) called for the use of objective scoring procedures for forensic DNA evidence, and declared that 'it is not permissible to decide which features of an evidence sample to count and which to discount on the basis of a comparison with a suspect sample, because this can bias one's interpretation'. But the NRC's recommendations in this area have not been followed by forensic laboratories.

Most laboratories score autorads with the help of computer-assisted imaging devices. But these devices, as currently used, leave the placement of bands, and the ultimate determination of whether bands are present, within the analyst's discretion. The analyst may cause the computer to add or delete bands when scoring a particular lane.[1] The analyst may also vary the system's threshold for detecting bands across different parts of the autorad, which may affect the system's scoring of bands. Hence, the use of computer-assisted imaging devices does not necessarily resolve concerns about examiner bias.

Examiner bias can lead analysts to overestimate the strength of DNA evidence. The analyst may, for example, infer that a discrepancy between two DNA profiles on one probe *must* be an artifact (rather than a true genetic difference) because there is a match on the other probes or, worse yet, because other evidence in the case suggests the two profiles have a common source. The author heard one forensic analyst defend the scoring of an ambiguous band (a judgment that incriminated the defendant in a rape case) by saying 'I must be right, they found the victim's purse in [the defendant's] apartment'. Inferential bootstrapping of this sort can convert problematic results into an apparently damning incrimination. The consistency (or inconsistency) of the DNA evidence with other evidence in the case is a matter that should be taken into account by the trier-of-fact (i.e., the jury) in a criminal trial, not by an analyst who is providing a putatively objective, independent interpretation of the DNA test.

Laboratory error

The possibility of laboratory error also affects the value of DNA evidence for proving that two samples have a common source. In forensic laboratories, samples from a given case are typically processed together, in a batch, through a sequence of procedures that require manual transfers of genetic material from container to container. Inadvertent switching, mixing or cross-contamination of samples can cause false matches (Thompson & Ford, 1989). For example, DNA from one sample may inadvertently be mixed with another sample at a number of stages: when the samples are collected, during DNA extraction, when enzymes and buffers are added to samples for restriction digestion, or when samples are loaded on gels (due to failure to use a fresh micropipet tip, spillage, or inaccurate sample placement) (Thompson & Ford, 1991).

False matches resulting from sample handling errors do not necessarily cause false incrimination or false exclusions, although they have that potential. In some cases, these errors produce improbable results that signal the possibility of an error. For example, Thompson and Ford (1991, p. 143) document a rape case in which a forensic laboratory found a match between the DNA profiles of the suspect and the victim. In other cases, however, a false match may be difficult or impossible to distinguish from a valid result. For example, if DNA of an innocent suspect was inadvertently mixed with an evidentiary sample from the crime scene, the appearance of the suspect's alleles in the evidentiary sample would be highly incriminating. Errors of this type would, in some instances, be signaled by the appearance of unexplained 'extra bands' (those of the true perpetrator). But these tell-tale indicators of a

[1] Most laboratories do not document manual overrides of the computer's initial scoring of bands. One exception is Cellmark Diagnostics, which requires analysts to record the number of bands added and deleted in each lane in the case notes. In a group of approximately 15 Cellmark cases reviewed by the author, additions and deletions were common. In almost every case there were a few; some cases had more than a dozen. Whether these cases are representative of Cellmark casework, or that of other laboratories is unclear. Manual overrides have also been noted to occur in FBI casework (Thompson, 1993, p.40), although their frequency is unknown.

The author does not mean to imply that these additions and deletions are improper. They may be necessary and appropriate given the limitations of computer-imaging technology. But the possibility of these overrides would seem to belie any claim that use of computer-scoring eliminates subjective elements from the process of interpreting autorads.

sample handling error might easily be missed or mistakenly attributed to other causes.[2]

Existing data on the rate of laboratory errors are often misinterpreted. The most frequently discussed study was conducted by the California Association of Crime Laboratory Directors (CACLD) (Graves & Kuo, 1989). In the first round of the study, the CACLD submitted approximately 50 simulated forensic samples (blood or semen stains) to each of three forensic DNA laboratories. The laboratories analyzed and compared the samples, and reported which ones 'matched' (i.e., had indistinguishable DNA patterns). In the second round of the study, which essentially replicated the first, a second set of approximately 50 samples was analyzed and compared.

One of the laboratories, Cellmark Diagnostics, reported two false matches, one in each round. In each instance, a sample was contaminated with DNA from another sample that was processed in the same batch, causing an apparent but erroneous match (Thompson & Ford, 1991). Cellmark had processed the samples in 16 batches. Thus, out of approximately 100 samples, two were contaminated with sufficient DNA from another sample to cause false matches. Another way of looking at the matter is that sample handling errors produced false matches among samples in two of the 16 batches that Cellmark processed.

Based on the CACLD study, Devlin, Risch and Roeder (1994) reach the surprising conclusion that 'the error rate for this lab [Cellmark] would be $1/1225 < .0008\ldots$' They argue that because there was only one false match in each round, and the study required laboratories to make pairwise comparisons among 50 samples in each round (i.e., 1255 comparisons), the error rate is 1 in 1255. This conclusion is misleading in several respects.

First, it misrepresents the results of the CACLD study. According to the CACLD's report (Graves & Kuo, 1989), in the first round Cellmark received 49 simulated forensic samples and obtained what is considered an interpretable DNA print from 44 of them. These 44 samples had originated in 21 different individuals; the number of samples from each individual ranged from one to six. Of the 946 possible pairwise comparisons among these samples, 36 were between samples from the same individual and 910 were between samples from different individuals. Cellmark's mistyping of one sample, due to cross-contamination, caused that sample to be mistakenly matched with two samples from a different individual. Thus, two of the comparisons produce false matches. Cellmark also erred with respect to another sample, failing to find a match between it and two others from the same person (see Graves & Kuo, 1989, p. 4, Sample 76). Thus, two of the comparisons produced false conclusions of non-match. If the error rate is stated in the manner preferred by Devlin, Risch and Roeder (1994), the the overall error rate is $4/946 = .004$ and the rate of false matches is and $2/946 = .002$.

This characterization of Cellmark's error rate is less than ideal, however, because the ratio of the number of errors to the number of comparisons is affected by the base rate of matches and non-matches in a particular study, which may be contrived (as it was in the CACLD study) and unrealistic. A more conventional way to describe the laboratory's performance would be to report the rate of false positives and false negatives. The false positive rate of a test is most usefully stated as the ratio of false positives to the sum of true positives and false positives (Hart, Webster & Menzies, 1992, p. 698). Hence, Cellmark's false positive rate in the first round of the CACLD study was $2/36 = .055$, or 5.5 percent. Correspondingly, Cellmark's false negative rate was $2/825 = .0024$, or 0.24 percent.

These estimates of error rate are still quite misleading, however, because they rest on the mistaken premise that the rate of errors is a function of the number of comparisons made among DNA prints in the study. In fact, the false matches arose from sample handling errors during the batch processing of samples and not from errors in comparing the resulting DNA prints. Hence, the number of comparisons that happened to have been made among DNA prints in the CACLD study is irrelevant to the likelihood that the laboratory would falsely match two samples in casework.

To illustrate this point, let us assume that Cellmark's performance in the CACLD study accurately reflects its performance in casework. Let us further assume that in Cellmark's next case it processes six

[2] Forensic analysts often find it difficult to distinguish extra bands caused by the presence of a second DNA pattern in a sample from extra bands caused by technical artifacts, such as partial digestion or star activity (Thompson & Ford, 1991). Moreover, examiner bias would favor discounting the unexplained 'extra bands' as uninformative artifacts. Alternatively, the extra bands might be attributed to a second person, the presence of whose DNA is consistent with the suspect's guilt. In a rape case, for example, the 'extra bands' in an evidentiary sample might be thought those of a husband or boyfriend of the victim. Although the proper procedure, in such cases, would be to confirm this possibility by testing the putative second contributor (see NRC Report, p. 58–59), this procedure is not always followed (Thompson & Ford, 1991).

forensic samples in a batch: one sample from the suspect, one from the victim and four evidentiary samples presumed to contain DNA of the perpetrator. What is the probability that Cellmark will falsely incriminate the suspect?

Based on the CACLD study, we would expect there to be one chance in eight that one of the samples within a batch will be contaminated with DNA of another. If we assume each of the six samples is equally likely to contaminate any of the other five, there are 30 equally likely ways that cross-contamination could occur. Of course, many of those ways would not incriminate the suspect. But contamination of any of the four evidentiary samples with the suspect's DNA would definitely be incriminating – so let us assume that four of the 30 ways cross-contamination could occur would be incriminating. Under these assumptions, the probability of a false incrimination would appear to be $1/8 \times 4/30 = .016$.[3]

Under this analysis, the probability of a false incrimination would vary somewhat from case to case depending on the number of samples in a case and the number of ways a false incrimination might arise from cross-contamination within batches of samples. Nevertheless, if Cellmark's performance on the CACLD study reflects its performance in casework, it is difficult to see how the probability of a false incrimination could possibly be as low as Devlin, Risch and Roeder (1994) suggest.

Whether Cellmark's performance on the CACLD study was typical of other laboratories (or even of Cellmark) is difficult to judge. However, another laboratory that participated in the CACLD study, Forensic Science Associates (which used a PCR-based test examining HLA DQ-alpha alleles), also had an 'incorrect match' due to 'sample manipulation or operator error' (Graves & Kuo, 1989). The third laboratory, Lifecodes Corporation, had no false matches on the CACLD study, but has been reported to have made several sample handling errors in casework (NRC Report, p. 88; Thompson & Ford, 1991). Hence, there is reason to believe that laboratory error is a significant factor affecting the value of DNA evidence (Koehler, 1993a; 1993b).

[3] Alternatively, we might assume that each sample had a probability of .02 of being contaminated by DNA of another sample in the same batch. Because there are four evidentiary samples, the probability is approximately $4 \times .02 = .08$ that one of the evidentiary samples would receive contaminating DNA. If each sample is equally likely to contaminate any other, the probability that the contaminating DNA would be from the suspect's sample (thereby creating a false match that incriminates the suspect) would be $1/5 \times .08 = .016$.

Case studies

To illustrate and elaborate the points made thus far, let us consider the DNA evidence in three actual forensic cases. In each instance, the laboratory results were made available to the author by defense counsel. (Copies of the laboratory reports are available from the author).

Case 1. Figure 1 shows the autorads of a case that came to trial in 1989 in which a forensic laboratory declared a match between the DNA print of the defendant (MP) and a print of semen taken from a vaginal swab (VS) of a murder victim (RM). Although bands matching those of the defendant appear in the vaginal swab, the vaginal swab also contains 'extra bands' matching neither the defendant nor the victim. How important is this discrepancy?

The laboratory analyst apparently concluded that the extra bands were unimportant and ignored them; the laboratory lab report does not even mention the extra bands. The report says that the defendant's DNA matches the evidence and that the frequency of the matching characteristics is about 1 in 30 billion. The report gives no indication of any technical problem that might affect the value of the evidence. The analyst probably assumed that the extra bands were artifacts – that is, the result of a minor technical problem in the assay. Artifactual extra bands are common in casework (Thompson & Ford, 1991).

An independent expert (Simon Ford), who reviewed the autorads at the request of defense counsel, had a different interpretation of the extra bands. Based on the appearance and location of the bands, Ford thought they were unlikely to be technical artifacts and were probably the DNA of a second individual. He thought that the laboratory's failure to rule out this possibility significantly undermined the value of the evidence for incriminating the defendant.

How much difference does it make whether the extra bands are due to an artifact (as the laboratory analyst apparently believed) or a second person's DNA (as Ford suspected)? One might think that Ford's concern is trivial – the existence of a second semen contributor does nothing to exculpate the defendant, particularly when, as in this case, defendant had resolutely insisted that he never had intercourse with the victim. From this perspective, Ford is making a big deal over almost nothing (a charge that has repeatedly been leveled against defense experts) (see Moenssens, 1990).

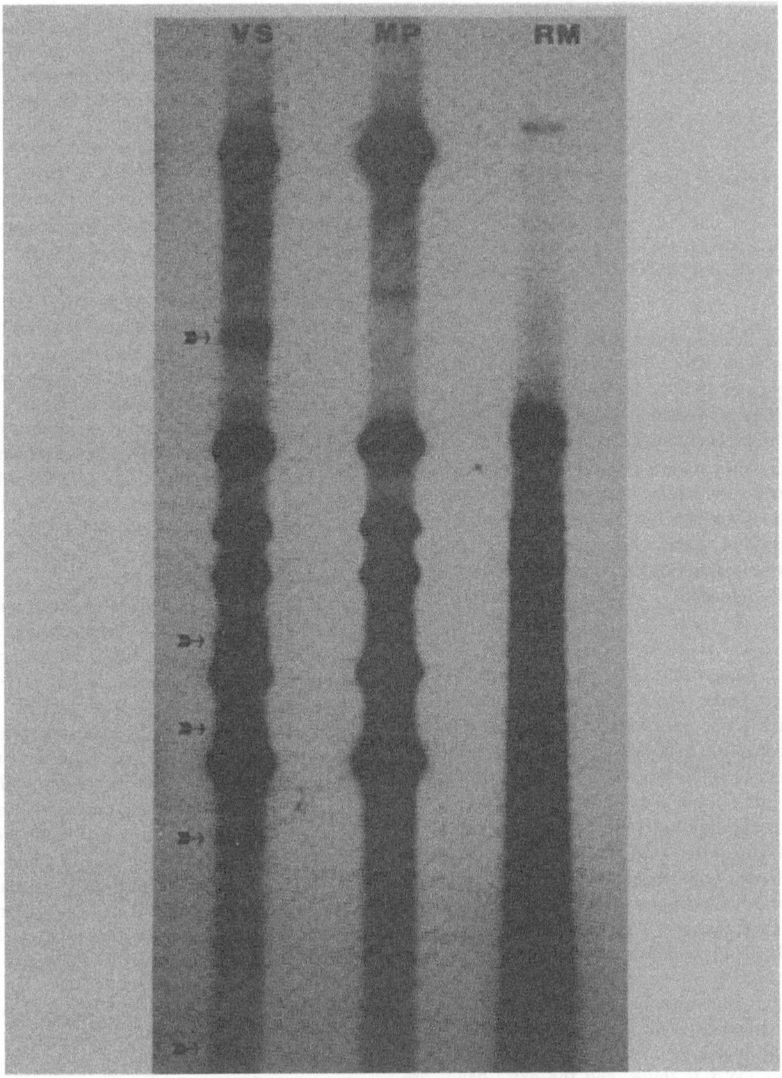

Fig. 1. DNA prints on vaginal swab (VS), defendant (MP) and rape victim (RM) in Case 1. This membrane was hybridized with a single-locus probe cocktail consisting of probes MS1, MS31, MS43 and g3. Arrows indicate 'extra bands' that the forensic laboratory failed to report.

But a careful analysis indicates that Ford's concerns were well grounded.

Let us begin with the two competing hypotheses in the case:

H1: Defendant is the source of the DNA in the vaginal swab

H2: Defendant is not the source of the DNA in the vaginal swab

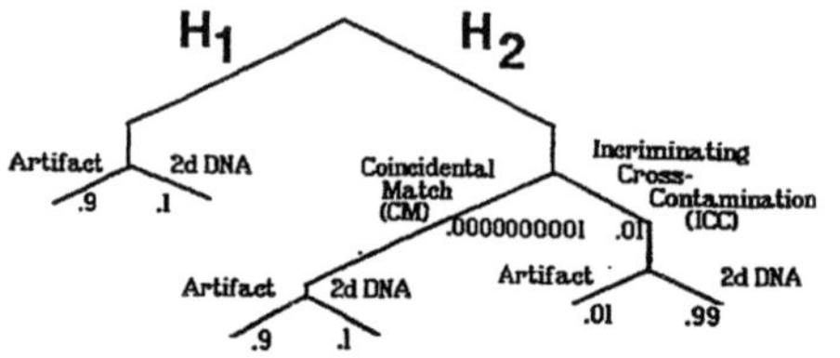

Fig. 2. Model of relationship between hypotheses and conditioning events, showing estimated conditional probabilities (Case 1).

What we must consider is how our belief in these hypotheses should be affected (conditioned) by the event in dispute – that is, by whether the extra bands are an artifact or a second DNA. Let us define the conditioning events precisely:

Artifact – the evidentiary sample contains DNA of a single person whose pattern matches defendant; the extra bands are the result of a technical artifact

Second DNA – the evidentiary sample contains DNA of two individuals, one whose DNA pattern matches defendant, the other whose pattern does not.

The analysis involves three steps. First, a simple model is constructed of how the two conditioning events might occur – that is, how there might come to be an artifact or a second DNA. Second, the conditional probability of these events under H1 and H2 is estimated. Finally, the conditional probabilities are combined (multiplied) to determine the likelihood ratio under each conditioning event. Comparison of the two likelihood ratios will reveal how much the value of the DNA evidence varies depending on which interpretation of the extra bands (artifacts or second DNA) is correct. Hence, the analysis will reveal how much the value of the DNA evidence depends on experts' subjective judgments. A model of how the conditioning events might occur is represented in Figure 2.

If the defendant is the source of the semen (H1), then there are matching bands in the evidentiary sample because his DNA is there, and the extra bands are either an artifact or a second DNA. If the defendant is guilty, then the extra bands are much more likely to be an artifact, as the laboratory analyst assumed, than a second DNA, so let us arbitrarily assign probabilities of .9 and .1 to these events.[4]

If the defendant is *not* guilty (H2), we must first come up with an explanation for the matching bands. One possibility is a coincidental match (CM) between the DNA profiles of the defendant and the real rapist. We will assume the likelihood of such a coincidence is low, and assign a probability of 0.000000001.[5]

An alternative possibility is cross-contamination of the evidentiary sample with the defendant's DNA, an event we call an incriminating cross-contamination (ICC). This was Ford's major concern. How likely is such an event? Let us assume that the probability of a given sample being cross-contaminated by another that is being processed at the same time is 0.02 (as suggested by Cellmark's performance in the CACLD study). In this case, there were two other samples being processed with the evidentiary sample – the defendant's and the victim's – and thus two ways the evidence could have been cross-contaminated, only one of which (by the defendant's sample) would be incriminating. So the probability of an incriminating cross-contamination would be $0.02/2 = 0.01$. Again, if readers' view of the world differs, they may try their own number.

If there was a coincidental match with the rapist, the relative likelihood of an artifact and a second DNA as a cause of the extra bands would presumably be the same as under H1 because the processes that produce artifactual extra bands are unlikely to depend on whether the rapist was the defendant or someone else with the same DNA pattern. On the other hand, if the match was the result of cross-contamination, the the extra bands are almost certainly those of the true rapist, whose DNA would be present because he was the source of the semen. So the likelihood of a second DNA is very high (let us assign it a probability of .99). The extra bands could be due to an artifact, rather than a second DNA, only if the true rapist's DNA was somehow lost from the sample, which seems very unlikely (let us assign it a probability of .01).

It is now possible to estimate the likelihood ratios for each conditioning event and to see how they vary depending on which interpretation of the extra bands is correct. If the extra bands are due to an artifact, as

[4] This conclusion depends partly on the author's understanding of other evidence in the case concerning the relationship between the victim and the defendant, the victim's background, and so on, that suggests a low *a priori* likelihood of a second sexual partner. Readers are encouraged to plug in different numbers if their view of the problem is different than ours, or if they wish to explore the implications of the analysis for different problems.

[5] The presence of extra bands in the evidentiary sample would make the probability of a coincidental match higher than the frequency of the defendant's DNA profile because the extra bands in the evidentiary sample would increase the number of ways a coincidental match would occur.

the laboratory analyst thought, then the likelihood ratio (LR) of the DNA evidence is specified by the following formula:

$$LR = \frac{p(\text{Artifact}/H_1)}{p(\text{Artifact}/CM)p(CM/H_2) + p(\text{Artifact}/ICC)p(ICC/H_2)}$$

Substituting the conditional probabilities we estimated earlier, we find that the value of the likelihood ratio is

$$LR = \frac{.9}{(.9 \times .0000000001) + (.01 \times .01)} \cong 9000$$

This likelihood ratio is not nearly as large as the reciprocal of the probability of a coincidental match (30 billion) because it takes into account the possibility of a laboratory error. On the other hand, it is not nearly as low as the reciprocal of the probability of an incriminating cross-contamination ($1/0.01 = 100$) because the observed data (artifact) is more diagnostic of H1 than H2.

Although the possibility of a laboratory error degrades the value of the DNA evidence dramatically (from a likelihood ratio of 30 billion to 9000), the DNA evidence is nevertheless powerfully incriminating *if* the analyst is right about the source of the extra bands. After hearing the DNA evidence, one's estimate of the odds of guilt should increase by a factor of 9000 if, in fact, the extra bands are due to an artifact.

On the other hand, if the extra bands are due to a second DNA, as Simon Ford thought, then the likelihood ratio is:

$$LR = \frac{p(\text{Second DNA}/H_1)}{p(\text{SecondDNA}/CM)p(CM/H_2) + p(\text{SecondDNA}/ICC)p(ICC/H_2)}$$

Substituting the conditional probabilities estimated earlier, we find that the value of the likelihood ratio is

$$LR = \frac{.1}{(.1 \times .0000000001) + (.99 \times .01)} \cong 10.$$

This likelihood ratio is even lower than the reciprocal of the laboratory error rate because a second DNA in the evidentiary sample is diagnostic of a laboratory error.

The analysis makes clear, then, that the value of the DNA evidence, in a case such as this, varies enormously depending on which interpretation of the extra bands is correct. One would expect similar variations whenever an analyst must interpret extra bands that are potentially diagnostic of incriminating cross-contamination. Despite initial appearances to the contrary, the subjective judgment of the analyst regarding the source of the extra bands is crucially important. If Ford's interpretation is correct, the evidence has only a tiny fraction of the value that one might reasonably assume it had on the basis of the laboratory report.

As it turned out, Ford was right about the source of the extra bands in this case. After consulting with Ford, defense counsel insisted that a second individual, MM, be tested.[6] As Figure 3 shows, and the laboratory later acknowledged, MM appeared to match the unreported 'extra' bands.

The case thus illustrates quite dramatically the importance of accounting for extra bands and illustrates the severe consequences of failure to do so. Based on the initial laboratory report, a trier-of-fact might reasonably have concluded that the likelihood ratio for the DNA evidence was on the order of 30 billion. In fact, our analysis suggests it was approximately 10. In other words, a person who relied upon the laboratory report might well have inferred that the DNA evidence in this case was three billion times more probative than is indicated by a more complete analysis of its value.

Case 2. Table 1 shows the band measurements of two DNA prints that were declared to match in another forensic DNA case that came to trial in 1993. Again there is a discrepancy in the number of bands, but in this case the discrepancy occurs only for two probes. On probe MS-1, the defendant has a high molecular weight band (16,166 base pairs) which was not observed in the evidentiary pattern. This discrepancy might well have been due to degradation of the DNA in the evidentiary stain, which is known to cause selective loss of high molecular weight bands. More troubling is the discrepancy on probe G3, where the defendant has an 'extra band' (4900 base pairs) that is not observed in the evidentiary

[6] The defense theory was that MM committed the rape and murder, but according to defense counsel MM was not taken seriously as a suspect in light of the report that appeared to link the semen from the vaginal swab only to the defendant.

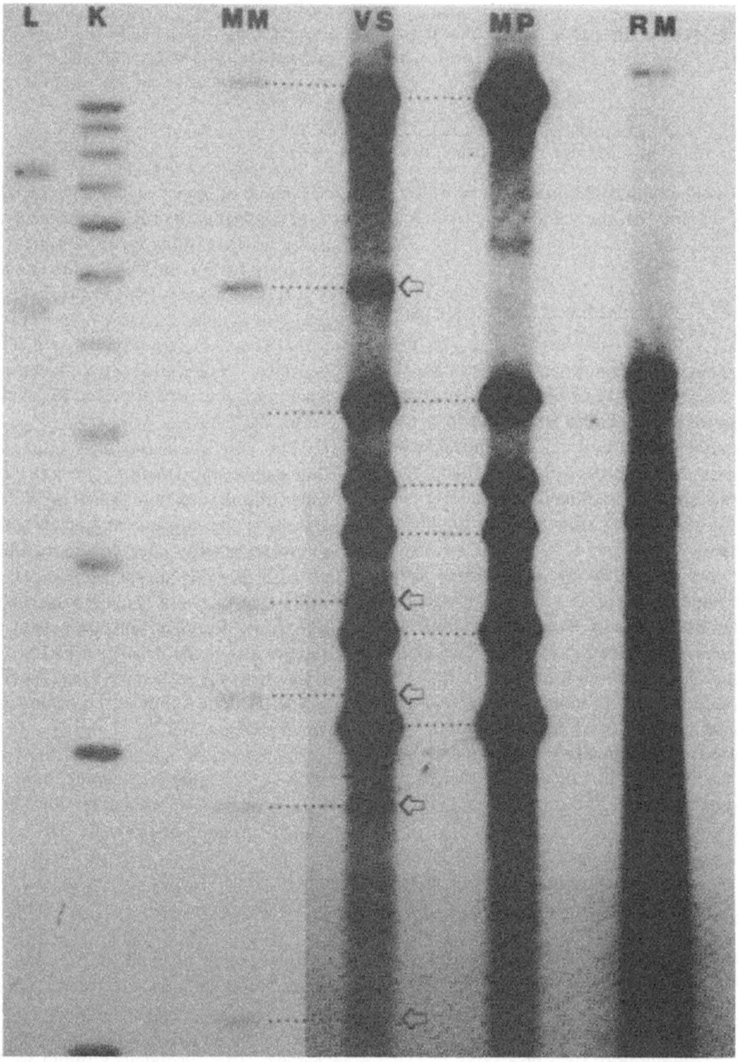

Fig. 3. Previously unreported 'extra bands' in vaginal swab (VS) in Case #1 are shown to match DNA pattern of a second individual (MM). Lanes L and K contain molecular weight markers. As with Figure 1, this membrane was hybridized with a single-locus probe cocktail consisting of probes MS1, MS31, MS43 and g3. Arrows indicate the controversial 'extra bands'.

pattern. This 'third band' was fainter than the other two for this probe, but quite distinct; and it appeared in two different samples taken from defendant but in none of several evidentiary samples that were processed.

The defense counsel gave the author permission to review his files on the case, which included three different experts' interpretations of the extra band on probe G3. One expert, a professor at a major medical

Table 1. Comparison of band measurements (in base pairs) in two DNA profiles (Case 2).

Probe	Evidentiary stain	Defendant
MS-1		16166
	5060	5030
MS-31	5710	5670
	3400	3400
MS-43	5710	5670
	3530	3550
G3	9740	9640
	5060	5030
		[4900]

school who has expertise in DNA biochemistry and extensive experience with RFLP analysis, concluded that the extra band was exculpatory, indicating that the discrepancy on probe G3 would, in his judgment, be 'very, very unlikely' if the samples were from the same source. A second expert, who is a professor of biology at a major university and has expertise in molecular genetics, and who has testified as an expert for both the prosecution and defense in cases involving DNA evidence, found the extra band troubling but not fully exculpatory. He said he would give no weight to the probe G3 and would rely solely on other probes in evaluating the DNA evidence. A third interpretation of the extra band is found in the laboratory report, which states: 'One additional band was observed in the DNA banding pattern of [the suspect]. The origin of this band has not been determined, but does not affect the ability to interpret the results'.

In this case, as in the previous one, a key issue is whether the difference among experts in their interpretation of the extra band is significant with regard to the value of the DNA evidence. Someone might think that a match over four or more probes is such powerful evidence of guilt that a minor inconsistency in the number or position of bands is of trivial importance: it is extremely unlikely that an innocent person would match so well, so one may safely ignore all concerns arising from imperfections in the match. A careful analysis shows, however, that the experts' disagreement is of crucial importance.

We begin again with two competing hypotheses in the case:

H1: Defendant is the source of the DNA in the evidentiary stain

H2: Defendant is *not* the source of the DNA in the evidentiary stain

What we must consider is how our belief in these hypotheses should be affected (conditioned) by evidence of a four-probe 'match' in this case, and how our beliefs will vary depending on which expert's assessment of the discrepancy on probe G3 we choose to credit.

We will assume that the four probes are independent. Hence, the likelihood ratio for the DNA evidence is the product of the likelihood ratios for each of the four probes. We will assume that the likelihood ratio for each probe, other than G3, is approximately the reciprocal of the frequency of the genotype identified by the probe. We will test how sensitive the value of the evidence is to the experts' interpretation of G3 by using three different values for the likelihood ratio of G3 that we believe to correspond with the experts' assessments. For Expert #1, who said the result on G3 was exculpatory and very, very unlikely if the samples have the same source, we will use a likelihood ratio of .001; for Expert 2, who said he would simply ignore probe G3, we will use a likelihood ratio of 1; for the laboratory's experts, who said the extra bands had no effect on interpretation of the results, we will use the reciprocal of the frequency of the G3 genotype. We will use two sets of genotype frequencies: those resulting from the application of the National Research Council's Modified Ceiling Principle, which produces a minimum genotype frequency of .02 (and therefore a maximum likelihood ratio of 50), and the more extreme frequencies resulting from the laboratory's own procedure for estimating genotype frequency.[7]

The results of this analysis are shown in Table 2. The likelihood ratio for a four probe match varies dramatically depending both on which genotype frequencies are used and on which expert's view of the extra band one chooses to credit. If expert 1 is right, and the NRC's frequency estimates are used, the likelihood ratio is 12.5; if the laboratory is right about the extra bands, and its own genotype frequencies are used, the likelihood ratio is 100 million.

[7] For convenience, round numbers that approximate the laboratory's typical genotype frequency estimates for these probes are used.

Table 2. Likelihood ratio for each probe in a four probe comparison under varying assumptions about genotype frequency and varying interpretations of the value of the match on probe G3 (Case 2).

Interpretation of probe G3	Probe G3		MS-43		MS-31		MS-1		Overall
Expert 1's interpretation									
NRC genotype freq.	.001	×	50	×	50	×	5	=	12.5
Lab's genotype freq.	.001	×	500	×	200	×	20	=	2000
Expert 2's interpretation									
NRC genotype freq.	1	×	50	×	50	×	5	=	12,500
Lab's genotype freq.	1	×	500	×	200	×	20	=	2,000,000
Lab's interpretation									
NRC genotype freq.	50	×	50	×	50	×	5	=	625,000
Lab's genotype freq.	500	×	500	×	200	×	20	=	100,000,000

This analysis holds several important lessons. First, the subjective judgment of experts regarding the extra band is of great importance. Under our assumptions, the likelihood ratio for the four probe match varies by a factor of 50,000 depending on whether Expert 1 or the laboratory is correct about the extra band. By comparison, the likelihood ratio varies only by a factor of 160 depending on whether the laboratory's or the NRC's genotype frequencies are used. For a case such as this, then, it is far more important to know which expert to trust regarding the extra band than to know which genotype frequencies are accurate.

The choice of genotype frequencies is, nevertheless, quite important if Expert 1 happens to be correct. If the laboratory's frequency estimates are chosen, the likelihood ratio is 2000 even if Expert 1 is right. In other words, under the laboratory's frequency estimates, the test has probative value to spare and is still quite powerful in spite of the problem on probe G3.[8] On the other hand, if the NRC's frequency estimates are chosen, and Expert 1 is correct, then the true likelihood ratio for the evidence is rather modest. Indeed, in a case with a match on only three probes, one of which produced a result like G3, the likelihood ratio could conceivably drop below one. That is, the DNA evidence might be exculpatory notwithstanding a three probe match.

Case 3. A woman was abducted and raped by two men. A commercial laboratory developed and compared DNA prints (via RFLP analysis of VNTRs) of five samples: an evidentiary sample consisting of vaginal aspirate from the victim (E), blood samples from two suspects (S1 and S2), who were half-brothers, a blood sample from the victim (V), and a blood sample from the victim's boyfriend (B), with whom she had intercourse before the rape. The band measurements reported by the laboratory for the samples are presented in Table 3. The case came to trial in 1994.

The laboratory report stated that there was a match between S1 and E with respect to six bands; the frequency of the set of six matching bands was estimated to be 1 in 300,000 in the relevant population (Western Hispanics). The report also stated that there was a match between S2 and E on seven bands; the frequency of the matching bands was estimated to be 1 in 1600 among Western Hispanics. The reason the frequency estimate for S2 is so much higher is that the laboratory ignored the three-bands produced by probe MS31 when making the calculation.[9]

[8] Of course the evidence is not nearly as powerful as the trier-of-fact might be led to believe based on the information in the laboratory report, which implies that the likelihood ratio is 100 million or more.

[9] The laboratory does not explain the decision to leave MS31 out of the calculation. Presumably the analyst either lacked adequate data on the frequency of the three-banded pattern or were uncertain how to apply the standard Hardy-Weinberg calculations to such a pattern.

Table 3. Band measurements in five DNA profiles (Case 3).

Probe	Suspect 1 (S1)	Suspect 2 (S2)	Vaginal aspirate (E)	Victim (V)	Victim's boyfriend
MS1	10078	5589	5589	4803	ns
	5589		4779	4226	
			4206		
MS31	5573	6599	7126	7273	ns
	4266	6427	6557	6605	
		5589	6386		
			5573		
MS43	10078	10276	10076	14133	ns
			8745	8885	
G3	7690	6601	8144	11763	8250
	6601	3167	7753	1659	6108
			6578		
			3168		

ns - not sized (none of the boyfriend's unsized bands appeared, on visual inspection, to correspond with evidentiary bands)

The laboratory report played a pivotal role in the resolution of the case. Because the lawyers who read the report focused on the bottom-line numbers, they concluded that the DNA evidence was much more powerfully incriminating with respect to S1, than S2. They assumed, in other words, that the reported frequencies provided an index of the value of the DNA evidence. S1 was persuaded to accept a plea bargain after his lawyer assured him that his case was hopeless, although he privately insisted he was innocent. The lawyer for S2 persuaded the prosecutor to withdraw the 'weak' DNA evidence against him by threatening to mount a lengthy and expensive legal challenge to its admissibility; his case proceeded to trial without DNA evidence.

This result is troubling because a close examination of the autorads and laboratory notes suggests, contrary to the lawyers' assessment, that the DNA evidence against S2 is strong, while the evidence against S1 is quite weak, perhaps even uninformative. Because four of the six 'matching' bands that incriminated S1 were shared with S2, the laboratory analyst relied on only two bands to distinguish the hypothesis that both men contributed to the semen from the hypotheses that only S2 contributed.

The scoring of these two critical bands was problematic because both bands, if present, are extremely faint. Neither band can be seen on high quality copies of the autorads. The laboratory's computer scoring system did not detect any bands in the evidentiary sample that corresponded to S1's lower band on probe MS31. The analyst apparently scored this band by eye. (This band will hereafter be called 'band a'). The computer system also failed to detect a band corresponding to S1's upper band on probe g3 when scoring results of a hybridization with a cocktail of four probes (including g3). A band in the correct location (hereafter called 'band b') was detected by the computer system on a subsequent hybridization with g3 alone, but only when the sensitivity of the optical scanner was increased to its maximum level – a level at which the possibility of a false reading may be of concern. At this level of sensitivity, the computer detected six other bands in the evidentiary sample that the analyst chose to 'delete' on the assumption that they were artifacts. The basis for the analyst's decision to 'delete' the other six bands, while accepting the computer's scoring of band b, is not described in the laboratory notes.

It is also noteworthy that no band was detected which corresponded to S1's upper band on probe MS1. The laboratory report attributed the absence of this band to 'the small quantity of human DNA obtained from the [evidentiary sample]'. But its absence is troubling because MS1 is normally one of the more sensitive probes, and bands in the same size range were detected by probe MS43. Moreover, if there is so little

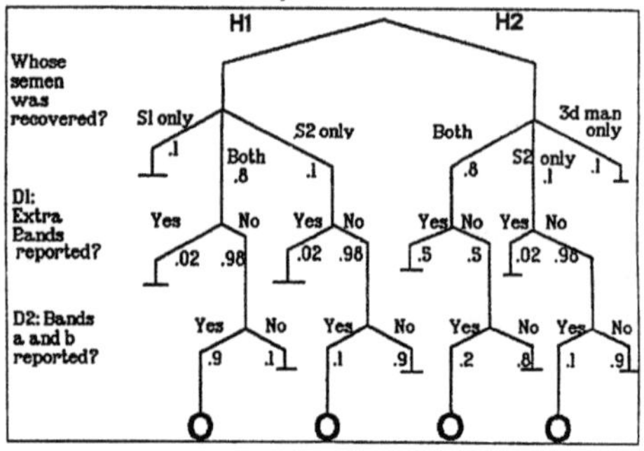

Fig. 4. Model of relationship between hypotheses and conditioning events, showing estimated conditional probabilities (Case 3).

DNA that some of this rapist's bands do not appear, then we cannot rule out the possibility that the victim was raped by S2 and a third man, who happened to have band a and band b, but whose other bands are too faint to see, or were dismissed as artifacts, or match those of S2, V or B.

In light of these uncertainties, what is the value of the DNA evidence for incriminating S1? To simplify matters, let us assume that the victim was indeed raped by two men, one of whom was S2. The problem, then, is to determine the value of the DNA evidence for distinguishing two hypotheses:

H1: S2 and S1 committed the rapes.

H2: S2 and a third man committed the rapes.

Belief in these hypotheses should be affected (conditioned) by two pieces of data:

D1: The analyst's report that no bands other than those corresponding to the patterns of S1, S2, V and B appeared in the evidentiary sample

D2: The analyst's report that bands a and b appeared in the evidentiary sample

The analysis offered here has three steps. First, as in Case 1, a model is constructed showing the various ways D1 and D2 might occur given H1 and H2. Second, conditional probabilities of the conditioning events under each hypothesis are estimated. Third, the conditional probabilities are combined (multiplied) in order to estimate the likelihood ratio for the conditioning events.

Figure 4 shows a model of the problem. When followed from top to bottom, the model shows the different chains of events that might lead to D1 and D2. The model has two main branches that reflect the competing hypothesis, H1 and H2, which are mutually exclusive and exhaustive.

Under each hypothesis there are three branches that reflect uncertainty about a key unknown event – whether the vaginal aspirate contains the DNA of both rapists or only one of the two. It is not uncommon, in cases involving two rapists, for the DNA of only one to be recovered. Because it is possible to imagine scenarios by which D1 and D2 would occur if only S2's DNA was recovered, this possibility must be considered. Let us tentatively assume that under both H1 and H2 the conditional probability that both rapists' DNA would be recovered is .80 and the probability that the DNA of only one rapist would be recovered is 0.10 for each.

Next, the model branches to show event D1 (no 'extra bands' were reported). The conditional probability of this event varies under the two hypotheses. Under H1 no 'extra bands' should be present; the only bands present in E will be those of S1, S2, V and B. Hence, a report of no extra bands is very likely regardless of whether semen of both rapists or only of S2 was recovered. There is, of course, a small chance that the analyst will mistake an artifact for an extra band, but examiner bias would tend to work against such a report because the analyst would expect no extra bands if the police theory of the case is correct. Let us assume the probability no extra bands would be reported under H1 is .98 regardless of whether DNA was recovered for both rapists or only S2.

Under H2, the probability that the analyst will report 'no extra bands' depends on whether both rapists' DNA or only S2's DNA is recovered. If both rapists' DNA is recovered, there should be 'extra bands' – i.e., those of the third man, in E. Hence, the probability that extra bands will be reported will be higher. But a report of extra bands is by no means certain. There is a fairly good chance that some of these bands would not be seen because they overlap with the bands of S2, V or B.[10] Given the faintness of bands a and b, the non-overlapping bands would presumably be very faint and easy to miss. Moreover, examiner bias would work against the report of such bands because their presence is inconsistent with the

[10] For example, S2, V, and B have, among them, seven different bands on probe MS31. Hence, there are 28 different genotypes a third man might have on MS31 that would be obscured completely by the bands of these other individuals. If the average frequency of each genotype were .02, then the probability the third man would have a distinct 'extra band' on MS31 would be $(1 - .02) = .98$ for each of 28 genotypes. The probability that he would have a distinct extra band that did not match any of the 28 genotypes would be only $.98^{28} = .57$.

police theory of the case. Consequently, let us assume only a 50-50 chance that such bands would be reported if DNA of both rapists is recovered. If only S2's DNA is recovered, there will be no extra bands, so let us assume the probability of a report of no extra bands would be 0.98, the same as under H1.

Finally, the model branches again to show event D2 (the report of bands a and b in the evidentiary sample). The conditional probability of this event varies. Under H1, if DNA of both rapists was recovered, then bands a and b would be present in E and there is a high likelihood that they would be reported. It seems unlikely that the analyst would miss these bands, if they were present, because the analyst would be looking for bands that matched those of S1. However, bands occasionally fail to appear where they should due to technical problems, hence the probability of D2 would be less than 1.00. Let us put it at 0.90.

Under H1, if DNA of only S2 is recovered then D2 would be less likely because bands a and b would not be present in E. But they might nevertheless be reported by error. Examiner bias would favor such an error because it would produce the expected result (incrimination of S1). In this case, the analyst reported bands a and b based on rather equivocal evidence of their presence. Given this low threshold for the reporting of these bands, the likelihood they would have been reported if not present could well be significant. Let us put that probability at 0.10.

Under H2, if DNA of both rapists was recovered, the probability of D2 might also be considerable. The frequency of 'band a' is approximately 3%, so there is a probability of approximately .06 that one of the third man's bands on probe MS31 would match 'band a'. The frequency of 'band b' is approximately 7%, so there is a probability of approximately .14 that one of the third man's bands on G3 would match 'band b'. Although the probability of the third man having both 'band a' and 'band b' would be only $.06 \times .14 = .0084$, the probability of his having at least one of the two bands would be approximately .19. If he did happen to have one of these two bands, it might appear to the analyst to confirm his guilt and thus create a particularly strong tendency for the analyst to 'see' and report the other band even if it was not present. For these reasons, the probability of D2 if both rapists' DNA was recovered would seem to be higher than the probability of D2 if only S2's DNA was recovered. Let us put the probability at 0.20. Put the probability of D2 if only S2's DNA was recovered at 0.1.

It is now possible to estimate the likelihood ratio that specifies the value of the joint occurrence of D1 and D2 for distinguishing H1 and H2. The likelihood ratio is the ratio of probability of the joint occurrence of D1 and D2 under H1 and under H2 – that is:

$$\mathrm{LR} = \frac{p(D1 \,\&\, D2/H1)}{p(D1 \,\&\, D2/H2)}.$$

Under the model, p(D1 & D2/H1) = p(D2/D1 & Both & H1)p(D1/Both & H1)p(Both/H1) + p(D2/D1 & S2 only & H1)p(D1/S2 only & H1)p(S2 only/H1). Substituting the conditional probability estimates discussed above, the numerator of the likelihood ratio is $(.9 \times .98 \times .8) + (.1 \times .98 \times .1) = .7156$. Correspondingly, p(D1 & D2/H2) = p(D2/D1 & Both & H2)p(D1/Both & H2)p(Both/H2) + p(D2/D1 & S2 only & H2)p(D1/S2 only & H2)p(S2 only/H2). Substituting the conditional probabilities discussed above, the denominator of the likelihood ratio is $(.2 \times .5 \times .8) + (.1 \times .98 \times .1) = .0898$. Hence, the likelihood ratio is $.7156/.0898 = 7.96$. Taking into account the uncertainty surrounding the interpretation of the autorads, then one's estimate of the odds of S1's guilt should increase by a factor of 8 after learning of the DNA evidence.

Obviously, the value of the DNA evidence against S1 is substantially lower than what one would assume if one thought that the likelihood ratio for DNA evidence is the reciprocal of the frequency of the matching genotypes. The likelihood ratio is low because, in this case, there is a significant chance that the incriminating evidence could have arisen if S1 was innocent.

Discussion

The role of subjective judgment

A major lesson to be learned from these cases is that subjective judgment can play a pivotal role in the interpretation of forensic DNA tests. The value of the DNA evidence in each of these cases balances on the razor's edge of a subjective judgment about whether critical bands are present or absent and about the source of these bands, i.e., whether they are true bands or artifacts.

Some courts have suggested that subjective judgment plays a minimal role in the interpretation of DNA evidence because 'ultimately the 'match' is confirmed or rejected through computer analysis, using wholly objective criteria' (*State v. Jobe,* 1992, p. 420). This statement reflects a misconception about the way computer scoring systems operate. These systems detect opacities in the evidentiary lanes of autorads and estimate their position relative to markers, but it is the analyst who decides which opacities to score as bands. These decisions are not based on objective criteria; they rest soley on the analyst's subjective judgment about what is and is not a convincing band. In Case 1, for example, the computer detected the tell-tale 'extra bands' in the evidence (which were diagnostic of an incriminating cross-contamination), but the analyst chose not to score them and failed even to report them. In Case 2, the computer detected the potentially exculpatory 'extra band' and the analyst scored it as a band, but dismissed it as unimportant. In Case 3, the computer failed to detect one critical band that the analyst scored anyway; at the level of sensitivity necessary to detect a second critical band, the computer detected six other opacities that the analyst chose not to score as bands.

Some commentators have suggested that the use of quantitative matching rules assures that subjective judgment cannot work against a suspect because '[a]s long as no visual match will be reported as a match unless confirmed by the quantitative matching rule, the imprecise, subjective phase serves as only a preliminary filter' (Kaye, 1994). This analysis, however, applies only to subjective judgments about whether the bands of a DNA print align closely enough to be called a match. Quantitative matching rules obviously do nothing to reduce the subjectivity of judgments made prior to the application of quantitative matching rules, such as whether a band is present and, if so, what its position is. The cases analyzed here clearly show that subjective judgments regarding the scoring of bands can work against the interests of a suspect, notwithstanding the subsequent application of quantitative matching rules.

This prospect is particularly worrisome because the analysts who make the crucial subjective interpretations do not do so 'blind'; they are free to make cross-lane comparisons, they know the identity of the samples and they often know the police theory of the case (Thompson & Ford, 1991). It seems improper that judgments of such crucial importance occur under circumstances so conducive to examiner bias. Further inquiry is surely needed to assess how serious this problem is and to weigh the dangers of examiner bias inherent in the current subjective scoring procedures against any arguments that might by offered for maintaining subjective scoring.

How common are ambiguous results?

One question that must be examined is how often forensic DNA tests produce results that are sufficiently ambiguous to make misinterpretation possible. If such results are extremely rare, the issue may be of little consequence. Unfortunately, little is known about how commonly problems like those discussed here arise in casework. No systematic reviews of random samples of casework have been reported. Certainly there are many cases in which the results are clear cut, interpretation is straightforward, and no one would disagree with the analyst's conclusions. On the other hand, it would be a mistake to assume (without evidence) that cases such as those analyzed here are unique, anomalous or even rare.

Commentators have discussed a number of actual cases in which the interpretation of DNA test results was problematic (Lander, 1989; Thompson & Ford, 1991; Shields, 1992; Thompson, 1993, p. 40, 48). Thompson and Ford (1991) found one or more 'potential sources of ambiguity' that 'affected the prints relied upon to incriminate the defendant' in 14 of 20 forensic cases they reviewed. For example, in 4 of 20 cases the laboratory invoked the theory of band shift in order to call a match where the difference in the measured sizes of 'matching bands' exceeded the laboratory's quantitative match standard. In 8 of the 20 cases a match was called despite the presence of extra bands, of uncertain origin, in one of the prints; in 5 of those cases the extra bands were not mentioned in the laboratory report. Although it is difficult to assess whether these cases are representative of casework in general, there is no obvious reason to think that they are not. (The cases had come to Ford's attention through his work as a consultant to lawyers and represented all of the cases on which he had complete records). Similar cases occasionally come to light when they lead to clashes of expert opinion in the courtroom (e.g., *People v. Keene,* 1992; *State v. Futch,* 1993). These examples in casework are numerous enough to raise legitimate concerns about subjective scoring procedures and to justify further inquiry into the issue.

Are there good reasons for allowing subjective scoring?

Another question that must be asked is whether there is adequate justification for the failure of forensic laboratories to adopt objective or blind scoring procedures. The National Research Council (1992, p. 53) has called for forensic laboratories to use 'an objective and quantitative procedure for identifying the pattern of a sample' and insists that '[p]atterns must be identified separately and independently in suspect and evidence samples'. Why are the forensic laboratories not following these recommendations? If objective scoring procedures are not practical, why not at least take steps to assure that the scoring is done 'blind'?

One argument that is sometimes offered in defense of subjective scoring is that criminal defendants who disagree with a laboratory's interpretation can always seek a second opinion. This argument is unpersuasive. Although problematic interpretations are sometimes uncovered by independent experts hired by the defendant, the great majority of criminal cases involving DNA evidence are resolved without any independent expert checking the work of the forensic laboratory.[11] To rely on the adversarial system to address the kind of interpretive problems discussed here is to rely on wishful thinking.

Another argument sometimes offered against blind scoring is that the analyst has a better chance of reaching the correct interpretation by taking into account all of the evidence. Those who accept this argument undoubtedly fail to appreciate the danger of bootstrapping critical judgments against other results. The danger is that DNA evidence will appear much stronger than it really is.

Understanding and explaining the value of DNA evidence

These cases show that the frequency of matching genotypes can be a misleading index of the value of DNA evidence. In each case, the reported rarity of the matching genotypes suggested (and probably was taken to mean) that the DNA evidence provided powerful proof of identity. Yet in each case the actual value of the DNA evidence was slight.

Some commentators have concluded that the likelihood ratio describing the value of DNA evidence is approximately 'the reciprocal of the population frequency calculated by the match/binning procedure' (Weir, 1992, p. 11657). Although sophisticated readers surely realize that this likelihood ratio is accurate only if one assumes that the laboratory is error-free and that the test results were unambiguous and correctly interpreted, they may not realize just how crucial these assumptions can be to the value of DNA evidence in particular cases. A person who accepted these assumptions with regard to Case 3 would have overestimated the likelihood ratio describing the value of the DNA evidence by a factor of 37,000; a person who accepted these assumptions with regard to Case 1 would have overestimated the likelihood ratio of the DNA evidence by a factor of 3 billion.

The possibility for striking discrepancies to exist between the actual value of the DNA evidence and the value suggested by frequency estimates has not received enough attention. Careful commentators have noted that the value of the likelihood ratio may be diminished if there is a possibility of laboratory error (e.g., Weir, 1992, p. 11658; Lempert, 1991; Hagerman, 1990). Indeed, having an accurate estimate of the laboratory error rate is probably far more important than having an accurate estimate of the frequency of multilocus genotypes because the former is likely to be much higher than the latter and therefore have a greater influence on the value of DNA evidence. But efforts to establish the rate of laboratory errors have, to date, received little attention relative to efforts to establish the frequency of multilocus genotypes. Even less attention has been paid to the potential of subjective interpretation to undermine the value of DNA evidence. Both matters require further investigation.

In this light, it is important to consider the accuracy and adequacy of the information typically provided to legal decision makers about the value of DNA evidence. In laboratory reports, and expert testimony, forensic analysts typically state whether samples 'match' and provide an estimate of the frequency of the matching genotypes. Their presentations seem designed to suggest, and sometimes state outright, that the frequency estimates provide an index of the value of the DNA evidence. Indeed, they sometimes make statements equating the frequency of the matching genotypes with the probability of the defendant's innocence (Koehler, 1993a; 1993b). To the extent that laboratory errors and interpretive ambiguities occur in casework, these characterizations of the value of DNA evidence are inadequate, misleading and potentially prejudicial. Further study of these issues is essential if we are to understand the value of DNA evidence and present it fairly in the courtroom.

[11] Case #3, for example, came to the author's attention only after the attorney for S1, who had consulted no experts, had convinced S1 to plead guilty. Moreover, the checking that does occur is sometimes cursory and uninformed. One expert (not Simon Ford) who reviewed the evidence in Case 1, for example, failed to appreciate that the extra bands in the evidentiary sample were diagnostic of an incriminating cross-contamination and, therefore, greatly reduced the value of the DNA evidence. This issue was not raised with the jury.

References

Devlin, B., N. Risch & K. Roeder, 1994. Comments on the statistical aspects of the NRC's report on DNA typing. J. Forensic Sci. 39: 28–40.

Graves, M.H. & M. Kuo, 1989. DNA: A blind trial study of three commercial testing laboratories. Presented at the meeting of the American Academy of Forensic Sciences, Las Vegas.

Hagerman, P.J., 1990. DNA typing in the forensic arena. Am. J. Hum. Genet. 47: 876–877.

Hart, S.D., C. Webster & R. Menzies, 1993. A note on portraying the accuracy of violence predictions. Law & Hum. Behav. 17: 695–700.

Kaye, D.H., 1994. DNA evidence: Probability, population genetics and the courts. Harvard J. Law & Technology 7: 101–172.

Koehler, J.J., 1993a. DNA matches and statistics: Important questions, surprising answers. Judicature 76: 222–229.

Koehler, J.J., 1993b. Error and exaggeration in the presentation of DNA evidence at trial. Jurimetrics 34: 21–35.

Lander, E., 1989. DNA fingerprinting on trial. Nature 339: 501–505.

Lempert, R., 1991. Some caveats concerning DNA as criminal identification evidence: With thanks to the reverend Bayes. Cardozo L. Rev. 13: 303–341.

Moenssens, A.A., 1990. DNA evidence and its critics–How valid are the challenges? Jurimetrics 31: 87–108.

National Research Council, 1992. DNA Technology in Forensic Science. National Academy Press, Washington, D.C.

Nisbett, R.E. & L. Ross, 1980. Human inference: Strategies and shortcomings of social judgment. Prentice-Hall, Englewood Cliffs, New Jersey.

People v. Castro, 545 N.Y.S.2d 985 (N.Y.Sup.Ct. 1989).

People v. Keene, 591 N.Y.S.2d 733 (N.Y.Sup.Ct. 1992).

Shields, W.M., 1992. Forensic DNA typing as evidence in criminal proceedings: Some problems and potential solutions, pp. 1–50 in Proceedings from the Third International Symposium on Human Identification. Promega Corp., Madison, Wisconsin.

State v. Futch, 860 p.2d 264 (Ore. 1993).

State v. Jobe, 486 N.W.2d 407 (Minn. 1992).

Technical Working Group on DNA Analysis Methods (TWIGDAM), 1990. Statement of the Working Group on Statistical Standards for DNA Analysis. Crime Lab. Dig. 17(3): 53–58.

Thompson, W.C. & S. Ford, 1989. DNA typing: Acceptance and weight of the new genetic identification tests. Virginia L. Rev. 75: 45–108.

Thompson, W.C. & S. Ford, 1991. The meaning of a match: Sources of ambiguity in the interpretation of DNA prints, pp. 93–152 in Forensic DNA Technology, edited by M. Farley & J. Harrington, Lewis Publishers, Chelsea, Michigan.

Thompson, W.C., 1993. Evaluating the admissibility of new genetic identification tests: Lessons from the "DNA War". J. Crim. Law & Criminology, 84: 701–781.

Weir, B.S., 1992. Population genetics in the forensic DNA debate. Proc. Natl. Acad. Sci. 89: 11654–11659.

Editor's comments

The author treats the timely and important issue of laboratory error. Readers will need to read the paper by Lempert in this volume for an alternative interpretation of the 1989 proficiency testing of Cellmark diagnostics.

B. S. Weir (ed.), Human Identification: The Use of DNA Markers, 169–178, 1995.

Exact tests for association between alleles at arbitrary numbers of loci

D. Zaykin[1], L. Zhivotovsky[2] & B.S. Weir[1]*
[1]*Program in Statistical Genetics, Department of Statistics, North Carolina State University, Raleigh NC 27695-8203, USA*
[2]*Institute of Gene Biology, Russian Academy of Sciences 34/5 Vavilov Street, Moscow 117334, Russia*
**Author for correspondence*

Received 1 July, 1994 Accepted 1 December, 1994

Key words: exact tests, allelic association, Hardy-Weinberg, linkage disequilibrium

Abstract

Associations between allelic frequencies, within and between loci, can be tested for with an exact test. The probability of the set of multi-locus genotypes in a sample, conditional on the allelic counts, is calculated from multinomial theory under the hypothesis of no association. Alleles are then permuted and the conditional probability calculated for the permuted genotypic array. The proportion of arrays no more probable than the original sample provides the significance level for the test. An algorithm is provided for counting genotypes efficiently in the arrays, and the powers of the test presented for various kinds of association. The powers for the case when associations are generated by admixture of several populations suggest that exact tests are capable of detecting levels of association that would affect forensic calculations to a significant extent.

Introduction

In the absence of evolutionary forces such as drift, selection, migration or mutations, genotypic frequencies are expected to be given by the products of corresponding allelic frequencies. Even if these forces are known to be present, however, it may be that genotypic frequencies are very close to the allelic frequency products. There are situations when it is convenient to be able to invoke this 'product rule' for multilocus genotypes, as in the use of genetic profiles for human identification. A specific multilocus genotype is unlikely to have been seen in samples collected for the purpose of estimating frequencies, even though all the constituent alleles are present, and then the product rule offers a means of providing an estimate. Of course, it is necessary first to test for consistency of genotype frequencies to products of allele frequencies and such tests are covered in this paper.

With many loci and many alleles per locus, there are very many possible associations among the frequencies of subsets of the alleles. Even for two alleles at two loci, for example, there are six pairs of genes, four triples and one set of four genes to be considered (Weir & Cockerham, 1989). If the only issue is whether allele frequencies can be used to construct genotype frequencies, all these associations are tested for simultaneously in a single test. This has the advantage of avoiding problems with multiple tests, and the power of the test sometimes increases with the number of loci and alleles per locus. If there is interest in some of the individual associations, then specific tests can be constructed, and will be discussed here.

The primary purpose of this paper is to examine the use of 'exact' tests, meaning tests based on the probabilities of sets of alleles conditional on observed counts of subsets of the alleles. In general these tests are expected to perform well and to avoid the problems faced by chi-square goodness-of-fit tests when expected numbers are small. Because of the large number of possible multi-allelic arrays, it is not possible to examine them all to compute significance levels for the tests. Samples of arrays are generated by permutation, following the suggestion of Guo and Thompson (1992).

General method

The general hypothesis is that there is no association among the frequencies of constituent genes of a genotype. For locus l, each of the two alleles received by an individual at that locus has probability p_{l_i} of being allelic type A_{l_i}. If $P_{1_i1_j,2_i2_j,\ldots,L_iL_j}$ is the population frequency of the genotype $A_{1_i}A_{1_j}A_{2_i}A_{2_j}\cdots A_{L_i}A_{L_j}$, then the hypothesis can be expressed as

$$P_{1_i1_j,2_i2_j,\ldots,L_iL_j} = 2^H \prod_l p_{l_i}p_{l_j} \tag{1}$$

where H is the number of loci that are heterozygous. When only one locus is being considered and $L = 1$, Equation 1 is just the Hardy-Weinberg law. In other cases, rejection of the hypothesis does not indicate whether it is allelic frequencies within or between loci that are associated.

The development of a testing strategy is based on the multinomial distribution, meaning that each member of a population is assumed to be equally likely to be sampled, and that there is the same probability for each sample member having a particular genotype. Sample sizes are therefore considered to be very much smaller than the population sizes. If **A** indicates a multilocus genotype, and n_{A_g} is the number of individuals of type $\mathbf{A}_g$ in a sample of size n from a population in which those genotypes have frequency P_{A_g}, then

$$\Pr(n_{A_1}, n_{A_2}, \cdots, n_{A_G}) = \frac{n!}{\prod_{g=1}^G n_{A_g}!}\prod_{g=1}^G (P_{A_g})^{n_{A_g}}$$

The quantity G is the number of different genotypes possible.

Under the null hypothesis of Equation 1, allelic counts n_{l_i} at locus l are multinomially distributed with sample size $2n$ and probabilities p_{l_i}. Furthermore, under the hypothesis these distributions are independent over loci. The joint probability of the sets of allelic counts $\{n_{l_i}\}$ is therefore

$$\Pr(\{n_{1_i}\}, \{n_{2_i}\}, \cdots, \{n_{L_i}\}) = \prod_{l=1}^L \left(\frac{(2n)!}{\prod_i n_{l_i}!} \prod_i (p_{l_i})^{n_{l_i}} \right)$$

and the probability of the genotypic counts conditional on the allelic counts is

$$\Pr(\{n_{A_g}\} \mid \{n_{l_i}\}) = \frac{n!}{\prod_{g=1}^G n_{A_g}!}\prod_{g=1}^G (P_{A_g})^{n_{A_g}} \prod_{l=1}^L \frac{\prod_i n_{l_i}!}{(2n)!\prod_i (p_{l_i})^{n_{l_i}}!}$$

Under the null hypothesis of complete independence of allele frequencies

$$\Pr(\{n_{A_g}\} \mid \{n_{l_i}\}) = \frac{n!\prod_{g=1}^G 2^{n_{A_g}H_g}}{\prod_{g=1}^G n_{A_g}!}\prod_{l=1}^L \frac{\prod_i n_{l_i}!}{(2n)!} \tag{2}$$

where count H_g is the number of heterozygous loci in genotype $\mathbf{A}_g$, of which there are n_{A_g} copies. Note that the unknown allelic frequencies p_{l_i} have cancelled out of this expression.

Genotypic arrays $\{n_{A_g}\}$, generated by permutation, with conditional probabilities equal to or less than that of the observed sample array contribute to the probability with which the null hypothesis would be rejected if it was true. This is the significance level, or p-value. In general, it is not feasible to calculate this quantity exactly because of the prohibitively large number of genotypic count array for a given array of allelic counts.

Gail and Mantel (1977) discussed methods for determining the numbers of two- and three-dimensional contingency tables with fixed marginals. Another approximate method for setting a lower bound on the number will suffice to show that the number is indeed prohibitive. If P_1 is the largest value of all the array probabilities, then the number of arrays must be at least $1+(1-P_1)/P_1 = 1/P_1$. This would be the actual number if all arrays had probability P_1. Similarly, if P_1 is the smallest of the T largest probabilities, the number of arrays must be at least $T + (1 - \sum_{i=1}^T P_i)/P_1$. For cases when all these P's are small, the lower bound is given essentially by $(1 - \sum_{i=2}^T P_i)/P_1$. Since the ordering of all possible P's is unknown, this bound is calculated by finding the set of T largest P's for a large number of permuted arrays. Having $T > 1$ protects against under-estimation. The procedure was applied to STR data from a sample of 182 people typed at loci with 6, 6, 9 and 14 alleles. With $T = 1$ and 65,000 permutations, the lower bound was estimated to be of the order of 10^{748} in each of three separate determinations. Obviously, it is not possible to examine all possible arrays.

Instead of identifying all arrays with lower conditional probabilities than the sample, a set of arrays is generated randomly by permuting those alleles hypothesized to be independent. For the hypothesis in Equation 1, this means that alleles are permuted among individuals within loci, and independent permutations performed for each locus. The proportion of permuted arrays as probable or less probable than the sample forms an estimate of the significance level. If the true significance level is α, then with probability 0.95 the estimate will be within δ of that value after m permutations if $m \approx 4\alpha(1-\alpha)/\delta^2 \leq 1/\delta^2$. Hence, with 95% probability, 10,000 permutations give an estimate accurate to two decimal places. Further discussion of this approach was given by Guo and Thompson (1992).

Applying the permutation method to estimate significance levels can be performed very efficiently. As all the arrays have the same allelic counts, for purposes of comparison it is necessary to compute only

$$P_s = \prod_g \frac{2^{n_{A_g} H_g}}{n_{A_g}!} \tag{3}$$

although it is the logarithm of P_s that is computed in practice, using an alogorithm of Press *et al.* (1988). Furthermore, it is not necessary to step through all possible multilocus genotypes, since only those with non-zero counts contribute to Equation 3. From now on, this will be indicated by $g \in z$, meaning that only those g values for which $n_{A_g} > 0$ are considered. The computer storage requirements therefore depend on the sample size rather than the number of possible genotypes.

Algorithm

The greatest saving in computing time is made in the way of counting the number of times each genotype appears in one of the genotypic arrays generated by permuting alleles. A naïve way would be to assign each genotype a numerical identifier, sort the identifiers and then count how many times each one occurs. For locus **B** with alleles numbered 1 to n_B, a possible identifier s_B for genotype B_iB_j, $j \leq i$ is

$$s_B = \frac{i(i-1)}{2} + j$$

Values of this quantity range from 1 for B_1B_1 to $S_B = n_B(n_B+1)/2$ for $B_{n_B}B_{n_B}$. If there is a second locus **C** with n_C alleles, then the identifier s_C ranging from 1 to S_C for genotypes at that locus can be defined similarly, and the identifier for two-locus genotypes is defined by

$$s_{BC} = S_C(s_B - 1) + s_C$$

which ranges from 1 for $B_1B_1C_1C_1$ to S_BS_C for $B_{n_B}B_{n_B}C_{n_C}C_{n_C}$. The extension to multiple loci is straightforward. If necessary, the genotype can be recovered from identifier. In the two-locus case, for example,

$$s_C = \frac{i_C(i_C-1)}{2} + j_C = s_{BC}(\text{mod} S_C)$$

$$s_B = \frac{i_B(i_B-1)}{2} + j_B = \frac{s_{BC} - s_C}{S_C} + 1$$

and the one-locus genotypes can be recovered from the one-locus identifiers by

$$i = 1 + \text{integer part of}[(\sqrt{8s_B+1} - 1)/2]$$
$$j = s_B - i(i-1)/2$$

The problem with this approach is the need to sort as many identifiers as there are individuals in the sample. The binary search tree method now described is very much faster.

The data set **D** has an element for every individual in the sample. A tree is constructed with nodes for every distinct genotype in **D**. Each element of **D** is placed on the tree, either at an existing node or at a new node, by comparing its identifier to the identifiers for the previously placed elements. This placement procedure begins at the root node of the tree, and at each node, the tree is followed in one of two directions depending on whether the identifier is greater than or less than the identifier for that node.

The algorithm is as follows:

1. Build the binary search tree
   - **a.** Set counter $i = 1$. Take the element of **D** and insert it at the root node as the node identifier (ID). Set the root internal counter C (the number of times that genotype occurs in the sample) to 1 and the number of nodes NN to 1.
   - **b.** Increment i. Take the ith element r_i from **D** and recursively traverse the tree. If r_i is equal to the

ID from some existing node, taking into account that for each locus genotype AA' is equal to $A'A$, increment the internal counter C at that node by 1. Otherwise, if any outer node (the top) is reached without finding an equal ID, insert r_i into the tree as a new node, setting the node for that ID to r_i, setting its count C to 1, and incrementing NN by 1.

c. If $i < n$, repeat the previous step.

2. Calculate $\ln P_s$ in Equation 3.

a. Set $\ln P_s = 0$.

b. For each node j in the tree, calculate $C_j H_j \ln 2 - \ln(C_j!)$ and add this to $\ln P_s$.

c. If **D** is the original sample, set $P_O = P_s$ and set $K = 0$. If **D** is a permuted array, increment K by 1 if $\ln P_s \leq \ln P_O$.

3. Permutation stage.

a. For each locus, randomly permute the $2n$ alleles in **D** for that locus. Return to the binary search tree step. Do this step while the number of permutations is less than the required number NR.

4. The estimated significance level is K/NR.

The binary search tree method for storing and retrieving the numbers of each multilocus genotype in a sample performs best when the genotypes are not sorted. This will certainly be the case for the permuted arrays. In the worst case, when the genotypes have been sorted, the method degenerates to a sequential search. Since the genotypes in the tree are stored in sorted order, it is guaranteed that no parts of the tree other than the current sub-tree can contain the particular genotype being sought. This means that only half the remainder of the tree needs to be considered after each comparison. The maximum number of nodes in the tree cannot exceed the sample size, so that the average branch length remains the same as the numbers of loci and alleles increase and the time to complete the algorithm is therefore affected very little by these two numbers. The most time-consuming part of the algorithm is the permutation stage. Its speed depends on the sample size and number of loci, but not the total number of alleles per locus.

The binary search tree method has been programmed in C++, taking advantage of features of that language. It may be helpful, however, to illustrate the nature of the algorithm using the more primitive genotype-identifiers described above. Suppose a sample of 11 multilocus genotypes has been reduced to integer identifiers by the method described above, and these identifiers are 24, 8, 37, 95, 24, 6, 28, 25, 23, 94, 37. The root node of the tree is 24, and at that stage $C = 1$, $NN = 1$. The tree can be drawn under the rule that the direction of travel is up to the left for identifiers smaller than that at the current node and up to the right for larger identifiers. The resulting tree is shown in Figure 1. A 12th genotype with identifier 25 would be located after only three steps: 25 is greater than 24, less than 37, and less than 28.

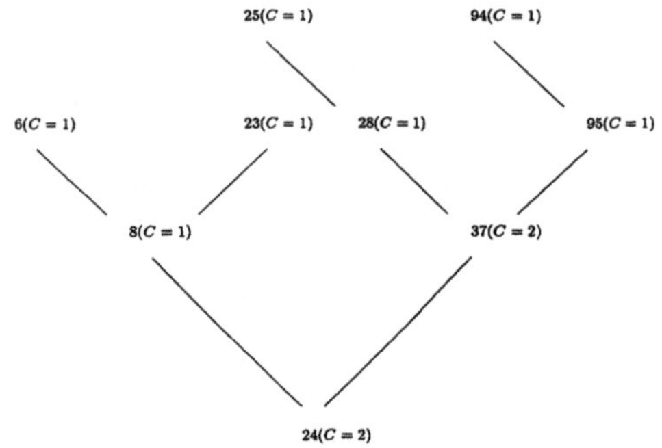

Fig. 1. Binary search tree.

Other tests

More specific tests can be performed in order the characterize associations among subsets of the alleles within genotypes, or to increase the power of detecting specific associations. Details will now be given for tests on two-locus data.

Products of one-locus frequencies

There may be interest in whether two-locus genotypic frequencies can be represented as products of one-locus frequencies without assuming Hardy-Weinberg equilibrium. The null hypothesis can be written as

$$P_{A_iA_jB_kB_l} = P_{A_iA_j} P_{B_kB_l} \quad (4)$$

In the human identification case, such a situation might be appropriate when there was evidence of Hardy-Weinberg disequilibrium at each locus, and two-locus genotypic frequencies were to be constructed as products of observed one-locus genotypic frequencies.

The sample counts can be written in a two-way table with row totals being **A**-locus genotype counts n_{ij} and column totals being **B**-locus counts n_{kl}. The table cell entries are the two-locus counts n_{ijkl}. The probability of the two-locus array conditional on the two one-locus arrays, under the null hypothesis in Equation 4, is

$$\Pr(\{n_{ijkl}\}|\{n_{A_{ij}}\},\{n_{B_{kl}}\}) = \frac{\prod_{i,j} n_{A_{ij}}! \prod_{k,l} n_{B_{kl}}!}{n! \prod_{i,j} \prod_{k,l} n_{ijkl}!}$$

In this notation, the subscripts i, j range over all **A**-genotypes in the sample and so include each of the rows in the table, and the subscripts k, l range over all **B**-genotypes and so include all the columns in the table. In estimating the significance level for the exact test of Equation 4, the only quantity to be calculated is $P_s = 1/\prod_{g \in z} n_g!$, where g indexes each of the two-locus genotypes with a non-zero count. Permutation proceeds by keeping the one-locus genotypes intact and permuting these genotypes among individuals at one of the two loci.

Products of genotypic frequencies at one locus and two allele frequencies at the other locus

To see if the two-locus genotypic frequencies $P_{A_iA_jB_kB_l}$ can be represented as the product of the genotype frequency $P_{A_iA_j}$ at one locus and the product of two allele frequencies $p_{B_k}p_{B_l}$ at the other locus, the hypothesis is

$$P_{A_iA_jB_kB_l} = 2^{H_{kl}} P_{A_iA_j} p_{B_k} p_{B_l} \qquad (5)$$

where H_{kl} is 1 if $k \neq l$ and is 0 if $k = l$. A human identification setting for this test may be when one of the two loci shows departures from Hardy-Weinberg frequencies.

The conditional probability of a sample under the hypothesis in Equation 5 is

$$\Pr(\{n_{ijkl}\}|\{n_{A_{ij}}\},\{n_{B_k}\}) = \frac{2^{H_B} \prod_{i,j} n_{A_{ij}}! \prod_k n_{B_k}!}{(2n)! \prod_{i,j} \prod_{k,l} n_{ijkl}!}$$

where H_B is the number of individuals in the sample that are heterozygous at locus **B**. Estimating the significance level in this case requires comparisons among values of $2^{H_B}/\prod_{g \in z} n_g!$, where n_g is still the number of occurrences of the gth two-locus genotype. Permutation proceeds by holding the **A**-locus genotypes intact and shuffling all $2n$ **B**-locus alleles among individuals.

Discarding some genotypes

There are systems of genetic markers, such as VNTRs, where there is difficulty in assigning genotypes unambiguously (Weir, 1992). A single band on an electrophoretic gel may represent a homozygote, a heterozygote for two alleles with similar copy numbers, or a heterozygote with an allele giving a band outside the scoring regions of the gel. It is only individuals scored as heterozygotes for which there is no ambiguity and for which tests of association are required. Only the heterozygotes in the original sample are used, and only permuted arrays where all resulting individuals are heterozygous are used. Since every individual in every array is heterozygous, $H_g = L$ for every genotype A_g, and the conditional probability in Equation 3 for L loci becomes

$$P_s = \prod_{g \in z} \frac{2^{Ln_{A_g}}}{n_{A_g}!}$$

Using goodness of fit measures

Hypothesis have been tested here by determining the proportion of a large number of permuted arrays that have a conditional probability no larger than that of the sample. Instead of using conditional probabilities, arrays could be compared to the hypothesized or expected array on the basis of goodness of fit statistics such as chi-square. The significance level would be the proportion of arrays with a larger statistic than that of the sample.

For multilocus genotype $\mathbf{A}_g$, the observed count is n_g and the expected count under the hypothesis of interest is written as e_j. The chi-square statistic is the usual

$$X^2 = \sum_g \frac{(n_g - e_g)^2}{e_g}$$

It is necessary to sum only over the observed genotypes, and use can be made of the fact that, since $\sum_g n_g = \sum_g e_g = n$, the statistic is $[\sum_{g \in z}(n_g^2/e_g)] - n$. It has been suggested (Anscombe, 1981) that Karl Pearson may have invented this statistic as an approximation for the exact test only because of the lack of computing resources at that time.

Table 1. Sample two-locus array. Marginal totals are one-locus counts.

0	0	0	0	0	0	0	0	0	0	1	0	0	0	0	0	0	0	0	0	1
1	0	0	1	0	0	0	0	0	0	0	0	0	0	0	0	0	0	0	0	2
0	0	0	0	1	0	0	0	0	0	0	0	0	0	0	0	0	0	0	0	1
0	1	0	0	0	0	0	0	0	0	0	1	0	0	0	0	0	0	0	0	2
0	0	0	0	0	0	0	0	0	0	0	0	0	1	0	0	0	0	0	0	1
0	0	0	0	0	0	0	0	0	0	0	0	0	0	0	0	0	0	1	0	1
1	0	0	1	1	0	0	0	0	0	0	1	0	0	0	0	0	0	0	0	4
0	0	0	0	0	0	0	1	1	0	0	0	0	0	0	0	0	0	0	0	2
0	0	0	0	0	0	1	0	0	0	0	0	0	0	0	0	0	0	0	0	1
0	0	0	0	0	0	0	0	0	0	0	1	0	0	0	0	1	0	0	0	2
0	0	0	0	0	1	0	0	0	0	0	0	0	0	0	0	0	0	0	0	1
1	1	0	1	0	0	0	0	0	0	0	1	0	1	0	1	0	1	0	0	7
1	0	0	0	0	0	0	0	0	0	0	1	0	0	0	0	1	0	0	1	4
0	0	0	0	0	0	0	0	0	0	0	0	1	0	0	0	0	0	0	0	1
0	0	0	0	0	0	0	0	0	1	0	0	0	0	0	0	0	0	0	0	1
1	0	0	0	0	0	0	0	0	0	0	0	0	0	0	0	0	0	0	0	1
0	0	1	0	0	0	0	0	0	0	0	0	0	0	0	0	0	0	0	0	1
0	0	0	0	0	0	0	0	0	0	0	0	0	0	0	0	0	1	1	0	2
0	0	0	0	0	0	0	0	0	0	0	0	0	0	0	1	0	0	0	0	1
0	0	0	1	0	0	0	0	0	0	0	0	0	0	1	0	0	0	0	0	2
5	2	1	4	2	1	1	1	1	1	1	5	1	2	1	2	2	2	2	1	38

The difference between goodness of fit tests and tests based on conditional probabilities was explored in some detail by Maiste (1993). The difference in the present case is greatest when there are so many different genotypes possible that each one is likely to be unique in a sample. Certainly this is the situation for forensic databases once five or more loci are scored. If each n_g equals 1, the relevant part of the chi-square test statistic is the sum of reciprocals of expected counts, each of which will be less than 1. The test statistic tends to increase with the number of possible genotypes. The conditional probability, by contrast, depends on the reciprocal of the product of factorial counts, each of which is 1, and this product does not change with the number of possible genotypes. As an extreme example, consider the 20×20 array in Table 1 which represents a sample of 38 two-locus genotype counts. The row and column totals are the one-locus counts. The sample size is much less than the 400 possible two-locus genotypes, and each two-locus genotype seen is unique in the sample. Other arrays, generated by permuting the genotypes at one locus holding both sets of one-locus counts constant, must have the same (when all two-locus counts remain at 0 or 1) or smaller (when some counts are greater than 1) conditional probability so that the significance level for the conditional probability test of no association is 1. However, the sampled individuals fall into cells with low expected frequencies and this is reflected by the goodness of fit test. Based on 17,000 shuffled arrays with the same marginals, the significance level for the X^2 test statistic was found to be 0.01. The same type of difference can happen in sparse arrays when each sampled individual is not unique.

Numerical results

The performance of the various test statistics and strategies has been investigated by applying them to simulated data sets.

Two loci

Cockerham and Weir (1973) expressed two-locus genotypic frequencies in terms of allelic frequencies and a set of disequilibrium coefficients. For example

$$\begin{aligned} P_{AABB} &= p_A^2 p_B^2 + 2p_A D_{ABB} + 2p_B D_{AAB} \\ &\quad + 2p_A p_B \Delta_{AB} + p_A^2 D_B + p_B^2 D_A \\ &\quad + D_{AB}^2 + D_{A/B}^2 + D_A D_B \end{aligned}$$

Table 2. Empirical powers of four tests for two loci with $m = 2$ or $m = 4$ equally frequent alleles when significance level is 0.05.

Nonzero disequilibria	m	P_1	P_2	P_3	P_4
None	2	0.049	0.052	0.051	0.056
	4	0.060	0.048	0.061	0.060
D_A	2	0.462	0.484	0.058	0.051
	4	0.951	0.976	0.041	0.056
$D_A + D_B$	2	0.778	0.486	0.498	0.065
	4	1.000	1.000	0.995	0.051
D_{AB}	2	0.708	0.744	0.742	0.783
	4	0.150	0.216	0.218	0.239
Δ_{AB}	2	1.000	1.000	1.000	1.000
	4	0.382	0.455	0.491	0.545
D_{AAB}	2	0.980	0.990	0.970	0.980
	4	0.275	0.351	0.294	0.324
$D_{AAB} + D_{ABB}$	2	1.000	1.000	1.000	1.000
	4	0.555	0.631	0.625	0.724
Δ_{AABB}	2	0.676	0.714	0.714	0.749
	4	0.204	0.266	0.212	0.259

For either two or four equally frequent alleles, genotypic frequencies were constructed with each disequilibrium coefficient in turn set to one-quarter of its maximum value, and other disequilibria being zero. Samples were taken from populations with these genotypic frequencies. For locus **A**, for example, with two equally frequent alleles, the coefficient D_A is bounded by ± 0.25 so that a population would be constructed with $D_A = 0.0625$. One thousand replicate samples of size 100 were drawn from each population. For each sample and each test, significance levels were computed from 10,000 permuted arrays. Although this number of permutations corresponds to the procedure used to analyze real data sets, it is not really necessary for simulation studies such as this. Oden (1991) and D.D. Boos (personal communication) have shown that the number of permutations can be set to a number very much smaller than the number of simulated data sets.

The entries in Table 2 are the proportions of the 1,000 simulated populations in which the significance level was estimated to be less than or equal to 0.05, and so are the powers of the tests for a 5% significance level. These powers are labelled P_1, P_2, P_3, P_4 for the four hypotheses considered:

- P_1: two-locus genotype frequency is product of four allele frequencies
- P_2: two-locus genotype frequency is product of two **A** allele frequencies and **B** genotype frequency
- P_3: two-locus genotype frequency is product of **A** genotype frequency and two **B** allele frequencies
- P_4: two-locus genotype frequency is product of two one-locus genotype frequencies

Table 2 shows that the test for overall association, as indicated by P_1, is working well in all situations, and is most sensitive to departures from Hardy-Weinberg disequilibrium. The power for detecting Hardy-Weinberg disequilibrium is greater for four than for two alleles. As the number of alleles increases beyond four, however, it has been found that power decreases when only one of the two loci has departures from Hardy-Weinberg. For the tests where genotype frequencies at one locus are held constant, the values of P_2 and P_3 show that Hardy-Weinberg disequilibrium at the other locus can be detected. The test characterized by P_4 is the most powerful of the conditional probability tests but, of course, does not detect departures from Hardy-Weinberg disequilibrium. Apart from the Hardy-Weinberg tests, the powers of tests for four-allele data are less than those for two-allele data. Larger sample sizes are needed to detect associations, although the loss of power with number of alleles varies among the tests.

Powers for the two-allele case can be verified by the method described by Fu and Arnold (1992). They discussed tests for 2 × 2 tables, and showed how powers could be calculated by restricting attention to those tables that make a significant contribution to power. For hypothesis P_1 when only D_{AB} is non-zero, their method can be used for the table of four gametes $A_1B_1, A_1B_2, A_2B_1, A_2B_2$ with frequencies $p_{A_1}p_{B_1} + D_{AB}, p_{A_1}p_{B_2} - D_{AB}, p_{A_2}p_{B_1} - D_{AB}, p_{A_2}p_{B_2} + D_{AB}$. Values very similar to those in Table 2 are obtained. The method should be able to be extended to other situations covered in the table.

It should be noted that the power values shown in Table 2 are, to some extent, dependent on the method of simulating data sets. In real populations, where evolutionary forces may create complex patterns of association, the powers may well be higher. For example, P_4 could be high because of the non-zero values of several of the disequilibrium coefficients, whereas only one coefficient at a time was allowed to be non-zero in the simulations.

Table 3. Empirical powers (with standard deviations) for exact multilocus test with $m = 5$ or $m = 10$ equally frequent alleles per locus, when significance level is 0.05.

		Number of loci									
θ	m	1	2	3	4	10	15	25	50	75	100
0	5	0.050	0.048	0.052	0.044	0.052	0.056	0.047	0.051	0.047	0.049
		(.01)	(.01)	(.01)	(.01)	(.02)	(.01)	(.01)	(.01)	(.01)	(.01)
	10	0.050	0.049	0.051	0.052	0.051	0.048	0.050	0.050	0.050	0.049
		(.01)	(.01)	(.01)	(.01)	(.01)	(.01)	(.01)	(.01)	(.01)	(.02)
0.005	5	0.053	0.062	0.074	0.089	0.098	0.147	0.145	0.244	0.275	0.354
		(.01)	(.01)	(.01)	(.01)	(.02)	(.02)	(.02)	(.03)	(.02)	(.03)
	10	0.065	0.067	0.097	0.126	0.129	0.192	0.240	0.327	0.493	0.600
		(.01)	(.01)	(.01)	(.01)	(.01)	(.02)	(.02)	(.02)	(.02)	(.02)
0.007	5	0.055	0.082	0.098	0.096	0.127	0.197	0.226	0.370	0.459	0.561
		(.01)	(.01)	(.01)	(.01)	(.01)	(.02)	(.01)	(.03)	(.03)	(.02)
	10	0.067	0.097	0.106	0.125	0.225	0.274	0.360	0.584	0.740	0.823
		(.01)	(.01)	(.01)	(.01)	(.02)	(.02)	(.02)	(.03)	(.04)	(.03)
0.010	5	0.065	0.078	0.104	0.119	0.217	0.270	0.342	0.519	0.682	0.809
		(.01)	(.01)	(.01)	(.01)	(.03)	(.02)	(.03)	(.02)	(.03)	(.02)
	10	0.068	0.104	0.141	0.193	0.298	0.411	0.583	0.845	0.942	0.986
		(.01)	(.01)	(.01)	(.02)	(.02)	(.02)	(.02)	(.02)	(.01)	(.01)
0.050	5	0.148	0.323	0.661	0.770	0.989	1.000	1.000	1.000	1.000	1.000
		(.02)	(.02)	(.02)	(.02)	(.01)	(.00)	(.00)	(.00)	(.00)	(.00)
	10	0.237	0.774	0.894	0.969	0.999	1.000	1.000	1.000	1.000	1.000
		(.01)	(.02)	(.01)	(.01)	(.00)	(.00)	(.00)	(.00)	(.00)	(.00)
0.100	5	0.434	0.856	0.996	1.000	1.000	1.000	1.000	1.000	1.000	1.000
		(.02)	(.02)	(.01)	(.00)	(.00)	(.00)	(.00)	(.00)	(.00)	(.00)
	10	0.757	0.998	1.000	1.000	1.000	1.000	1.000	1.000	1.000	1.000
		(.01)	(.00)	(.00)	(.00)	(.00)	(.00)	(.00)	(.00)	(.00)	(.00)

Many loci

For more than two loci, power was studied only for the test of the hypothesis in Equation 1 of no association between any of the alleles. Populations under the alternative hypothesis were simulated in a different way from previously, and in such a way to address the concerns of several authors (e.g. Lewontin & Hartl, 1991) that tests of association have low power as tests for population substructure.

Populations were constructed as amalgamations of several subpopulations that were subjected to drift, and were simulated by the coalescent process (Hudson, 1990) using a program written by P.O. Lewis. The degree of drift was specified by the coancestry coefficient $\theta \equiv F_{ST}$ (Weir, 1990) that measures the probability of any two allelic genes within a randomly mating population being identical in descent (relative to identity between populations), and serves as a measure of divergence of populations from an ancestral population. Apparent associations between alleles were created by constructing samples as composites of samples of size 20 from 10 such populations. These admixed samples serve to address the issue of subpopulation structure that has caused concern over forensic calculations (e.g. Nichols & Balding, 1991) and correspond to the situation where a population consists of a collection of somewhat distinct subpopulations, but a sample is taken from the population as a whole.

Only unlinked loci were simulated, and either 5 or 10 equally frequent alleles at 1 to 100 loci were used. The number of replicate simulated populations used to determine power values ranged from 500 to 1,000 and the number of permuted arrays used to determine the significance level for each replicate was 3,200. Power values in the case where $\theta = 0$ (no association) should be 0.05.

Table 4. Empirical power of exact tests with homozygotes excluded when significance level is 0.05.

	Number of 10-allele loci			
θ	1	2	3	4
0	0.052	0.047	0.005	0.000
0.01	0.062	0.050	0.006	0.002
0.05	0.104	0.065	0.019	0.007
0.10	0.334	0.292	0.083	0.014

When $\theta \neq 0$, power always increased with the number of loci, and with the number of alleles per locus. When $\theta = 0.05$, for example, the power shown in Table 3 is 0.969 for four loci and 10 alleles and 0.765 for 5 alleles, but it drops to 0.552 for 2 alleles (data not shown). The explanation is as follows. As the number of loci and alleles per locus increases, there is an increasing chance that each multilocus genotype in a sample becomes unique and the value of P_s in Equation 3 reduces to $\prod_{g \in z} 2^{H_g}$. Because of the Wahlund effect of reducing heterozygosity by amalgamating 10 populations to provide the original simulated samples, the P_s value is expected to be smaller for this sample than for a permuted array.

The same does not apply for tests on heterozygotes only since the number of heterozygotes is the same in the sample and the permuted arrays. Indeed, the entries in Table 4 show a decrease in power, and poor performance of the test when $\theta = 0$ for more than two loci. As Equation 3 tends to a value of 2^L, zero power when $\theta = 0$ suggests a very low chance of obtaining a sample with more than one copy of any four-locus genotype. When $\theta \neq 0$, power values are greater than zero, since then the samples contain genotypes homozygous at some loci and these individuals are discarded. This increases the chance of having some duplicate genotypes in the remaining data.

Discussion

Testing for associations between alleles, within and between loci, can be performed satisfactorily with exact tests conditional on allelic counts. Significance levels can be found by permutation procedures. These tests are an alternative to those based on normal statistics constructed as estimated disequilibrium coefficients divided by their estimated standard deviations (Weir and Cockerham, 1989). In general, Maiste (1993) found that conditional tests performed better than unconditional tests, of which the variance-based tests are an example

In the current forensic setting, any associations between neutral genetic markers are most likely to be due to population substructure. Comparing multi-locus genotypic frequencies with the products of corresponding allelic frequencies is likely to detect these associations with a power that increases with the number of alleles per locus and/or the numbers of loci. Specifically, samples of size as small as 100 individuals have high powers for detecting the associations accompanying θ values of 0.05 when several loci are used. This θ value is the one suggested by Nichols and Balding (1991) as being a upper bound on actual values in human populations. The fact that associations are generally not found when exact tests are applied (e.g. Evett *et al.*, 1995) suggests that θ is less than this upper bound.

Conversely, Table 3 shows that the power is low when θ is 0.01. What is the effect of not detecting this level of population substructure? The strength of the evidence of a matching genotype between a evidentiary sample and a person suspected of having contributed that sample can be measured as a likelihood ratio. This is the ratio of the probability of the matching genotypes conditional on the suspected person being the contributor to the probability when another person is the contributor. The ratio, termed the forensic index by Weir (1994) by analogy to the usual paternity index, changes very little for small θ values. For a heterozygous matching genotype, when both alleles have frequencies of 0.05, the one-locus index changes from 200 when $\theta = 0$ to 145 when $\theta = 0.01$ (Weir, 1994). The change is greater for smaller allelic frequencies or larger θ values, but these are less likely for the loci currently being used in human identification.

If the product rule is to be employed to estimate multi-locus genotypic frequencies, it is necessary to check for associations between the constituent allelic frequencies. If exact tests do not detect associations, it appears appropriate to base forensic calculations on the product rule. Associations due to population substructure can be accommodated by modifications of the type proposed by Evett *et al.* (1995) or Balding and Nichols (1994).

Acknowledgements

This work was supported in part by NIH grant GM43544. Helpful comments were provided by Dr M.A. Asmussen. The possibility of calculating exact powers was suggested by Dr J. Arnold, who also provided the Anscombe reference.

References

Anscombe, F.J., 1981. Computing in Statistical Science through APL. Springer-Verlag, New York.

Balding, D.J. & R.A. Nichols, 1994. DNA profile match probability calculation: how to allow for population stratification, relatedness, database selection and single bands. Forensic Sci. Int. 64:125–140.

Cockerham, C.C. & B.S. Weir, 1973. Descent measures for two loci with some applications. Theor. Pop. Biol. 4:300–330.

Evett, I.W., P.D. Gill, J.K. Scranage & B.S. Weir, 1995. Establishing the robustness of STR statistics for forensic applications (submitted)

Fu, Y.X. & J. Arnold, 1992. A table of exact sample sizes for use with Fisher's exact test for 2 × 2 tables. Biometrics 48:1103–1112.

Gail, M. & N. Mantel, 1977. Counting the number of $r \times c$ contingency tables with fixed marginals. JASA 72:859–862.

Guo, S-W. & E.A. Thompson, 1992. Performing the exact test of Hardy-Weinberg proportion for multiple alleles. Biometrics 48:361–372.

Hudson, R.R., 1990. Gene genealogies and the coalescent process. In D. Futuyma and J. Antonovics (Eds.) Oxford Surveys in Evolutionary Biology, pp 1–4.

Lewontin, R.C. & D.L. Hartl, 1991. Population genetics in forensic DNA typing. Science 254:1745–1750.

Maiste, P.J., 1993. Comparison of Statistical Tests for Independence at Genetic Loci with Many Alleles. Ph.D. Thesis, North Carolina State University, Raleigh, NC.

Nichols, R.A. & D.J. Balding, 1991. Effects of population structure on DNA fingerprint analysis in forensic science. Heredity 66:297–302.

Oden, N.L., 1991. Allocation of effort in Monte Carlo simulation for power of permutations tests. J. Am. Stat. Assoc. 86:1074–1076.

Press, W.H., S.A. Teukolsky, W.T. Vetteling & B.P. Flannery, 1988. Numerical Recipes in C. The Art of Scientific Computing. 2nd Edition. Cambridge Univ. Press, New York.

Weir, B.S., 1990. Genetic Data Analysis. Sinauer Associates, Sunderland, MA.

Weir, B.S., 1992. Independence of VNTR alleles defined as fixed bins. Genetics 130:873–887.

Weir, B.S., 1994. The effects of inbreeding on forensic calculations. Ann. Rev. Genet. 28:597–621.

Weir, B.S. & C.C. Cockerham, 1989. Complete characterization of disequilibrium at two loci. In M.W. Feldman (Ed.) Mathematical Evolutionary Theory, Princeton Univ. Press, Princeton, pp 86–110.

B. S. Weir (ed.), Human Identification: The Use of DNA Markers, 179–213, 1995.

A bibliography for the use of DNA in human identification

B.S. Weir
Program in Statistical Genetics, Department of Statistics, North Carolina State University, Raleigh NC 27695-8203, USA

Key words: DNA, PCR, STR, VNTR, RFLP, forensic science, paternity

Abstract

A bibliography of material relating to the use of DNA in human identification is presented. It includes bibliographies previously compiled by the author and other individuals, the DNA Legal Assistance Subunit of the FBI, and references from all the chapters in this volume.

Introduction

The following list is based on several previously compiled bibliographies, including that in the author's lecture notes 'Statistics and Population Genetics for Forensic Biology'. The DNA Legal Assistance Subunit, Investigative Law Unit, Federal Bureau of Investigation (Washington, DC 20533) kindly provided their May, 1994 bibliography. Dr. David Kaye of Arizona State University and Dr. John Marinopoulos of LaTrobe University also gave their bibliographies. For convenience, all the references of the papers in this volume, as well as the papers themselves, are included. Each of the authors of the papers in this volume also had the opportunity to add references to the bibliography. There is no guarantee of complete coverage, although the major papers of the past four years are likely to be included.

Papers supportive as well as critical of the use of DNA for human identification have been included, as well as most of the recent news items from *Science* and *Nature*. Inclusion of any material does not constitute an endorsement of the views expressed therein, and especial care needs to be given to critical views expressed in the early days of this application of DNA markers. Many of the issues raised by early critics have either been addressed or found to be of less concern than feared (see, for example the *Proceedings of the National Academy of Sciences, USA* review by Weir, 1992 and the *Statistical Science* review by Roeder, 1994).

Thanks are due to Dr. John Buckleton for his assistance in preparing this bibliography.

Introductory materials

Books

Aitken, C.G.G., 1995. Statistics and the Evaluation of Evidence for Forensic Scientists. Wiley, New York.
Aitken, C.G.G. & D.A. Stoney, (Editors), 1991. The Use of Statistics in Forensic Science. Ellis Horwood, Chichester, UK.
Ballantyne, J., G. Sensabaugh & J. Witkowski, (Editors), 1989. DNA Technology and Forensic Science, Cold Spring Harbor Laboratory Press, Cold Spring Harbor, New York. (See reviews: Chakraborty, R., 1991. Am. J. Hum. Genet. 48: 173–174; Reeve, E.C.R., 1990. Genet. Res. 56: 277–280).
Billings, P.R., (Editor), 1992. DNA on Trial: Genetic Identification and Criminal Justice. Cold Spring Harbor Laboratory Press, Cold Spring Harbor, New York. (See reviews: Crow, J.F., 1993. Trends in Genetics 9: 331–332; Reeve, E.C.R., 1993. Genet. Res. 61: 233–237; Weir, B.S., 1993. Am. J. Hum. Genet. 53: 1158–1160).
Burke, T., G. Dolf, A.J. Jeffreys & R. Wolff, (Editors), 1991. DNA Fingerprinting: Approaches and Appli-

cations. Birkhäuser, Basel. (See review: Reeve, E.C.R., 1993. Genet. Res. 60: 235.)

Coleman, H. & E. Swenson, 1994. DNA in the Courtroom. A Trial Watcher's Guide. GeneLex, Seattle.

Eastel, S., N. McLeod & K. Reed, 1991. DNA Profiling: Principles, Pitfalls and Potential. Harwood, Chur, Switzerland.

Farley, M.A. & J.J. Harrington, (Editors), 1991. Forensic DNA Technology. Lewis, Chelsea, Michigan. (See review: Caskey, C.T., 1992. Am. J. Hum. Genet. 50: 1143–1144).

Home Office, 1988. DNA Profiling in DNA Immigration Casework. Home Office, London.

Kirby, L.T., 1990. DNA Fingerprinting. An Introduction. MacMillan, London.

Krawczak, M. & J. Schmidtke, 1994. DNA Fingerprinting; Series: Medical Perspective. Bios, Oxford. (See reviews: J.O. Falkinham, 1995. J. Heredity 86: 251. A. Reynolds, 1995. Am. J. Hum. Genet. 56: 816–817).

Lampton, C., 1991. DNA Fingerprinting. Impact Books.

National Research Council, 1992. DNA Technology in Forensic Science. National Academy Press, Washington, DC. (See reviews: Ayala, F.J., 1992. J. Mol. Evolution 35: 273; NDW, 1993. J. Forensic Sci. Soc. 33: 255–256; Reeve, E.C.R., 1993. Genet. Res. 60: 235–236).

Pena, S.D.J., R. Chakraborty, J.T. Epplen & A.J. Jeffreys, (Editors), 1993. DNA Fingerpringing: State of the Science. Basel. (See review: Reeve, E.C.R., 1994. Genet. Res. 63: 79–80).

Promega Corporation, 1989–94. Proceedings of the International Symposia on Human Identification. Promega, Madison, Wisconsin.

Robertson, J., A.M. Ross & L.A. Burgoyne, (Editors), 1990. DNA in Forensic Science: Theory, Techniques and Applications. Ellis Horwood, Sydney.

Science and Life Consultants Association, 1991. DNA Finger- Printing: Index of Developments and Progress with Guide for Rapid Research. A E Pubs. Assn.

US Congress, Office of Technology Assessment, 1990. Genetic Witness: Forensic Uses of DNA Tests, OTA-BA-438. US Government Printing Office, Washington, DC.

US Department of Justice, Federal Bureau of Investigation, 1991. Proceedings of the International Symposium on the Forensic Aspects of DNA Analysis. US Government Printing Office, Washington, DC.

Vernon, J. & V. Selinger, (Editors), 1990. DNA and Criminal Justice. Aust. Inst. Criminology, Canberra.

Walker, R.H. (ed.), 1982. Inclusion Probabilities in Parentage Testing. Am. Assoc. Blood Banks, Arlington, VA.

Wambaugh, J., 1989. The Blooding. William Morrow, New York.

Weir, B.S., (Editor), 1995. Human Identification. Kluwer, Dordrecht.

Weir, B.S. 1995. Genetic Data Analysis II. Sinauer, Sunderland, MA. (in press).

RFLP Based Technology

Baechtel, F.S., 1988. A primer on the methods used in the typing of DNA. Crime Laboratory Digest 15: 3.

Berghaus, G., B. Brinkmann, C. Rittner & M. Staak, (Editors), 1991. DNA-Technology and Its Forensic Application: Proceedings of an International Symposium. Springer-Verlag, Berlin.

Budowle, B., H.A. Deadman, R.S. Murch & F.S. Baechtel, 1988. An introduction to the methods of DNA analysis under investigation in the FBI laboratory. Crime Laboratory Digest 15: 8–21.

Caskey, C.T., 1991. Comments on DNA-based forensic analysis. Am. J. Hum. Genet. 49: 893–894.

Daiger, S.P., 1991. DNA fingerprinting. Am. J. Hum. Genet. 49: 897.

DuChesne, A., A. Rand & B. Brinkman, 1993. Forensic trace examination with DNA technology. Arch. Kriminol, 192: 87.

Eubanks, W., 1988. FBI Laboratory DNA Evidence Examination Policy. Crime Laboratory Digest 15: 114.

Farley, M.A. & J.J. Harrington, 1990. Forensic DNA Technology. Lewis, New York.

Herrera, R.J. & M.J. Tracey, 1994. DNA fingerprinting: Basic techniques, problems, and solutions. J. Crim. Justice, May–June 1992: 237.

Honma, M., T. Yoshii, I. Ishiyama, K. Mitani, R. Kominami & M. Muramatsu, 1989. Individual identification from semen by deoxyribonucleic acid (DNA) fingerprint technique. J. Forensic Sci. 34: 222.

International Society of Forensic Haemogenetics, 1992. Recommendations relating to the use of DNA polymorphisms. Forensic Sci. Int. 52: 125.

Jeffreys, A.J., 1993. 1992 William Allan Award Address. Am. J. Hum. Genet. 43: 1–5.

Jeffreys, A.J., N. Royle, I. Patel, J. Armour, A. Maclead, A. Collick, I. Gray, R. Neumann, M. Gibbs, M. Crosier, M. Hill, E. Signer & D. Monckton, 1991. Principles and recent advances in human DNA fingerprinting, in DNA Fingerprinting: Approaches and Applications, T. Burke, G. Dolf, A. Jeffreys & R. Wolff (eds) Birkhauser Verlag, Basel.

Jeffreys, A.J., V. Wilson & S.L. Thein, 1985. Hypervariable 'minisatellite' regions in human DNA. Nature 314: 67–73.

Jeffreys, A.J., V. Wilson & S.L. Thein, 1985. Individual specific 'fingerprints' of human DNA. Nature 316: 76–79.

Lee, H.C., 1990. DNA and Other Polymorphisms in Forensic Sciences. Adv. in Forensic Sci. 3. Mosby, New York.

Reilly, P.R., 1992. DNA banking. Am. J. Hum. Genet. 51: 1169.

Roberts, L., 1992. Science in court: a culture clash. Science 257: 732.

Rowe, W., 1989. DNA testing not ready for court?! A tale of two surveys. J. Forensic Sci. 34: 803.

Ross, A.M. & H.W.J. Harding, 1989. DNA Typing and Forensic Science. Forensic Sci. Int. 41: 197.

PCR Based Technology

Erlich, H.A., (Editor), 1989. PCR Technology. Principles and Applications for DNA Amplification. Stockton, New York.

Gyllensten, T.B. & H.A. Erlich, 1988. Generation of single-stranded DNA by the polymerase chain reaction and its application to direct sequencing of the HLA-DQα Locus. Proc. Natl. Acad. Sci. USA 85: 7652–7656.

Gyllensten, T.B. & H.A. Erlich, 1989. Ancient roots for polymorphisms at the HLA-DQα locus in primates. Proc. Natl. Acad. Sci. USA 86: 9986.

International Society of Forensic Haemogenetics, 1992. Recommendations relating to the use of PCR based polymorphisms. Int. J. Leg. Med. 105: 63–64.

Technology (methods and procedures)

RFLP

Adams, D.E., 1988. Validation of the FBI procedure for DNA analysis. Crime Laboratory Digest 15: 106.

Adams, D.E., L.A. Presley, A.L. Baumstark, K.W. Hensley, A.L. Hill, K.S. Anoe, P.A. Campbell, C.M. McLaughlin, B. Budowle, A.M. Giusti, J.B. Smerick & F.S. Baechtel, 1991. Deoxyribonucleic acid (DNA) analysis by restriction fragment length polymorphisms of blood and other body fluid stains subjected to contamination and environmental insults. J. Forensic Sci. 36: 1284.

Agard, D.A., R.A. Steinberg & R.M. Stroud, 1981. Quantitative analysis of electrophoretograms: a mathematical approach to super-resolution. Anal. Biochem. 111: 257–268.

Bar, W., A. Kratzer, M. Machler & W. Schmid, 1988. Postmortem stability of DNA. Forensic Sci. Int. 39: 59.

Barnett, P.D., E.T. Blake, J. Super-Mihalovich, G. Harmor, L. Rawlinson & B. Wraxall, 1992. Discussion of 'Effects of presumptive test reagents on the ability to obtain restriction fragment length polymorphism (RFLP) patterns from human blood and semen stains'. J. Forensic Sci. 37: 369.

Brauner, P. & N. Galili, 1993. A condom – The link in a rape. J. Forensic Sci. 38: 1233.

Budowle, B., 1988. The RFLP technique. Crime Laboratory Digest 15: 97.

Budowle, B. & F.S. Baechtel, 1990. Modifications to improve the effectiveness of restriction fragment length polymorphism typing. Appl. Theoret. Electrophoresis 1: 181.

Budowle, B., D.E. Adams & R.C. Allen, 1991. Fragment-length polymorphisms for forensic science applications. Methods in Nucleic Acids Research 181: 182.

Budowle, B., J.S. Waye, G.G. Shutler & F.S. Baechtel, 1990. Hae III – A suitable restriction endonuclease for restriction fragment length polymorphism analysis of biological evidence samples. J. Forensic Sci. 35: 530.

Buffery, C., T. Catterick, M. Greenhalgh, S. Jones & J.R. Russell, 1991. Assessment of a video system for scanning DNA autoradiographs. Forensic Sci. Int. 49: 17.

Davies, A., 1991. The use of DNA profiling and behavioral science in the investigation of sex offenses. Med. Sci. Law 31: 95.

Decorte, R., R. Wu, P. Marynen & J.-J. Cassiman, 1994. Identification of internal variation in the pseudoautosomal VNTR DXYS17, with nonrandom distribution of the alleles on the X and Y chromosomes. Am. J. Hum. Genet. 54: 506–515.

Dubeau, L., L.A. Chandler, J.R. Grawlow, P.W. Nichols & P.A. Jones, 1986. Southern blot analysis of DNA extracted from formalin-fixed pathology specimens. Cancer Research 46: 2964.

Elder, J.K., A. Amos, E.M. Southern & G.A. Shippey, 1983. Measurement of DNA length by gel electrophoresis. I. Improved accuracy of mobility measurements using a digital microdensitometer and computer processing. Anal. Biochem. 128: 223–226.

Elder, J.K. & E.M. Southern, 1983. Measurement of DNA length by gel electrophoresis II: Comparison of methods for relating mobility to fragment length. Analytical Biochemistry 128: 227.

Eriksen, B., A. Bertelsen & O. Svensmark, 1992. Statistical analysis of the measurement errors in the determination of fragment length in DNA-RFLP analysis. Forensic Sci. Int. 52: 181.

Eriksen, B. & O. Svensmark, 1993. DNA-profiling of stains in criminal cases: Analysis of measurement errors and band-shift. Discussion of match criteria. Forensic Sci. Int. 61: 21.

Fadda, S., S. Pelotti & G. Pappalardo, 1991. A common disinfectant used in condom processing inhibits endonuclease digestion of sperm DNA. Int. J. Leg. Med. 104: 281.

Feinberg, A.P. & B. Vogelstein, 1987. A technique for radiolabelling DNA restriction endonuclease fragments to high specific activity. Anal. Biochem. 132: 6–13.

Fourney, R.M., 1992. Forensic reality and DNA typing. Proc. Int. Symp. Human Indent. p. 301. Promega.

Fowler, J.C.S., L.A. Burgoyne, A.C. Scott & H.W.J. Harding, 1988. Repetitive deoxyribonucleic acid (DNA) and human genome variation – A concise review relevant to forensic biology. J. Forensic Sci. 33: 1111. (see also: Waye, J.S., 1989. Discussion of 'Fowler, J.C.S., L.A. Burgoyne, A.C. Scott & H.W.J. Harding, 1988. Repetitive deoxyribonucleic acid (DNA) and human genome variation – A concise relevant to forensic biology'. J. Forensic Sci. 34: 1296. and: Fowler, J.C.S., 1989. Author's reply. J. Forensic Sci. 34: 1299).

Gill, P., A.J. Jeffreys & D.J. Werrett, 1985. Forensic applications of DNA 'Fingerprints'. Nature 318: 577–579.

Gill, P., J.E. Lygo, S.J. Fowler & D.J. Werrett, 1987. An evaluation of DNA fingerprinting for forensic purposes. Electrophoresis 8: 38–44.

Gill, P. & D.J. Werrett, 1987. Exclusion of a man charged with murder by DNA fingerprinting. Forensic Sci. Int. 35: 145.

Giusti, A., M. Baird, S. Pasquale, I. Balazs & J. Glassberg, 1986. Application of deoxyribonucleic acid (DNA) polymorphisms to the analysis of DNA recovered from sperm. J. Forensic Sci. 31: 409.

Giusti, A.M. & B. Budowle, 1992. Effect of storage conditions on restriction fragment length polymorphism (RFLP) analysis of deoxyribonucleic acid (DNA) bound to positively charged nylon membranes. J. Forensic Sci. 37: 597.

Gustafson, S., J.A. Proper, E.J.W. Bowie & S.S. Sommer, 1987. Parameters affecting the yield of DNA from human blood. Analyt. Biochem. 165: 294.

Hagerman, P.J., 1990. DNA typing in the forensic arena. Am. J. Hum. Genet. 47: 876 (See also: Higuchi, R., 1991. Human error in forensic DNA typing. Am. J. Hum. Genet. 48: 1215).

Haglund, W.D., D.T. Reay & S.L. Tepper, 1990. Identification of decomposed human remains by deoxyribonucleic acid (DNA) profiling. J. Forensic Sci. 35: 724.

Herrin, G., 1993. Probability of matching RFLP patterns from unrelated individuals. Am. J. Hum. Genet. 52: 491.

Hochmeister, M.N., 1992. Author's Reply. J. Forensic Sci. 37: 370.

Hochmeister, M.N., B. Budowle & F.S. Baechtel, 1991. Effects of presumptive test reagents on the ability to obtain restriction fragment length polymorphism (RFLP) patterns from human blood and semen stains. J. Forensic Sci. 36: 656.

Hopkins, B., J.E.N. Morten, J.C. Smith & A.F. Markham, 1989. The development of methods for the analysis of DNA extracted from forensic samples. J. Methods in Cell Biology 1: 96.

The American Society of Human Genetics, 1990. Individual identification by DNA analysis: points to consider. Am. J. Hum. Genet. 46: 31.

Jeffreys, A.J., 1987. Highly variable minisatellites and DNA fingerprints. Biochemical Soc. Trans. 15: 309.

Jeffreys, A.J., A. Macleod, K. Tamaki, D.L. Neil & D.G. Monckton, 1991. Minisatellite repeat coding as a digital approach to DNA typing. Nature 354: 204–209.

Jeffreys, A.J., V. Wilson & S.L. Thein, 1985. Hypervariable 'minisatellite' regions in human DNA. Nature 314: 67–73.

Jeffreys, A.J., V. Wilson & S.L. Thein, 1985. Individual– specific 'fingerprints' of human DNA. Nature 316: 76–79.

Johnson, P.H., M.J. Miller & L.J. Grossman, 1980. Electrophoresis of DNA in agarose gels. Analyt. Biochem. 102: 159.

Kanter, E., M. Baird, R. Shaler & I. Balazs, 1986. Analysis of restriction fragment length polymorphisms in deoxyribonucleic acid (DNA) receovered from dried bloodstains. J. Forensic Sci. 31: 403.

Laber, T.L., J.M. O'Connor, J.T. Iverson & J.A. Liberty, 1992. Evaluation of four deoxyribonucleic acid (DNA) extraction protocols for DNA yield and variation in restriction fragment length polymorphism (RFLP) sizes under varying gel conditions. J. Forensic Sci. 37: 404.

Lander, E.S., 1989. DNA fingerprinting on trial. Nature 339: 501.

Lee, H.C., E.M. Pagliaro, K.M. Berka, N.L. Folk, D.T. Anderson, G. Ruano, T.P. Keith, P. Phipps, G.L. Herrin, D.D. Garner & R.E. Gaensslen, 1991. Genetic markers in human bone: I. deoxyribonucleic acid (DNA) analysis. J. Forensic Sci. 36: 320.

Lee, H.C., E.M. Pagliaro, R.E. Gaensslen, K.M. Berka, T.P. Keith, G.N. Keith & D.D. Garner, 1991. DNA analysis in human bone tissue: RFLP typing. J. Forensic Sci. Soc. 31: 209.

Lewis, M.E., R.E, Kouri, D. Latorra, K.M. Berka, H.C. Lee & R.E. Gaensslen, 1990. Restriction fragment length polymorphism DNA analysis by the FBI laboratory protocol using a simple, convenient hardware system. J. Forensic Sci. 35: 1186.

McNally, L., M. Baird, K. McElfresh, A. Eisenberg & I. Balazs, 1990. Increased migration rate observed in DNA from evidentiary material precludes the use of sample mixing to resolve forensic cases of identity. Appl. Theoret. Electrophoresis 1: 267.

McNally, L., R.C. Shaler, M. Baird, I. Balazs, P. DeForest & L. Kobilinsky, 1989. Evaluation of deoxyribonucleic acid (DNA) isolated from human bloodstains and exposed to ultraviolet light, heat, humidity, and soil contamination. J. Forensic Sci. 34: 1059.

McNally, L., R.C. Shaler, M. Baird, I. Balazs, L. Kobilinksy & P. DeForest, 1989. The effects of environment and substrata on deoxyribonucleic acid (DNA): the use of casework samples from New York City. J. Forensic Sci. 34: 1070.

Monson, K.L., 1988. Semiautomated analysis of DNA autoradiograms, Crime Laboratory Digest 15: 104.

Monson, K.L. & B. Budowle, 1989. A system for semi-automated analysis of DNA autoradiograms. Proc. Int. Symp. Forensic Aspects of DNA Analysis. p. 127.

Moreno, R.F., F. Booth, S.M. Thomas & L.L. Tilzer, 1990. Enhanced conditions for DNA finger-printing with biotinylated M13 bacteriophage. J. Forensic Sci. 35: 831.

Murphy, P., J. Amos, N. Carpenter, R. Fenwick, M.E. Hodes, T. Kelly, K. Matteson, W. Seltzer, J.E. Spence, S. Thibodeau, V. Venne, V. Wilson & J. Zonana, 1992. Minimum qualifications for directors: DNA-based genetic-testing Laboratories. Am. J. Hum. Genet. 51: 910.

Nakamura, Y., M. Leppert, P. O'Connell, G.M. Lathrop, P. Cartwright, J.M. Lalouel & R. White, 1987. A primary genetic linkage map of distal chromosome 14q. Cytogenetics Cell Genet. 46: 668.

Nakamura, Y., M. Leppert, P. O'Connell, R. Wolff, T. Holm, M. Culver, C. Martin, E. Fujimoto, M. Hoff, E. Kumlin & R. White, 1987. Variable number of tandem repeat (VNTR) markers for human gene mapping. Science 235: 1616.

Nakamura, Y., P. O'Connell, M. Leppert, D. Barker, E. Wright, M. Skolnick, M. Lathrop, P. Cartwright, J.M. Lalouel & R. White, 1987. A primary genetic map of chromosome 17. Cytogenetics Cell Genet. 46: 668.

Odelberg, S.J., R. Plaetke, J.R. Eldridge, L. Ballard, P. O'Connell, Y. Nakamura, M. Leppert, J.M. Lalouel & R. White, 1989. Characterization of eight VNTR loci by agarose gel electrophoresis. Genomics 5: 915.

Polymeropolous, M.H., D.S. Rath, H. Xiao & C.R. Merril, 1991. Tetranucleotide repeat polymorphism at the human tyrosine hydrolase gene (TH). Nuc. Acids Res. 19: 3753.

Polymeropolous, M.H., D.S. Rath, H. Xiao & C.R. Merril, 1991. Tetranucleotide repeat polymorphism at the human c-fes/fps proto- oncogene (FES). Nuc. Acids Res. 19: 4018.

Polymeropolous, M.H., D.S. Rath, H. Xiao & C.R. Merril, 1991. Tetranucleotide repeat polymorphism at the human coagulation factor XIIII. A subunit gene (F13A1). Nuc. Acids Res. 19: 4036.

Rand, S., P. Wiegand & B. Brinkmann, 1991. Problems associated with the DNA analysis of stains. Int. J. Leg. Med. 104: 293.

Sambrook, J., E.F. Fritch & T. Maniatis, 1989. Molecular Cloning: A Laboratory Manual (2nd Edition), Cold Spring Harbor Laboratory Press, Cold Spring Harbor, NY.

Roewer, L., P. Nurnberg, E. Fuhrmann, M. Rose, O. Prokop & J.T. Epplen, 1990. Stain analysis using oligonucleotide probes specific for simple repetitive DNA sequences. Forensic Sci. Int. 47: 59.

Society for Forensic Haemogenetics, 1989. Recommendations concerning DNA polymorphisms. Forensic Sci. Int. 43: 109.

Southern, E.M., 1975. Detection of specific sequences among DNA fragments separated by gel electrophoresis. J. Mol. Biol. 98: 503–527.

Schacker, U., P.M. Schneider, B. Holtkamp, E. Bohnke, R. Fimmers, H.H. Sonneborn & C. Rittner, 1990. Isolation of the DNA minisatellite probe MZ 1.3 and its application to DNA 'Fingerprinting' analysis. Forensic Sci. Int. 44: 209.

Scheithauer, R. & H.J. Weisser, 1991. DNA profiling of bloodstains on linen pretreated with remedies used for cleaning and maintaining clothes. Int. J. Leg. Med. 104: 273.

Schneider, P.M., R. Fimmers, S. Woodroffe, D.J. Werrett, W. Bar, B. Brinkmann, E. Eriksen, S. Jones, A.D. Kloosterman, B. Mevag, V.L. Pascali, C. Rittner, H. Schmitter, J.A. Thomson & P. Gill, 1991. Report of a European collaborative exercise comparing DNA typing results using a single locus VNTR probe. Forensic Sci. Int. 49: 1.

Schwartz, T.R., E.A. Schwartz, L. Mieszerski & L. Kobilinsky, 1991. Characterization of deoxyribonucleic acid (DNA) obtained from teeth subjected to various environmental conditions. J. Forensic Sci. 36: 979.

Shapiro, M.M., 1991. Imprints on DNA fingerprints. Nature 353: 121. (See also: Evett I.W., 1991. Trivial Error. Nature 354: 114).

Smith, J.C., C.R. Newton, A. Alves, R. Anwar, D. Jenner & A.F. Markham, 1990. Highly polymorphic minisatallite DNA probes: Further evaluation for individual identification and paternity testing. J. Forensic Sci. Soc. 30: 3.

Southern, E.M., 1975. Detection of specific sequences among DNA fragments separated by gel electrophoresis. J. Mol. Bio. 98: 503.

International Society for Forensic Haemogenetics, 1993. Statement concerning the National Academy of Sciences 'Report on DNA Technology in Forensic Science'. Forensic Sci. Int. 59: 1.

Technical Working Group on DNA Analysis Methods (TWGDAM), 1990. Guidelines for a quality assurance program for DNA restriction fragment length polymorphism analysis. Crime Laboratory Digest 16: 40–59.

Technical Working Group on DNA Analysis Methods (TWGDAM), 1990. Guidelines for a proficiency testing program for DNA restriction fragment length polymorphism analysis. Crime Laboratory Digest 17: 50–60.

Technical Working Group on DNA Analysis Methods (TWGDAM), 1990. Statement of the Working Group on Statistical Standards for DNA Analysis. Crime Laboratory Digest 17: 53–58.

Technical Working Group on DNA Analysis Methods (TWGDAM) & California Association of Criminalists Ad Hoc Committee on DNA Quality Assurance, 1991. The Guidelines for a Quality Assurance Program for DNA Analysis. Crime Laboratory Digest 18: 44.

Van Eade, P.H., L. Henke, R. Fimmers, J. Henke & G.G. de Lange, 1991. Size calculation of restriction enzyme Hae III- generated fragments detected by probe YNH24 by comparison of data from two laboratories: the generation of fragment-size frequencies. Forensic Sci. Int. 49: 21.

Verbovaya, L.V. & P.L. Ivanov, 1991. 'Sexing' deoxyribonucleic acid (DNA) on DNA fingerprint gel: An internal control for DNA fingerprint evidence. J. Forensic Sci. 36: 991.

Walsh, D.J., A.C. Corey, R.W. Cotton, L. Forman, G.L. Herin, C.J. Word & D.D. Garner, 1992. Isolation of deoxyribonucleic acid (DNA) from saliva and forensic science samples containing saliva. J. Forensic Sci. 37: 387.

Waye, J.S., 1989. Discusion of 'Repetitive deoxyribonucleic acid and human genome variation – A concise review relevant to forensic biology'. J. Forensic Sci. 34: 1296 (see also: Fowler, J.C.S., 1989. Author's Reply. J. Forensic Sci. 34: 1299).

Waye, J.S., 1993. Prenatal paternity testing following sexual assault: A novel application of forensic DNA typing. Can. Soc. Sci. J. 26: 81.

Waye, J.S. & R.M. Fourney, 1990. Agarose gel electrophoresis of linear genomic DNA in the presence of ethidium bromide: band shifting and implications for forensic identity testing. Appl. Theoret. Electrophoresis 1: 193.

Waye, J.S., D. Michaud, J.H. Bowen & R.M. Fourney, 1991. Sensitive and specific quantification of human genomic deoxyribonucleic acid (DNA) in forensic science specimens: Casework examples. J. Forensic Sci. 36: 1198.

Waye, J.S., L.A. Presley, B. Budowle, G.G. Shutler & R.M. Fourney, 1989. A simple and sensitive method for quantifying human genomic DNA in forensic specimen extracts. BioTechniques 4: 852.

Webb, M.B.T., N.J. Williams & M.D. Sutton, 1993. Microbial DNA challenge studies of variable number tandem repeat (VNTR) probes used for DNA profiling analysis. J. Forensic Sci. 38: 1172.

Wong, Z., V. Wilson, I. Patel, S. Povey & A.J. Jeffreys, 1987. Characterization of a panel of highly variable minisatellites cloned from human DNA. Ann. Hum. Genet. 51: 269–288.

Wong, Z., V. Wilson, A.J. Jeffreys & S.L. Thein, 1986. Cloning a selected fragment from a human DNA 'fingerprint': isolation of an extremely polymorphic minisatellite. Nucleic Acids Research 14: 4605.

Wyman, A.R., J. Mulholland & D. Botstein, 1986. Oligonucleotide repeats involved in the highly polymorphic locus D14S1. Am. J. Hum. Genet. 39: A226.

Wyman, A.R. & R. White, 1990. A highly polymorphic locus in human DNA. Proc. Natl. Acad. Sci. USA 77: 6754–6758.

Yokoi, T., M. Naia, T. Odaira & K. Sagisaka, 1990. Hypervariable polymorphic VNTR loci for parentage testing and individual identification. Jpn. J. Hum. Genet. 35: 179.

Yokoi, T., T. Odaira, M. Nata & K. Sagisaka, 1990. Investigation of paternity establishing without the putative father using hypervariable DNA probes. Jpn. J. Hum. Genet. 35: 235.

Yokoi, T., T. Odaira, M. Nata, Y. Aoki & K. Sagisaka, 1991. Application of single-locus hypervariable region DNA probes to deficiency cases in paternity testing. Int. J. Leg. Med. 104: 117.

Yokoi, T. & K. Sagisaka, 1990. Haptoglobin typing of human bloodstains using a specific DNA probe. Forensic Sci. Int. 45: 39.

PCR

Akane, A., H. Shiono, K. Matsubara, H. Nakamura, M. Hasegawa & M. Kagawa, 1993. Purification of forensic specimens for the polymerase chain reaction (PCR) analysis. J. Forensic Sci. 38: 691.

Armour, J.A.L., P.C. Harris & A.J. Jeffreys, 1993. Allelic diversity at minisatellite MS205 (D16S309): evidence for polarized variability. Hum. Molecular Genet. 2: 1137.

Baechtel, F.S., J.B. Smerick, K.W. Presley & B. Budowle, 1993. Multigenerational amplification of a reference ladder for alleles at locus D1S80. J. Forensic Sci. 38: 1176.

Barros, F., A. Carracedo, M.V. Lareu & M.S. Rodriguez-Calvo, 1991. Electrophoretic human leukocyte antigen HLA-DQA1 DNA typing after polymerase chain reaction amplification. Electrophoresis 12: 1041.

Blake, E., J. Mihalovich, R. Higuchi, P.S. Walsh & H. Erlich, 1992. Polymerase chain reaction (PCR) amplification and human leukocyte antigen (HLA)-DQα oligonucleotide typing on biological evidence samples: casework experience. J. Forensic Sci. 37: 700.

Brown, K. & T. Brown, 1992. Amount of human DNA in old bones. Ancient DNA Newsletter 1: 18–19.

Budowle, B., R. Chakraborty, A.W. Guisti, A.J. Esienberg & R.C. Allen, 1991. Analysis of the VNTR Locus D1S80 by PCR followed by high resolution PAGE. Am. J. Hum. Genet. 48: 137–144.

Bugawan, T.L., R.K. Saiki, C.H. Levenson, R.M. Watson & H.A. Erlich, 1988. The use of non-radioactive oligonucleotide probes to analyze enzymatically amplified DNA for prenatal diagnosis and forensic HLA typing. Biotechnology 6: 943.

Comey, C.T. & B. Budowle, 1991. Validation studies on the analysis of the HLA DQα locus using the polymerase chain reaction. J. Forensic Sci. 36: 1633–1648.

Comey, C.T., B. Budowle, D.E. Adams, A.L. Baumstark, J.A. Lindsey & L.A. Presley, 1992. PCR amplification and typing of the HLA DQα gene in forensic samples. J. Forensic Sci. 38: 239.

Cooper, A., 1992. Removal of colourings, inhibitors of PCR, and the carrier effect of PCR contamination from ancient DNA samples. Ancient DNA Newsletter 1: 31–32.

Crouse, C.A., Vincek & B. Carabello, 1994. Analysis and interpretation of the DQα '1.1 Weak Signal' observed during the PCR typing method. J. Forensic Sci. 39: 41–51.

Dallapiccola, B. & G. Novelli, 1991. PCR DNA typing for forensics. Nature 354: 179.

Das, A., 1991. DNA fingerprinting in India. Nature 350: 387.

Eckert, K.A. & T.A. Kunkel, 1990. High fidelity DNA synthesis by the thermus aquaticus DNA polymerase. Nucleic Acids Research 18: 3739.

Edwards, A., A. Civitello, H.A. Hammond & C.T. Caskey, 1991. DNA typing and genetic mapping with trimeric and tetrameric tandem repeats. Am. J. Hum. Genet. 49: 746–756.

Erlich, H.A., D. Gelfand & J.J. Sninsky, 1991. Recent advances in the polymerase chain reaction. Science 252: 1643.

Erlich, H.A., R. Higuchi, K. Lichtenwalter, R. Reynolds & G. Sensabaugh, 1990. Reliability of the HLA-DQα PCR-based oligonucleotide typing system. J. Forensic Sci. 35: 1017.

Fregeau, C.J. & R.M. Fourney, 1993. DNA typing with fluorescently tagged short tandem repeats: a sensitive and accurate approach to human identification. Biotechniques 15: 100–119.

Gaensslen, R.E., K.M. Berka, D.A. Grosso, G. Ruano, E.M. Pagliaro, D. Messina & H.C. Lee, 1992. A polymerase chain reaction (PCR) method for sex and species determination with novel controls for deoxyribonucleic acid (DNA) template length. J. Forensic Sci. 37: 6.

Gaensslen, R.E. & H.C. Lee, 1990. Genetic markers in human bone tissue. Forensic Sci. Rev. 2: 125–146.

Ginther, C., L. Isseltarver & M.-C. King, 1992. Identifying individuals by sequencing DNA from teeth. Nature Genet. 2: 135–138.

Gray, M.R., 1992. Detection of DNA sequence polymorphisms in human genomic DNA by using denaturing gradient gel blots. Am. J. Hum. Genet. 50: 331–346.

Harrington, C.S., V. Dunaiski, K.E. Williams & C. Fowler, 1991. HLA DQα typing of forensic specimens by amplification restriction fragment polymorphism (ARFP) analysis. Forensic Sci. Int. 51: 147.

Helminen, P., A. Sajantila, V. Johnsson, M. Lukka, C. Ehnholm & L. Peltonen, 1992. Amplification of three hypervariable regions by polymerase chain reaction in paternity determinations: comparison with conventional methods and DNA fingerprinting. Mol. Cell Probes 6: 21–26.

Higuchi, R., C.H. von Beroldingen, G.F. Sensabaugh & H.A. Erlich, 1988. DNA typing from single hairs. Nature 332: 543.

Hochmeister, M.N., B. Budowle, U.V. Bore, U. Eggmann, C.T. Comey & R. Dirnhofer, 1991. Typing of DNA extracted from compact bone from human remains. J. Forensic Sci. 36: 1649.

Hochmeister, M.N., B. Budowle, J. Jung, U.V. Borer, C.T. Comey & R. Dirnhofer, 1991. PCR-based typing of DNA extracted from cigarette butts. Int. J. Legal Med. 104: 229.

Holland, M.H., D.L. Fisher, L.G. Mitchell, W.C. Rodriguez, J.S. Canik, C.B. Merril & V.W. Weedn, 1993. Mitochondrial DNA sequence analysis of human skeletal remains: identification of remains from the Vietnam war. J. Forensic Sci. 38: 542–553.

Hubert, R., J.L. Weber, K. Schmitt, L. Zhang & Arnheim, 1992. A new source of polymorphic DNA markers for sperm typing: analysis of microsatellite repeats in single cells. Am. J. Hum. Genet. 51: 985–991.

Ivey, J.N., B.A. Atchison & A.M. Georgalis, 1994. Assessment of PCR of the D17S30 locus for forensic identification. J. Forensic Sci. 39: 52.

Kimpton, C.P., D. Fisher, S. Watson, M. Adams, A. Urquhart, J.E. Lygo & P. Gill, 1994. Evaluation of an automated DNA profiling system employing multiplex amplification of four tetrameric STR loci. Int. J. Leg. Med. (In press).

Kimpton, C.P., P. Gill, A. Walton, A. Urquhart, E.S. Millican, S. Kwok & R. Higuchi, 1989. Avoiding false positives with PCR. Nature 339: 237.

Kimpton, C.P., A. Walton & P. Gill, 1992. A further tetranucleotide repeat polymorphism in the vWF gene. Hum. Mol. Genet. 1: 287.

Kurosaki, K., T. Matsushita & S. Ueda, 1993. Individual DNA identification from ancient human remains. Am. J. Hum. Genet. 53: 638–43.

Lee, H.C., E.M. Pagliaro, K.M. Berka, N.L. Folk, D.T. Anderson, G. Ruano, T.P. Keith, P. Phipps, G.L. Herrin Jr., D.D. Garner & R.E. Gaensslen, 1991. Genetic markers in human bone I: deoxyribonucleic acid (DNA) analysis. J. Forensic Sci. 36: 320.

Lee, J.C.-I. & C. Jan-Growth, 1992. ABO genotyping by polymerase chain reaction. J. Forensic Sci. 37: 1269–1275.

Neil, D.L. & A.J. Jeffreys, 1993. Digital DNA typing at a second hypervariable locus by minisatellite variant repeat mapping. Hum. Molecular Genet. 2: 1129.

Pascal, O., D. Aubert, E. Gilbert & J.P. Moisan, 1991. Sexing of forensic samples using PCR. Int. J. Legal Med. 104: 205.

Puers, C., H.A. Hammond, L. Jin, C.T. Caskey & J.W. Schumm, 1993. Identification of repeat sequence heterogeneity at the polymorphic short tandem repeat locus HUMTH01 [AATG]n and reassignment of alleles in population analysis by using a locus- specific allelic ladder. Am. J. Hum. Genet. 53: 953–958.

Reiss, J., M. Krawczak, M. Schloesser, M. Wagner & D.N. Cooper, 1990. The effect of replication errors in the mismatch analysis of PCR-amplified DNA. Nucleic Acids Research 18: 973.

Reynolds, R. & G. Sensabaugh, 1991. Analysis of genetic markers in forensic DNA samples using the polymerase chain reaction. Analytical Chemistry 63: 2.

Saiki, R.K., T.L. Bugawan, G.T. Horn, K.B. Mulis & H.A. Erlich, 1986. Analysis of enzymatically amplified beta-globin and HLA-DQα DNA with allele-specific oligonucleotide probes. Nature 324: 163.

Saiki, R.K., D.H. Gelfand, S. Stoffel, S.J. Scharf, R. Higuchi, G.T. Horn, K.B. Mullis & H.A. Erlich, 1988. Primer-directed enzymatic amplification of DNA with a thermostable DNA polymerase. Science 239: 487.

Saiki, R.K., S. Scharf, F. Faloona, K.B. Mullis, G.T. Horn, H.A. Erlich & N. Arnheim, 1985. Enzymatic amplification of beta- globin genomic sequences and restriction site analysis for diagnosis of sickle cell anemia. Science 230: 1350.

Saiki, R.K., P.S. Walsh, C.H. Levenson & H.A. Erlich, 1989. Genetic analysis of amplified DNA with immobilized sequence-specific oligonucleotide probes. Proc. Natl. Acad. Sci. USA. 86: 6230.

Sajantila, A., B. Budowle, M. Ström, V. Johnsson, M. Lukka, L. Peltonen & C. Ehnholm, 1992. PCR amplification of alleles at the D1S80 locus: comparison of a Finnish and a North American Caucasian population sample, and a forensic casework evaluation. Am. J. Hum. Genet. 50: 816.

Sajantila, A., M. Ström, B. Budowle, C. Ehnholm & L. Peltonen, 1991. The distribution of the HLA-DQα alleles and genotypes in the Finnish population as determined by the use of DNA amplification and allele specific oligonucleotides. Int. J. Legal Med. 104: 181–184.

Sajantila, A., M. Ström, B. Budowle, P.J. Karhunen & L. Peltonen, 1991. The polymerase chain reaction and post-mortem forensic identity testing: application of amplified D1S80 and HLA-DQα loci to the identification of fire victims. Forensic Sci. Int. 51: 123.

Salazar, M., J. Williamson & D.H. Bing, 1994. Genetic typing of the DqA1*4 alleles by restriction enzyme digestion of the PCR product obtained with the DQα Alpha Amplitype TM Kit. J. Forensic Sci. 39: 518–525.

Siebert, P.M. & M. Fukuda, 1987. Molecular cloning of human glycophorin B cDNA: nucleotide sequence and genomic relationship to glycophorin A. Proc. Natl. Acad. Sci. USA 84: 6735–6739.

Slightom, J.L., A.E. Blechl & O. Smithies, 1980. Human fetal G-gamma and A-gamma globin genes: complete nucleotide sequences suggest that DNA can be exchanged between these duplicated genes. Cell 21: 627–638.

Smith, B.C., D.L. Fisher, V.W. Weedn, G.R. Warnock & M.M. Holland, 1993. A systematic approach to the sampling of dental DNA. J. Forensic Sci. 38: 1194.

Tindall, K.R. & T.A. Kunkel, 1988. Fidelity of DNA synthesis by the thermus aquaticus DNA polymerase. Biochemistry 27: 6008.

Tsongalis, G.J., W.B. Coleman, G.L. Esch, G.J. Smith & D.G. Kaufman, 1993. Identification of human DNA in complex biological samples using the Alu polymerase chain reaction. J. Forensic Sci. 38: 961.

Tulley, G., K.M. Sullivan & P. Gill, 1993. Analysis of 6 VNTR loci by 'multiplex' PCR and automated fluorescent detection. Hum. Genet. 92: 554–562.

Uitterlinden, A.G., P.E. Slagboom, D.L. Knook & J. Ving, 1989. Two-dimensional DNA fingerprinting of human individuals. Proc. Natl. Acad. Sci. USA 86: 2742–2746.

Vigilant, L., R. Pennington, H. Harpending, T.D. Kocher & A.C. Wilson, 1989. Mitochondrial DNA sequences in single hairs from a southern African population. Proc. Natl. Acad. Sci. USA 86: 9350.

von Beroldingen, C.H., E.T. Blake, R. Higuchi, G.F. Sensabaugh & H.A. Erlich, 1989. Applications of PCR to the analysis of biological evidence, pp. 209–223 in PCR Technology-Principles and Applications for DNA Amplification, edited by H.A. Ehrlich. Stockton, New York.

Vuorio, A.F., A. Sajantila, T. Hamalainen, A.-C. Syvanen, C. Ehnholm & L. Peltonen, 1990. Amplification of the hypervariable region close to the apoliprotein B gene: application to forensic problems. Biochem. & Biophys. Research Communications 170: 616–620.

Walsh, P.S., N. Fildes, A.S. Louie & R. Higuchi, 1991. Report of the blind trial of the Cetus AmpliType HLA DQαforensic deoxyribonucleic acid (DNA) amplification and typing kit. J. Forensic Sci. 36: 1551.

Walsh, P.S., D.A. Metzger & R. Higuchi, 1991. Chelex 100 as a medium for simple extraction of DNA for PCR-based typing from forensic material BioTechniques 10: 506–513.

Waye, J.S., L.A. Presley, B. Budowle, G.G. Shutler & R.M. Fourney, 1989. A simple and sensitive method for quantifying human genomic DNA in forensic specimen extracts. BioTechniques 7: 852.

Weber, J.L. & P.E. May, 1989. Abundant class of human DNA polymorphism which can be typed using the polymerase chain reaction. Am. J. Hum. Genet. 44: 388–396.

Weber, J.L. & C. Wong, 1993. Mutation of human short tandem repeats. Hum. Mol. Genet. 2: 1123–1128.

Westwood, S.A. & D.J. Werrett, 1990. An evaluation of the polymerase chain reaction method for forensic applications. Forensic Sci. Int. 45: 201.

Woodroffe, S., J.E. Lygo & D.J. Lingard, 1991. Validation oof the Amplitype HLA-DQα forensic amplification and typing kit., Proc. European Forensic Science PCR Symp.

Yamamoto, T., C.G. Davis, M.S. Brown, W.J. Schneider, M.J. Casey, J.L. Goldstein & D.W. Russell, 1984. The human LDL receptor: A cystein-rich protein with multiple Alu sequences in its mRNA. Cell 39: 27–38.

Yang, F., J.L. Brune, S.L. Naylor, R.L. Apples & K.H. Naberhaus, 1985. Human group-specific Component (Gc) is a member of the albumin family. Proc. Natl. Acad. Sci. USA 82: 7994–7998.

Ziegle, J.S., Y. Su, P. Kevin, L.N. Corcoran, P.E. Maynard, L.B. Hoff, L.J. McBride, M.N. Kronick & S.R. Diehl, 1992. Application of automated DNA sizing technology for genotyping microsatellite loci. Genomics 14: 1026–1031.

Validation of methods and procedures

RFLP

Adams, D.E., L.A. Presley, A.L. Baumstark, K.W. Hensley, A.L. Hill, K.S. Anoe, P.A. Campbell, C.M. McLaughlin, B. Budowle, A.M. Giusti, J.B. Smerick & F.S. Baechtel, 1991. Deoxyribonucleic acid (DNA) analysis by restriction fragment length polymorphisms of blood and other body fluid stains subjected to contamination and environmental insults. J. Forensic Sci. 36: 1284.

Adams, D.E., 1988. Validation of the FBI procedure for DNA Analysis: A summary. Crime Laboratory Digest 15: 106.

Akane, A., K. Matsubara, H. Nakamura, 1994. Identification of the heme compound copurified with deoxyribonucleic acid (DNA) from bloodstains, a major inhibitor of polymerase chain reaction (PCR) amplification. J. Forensic Sci. 39: 362–372.

Bar, W., A. Kratzer, M. Machler & W. Schmid, 1988. Postmortem stability of DNA. Forensic Sci. Int. 39: 59.

Budowle, B. & R.C. Allen, 1987. Electrophoresis reliability: I. the contaminant issue. J. Forensic Sci. 32: 1537. Collaborative Testing Services, Inc., 1991. DNA Profiling. Report 91-15. Herndon, Virginia.

Budowle, B., F.S. Baechtel & D.E. Adams, 1991. Validation with regard to environmental insults of the RFLP procedure for forensic purposes. pp. 82–91 in M.A. Farley & J.J. Harrington, 1991, Forensic DNA Technology. Lewis, Chelsea, MI.

Cotton, R.W., L. Forman & C.J. Word, 1991. Research on DNA typing validated in the literature. Am. J. Hum. Genet. 49: 898.

Graves, M.H. & M. Kuo, 1989. DNA: A blind trial study of three commercial testing laboratories. Presented at the meeting of the American Academy of Forensic Sciences, Las Vegas.

Hochmeister, M.N., B. Budowle, U.V. Borer & R. Dinhofer, 1993. Effects of nonoxinol-9 on the ability to obtain DNA profiles from postcoital vaginal swabs. J. Forensic Sci. 38: 442–447.

Ludes, B., H. Pfitzinger & P. Mangin, 1993. DNA fingerprinting from tissues after variable post-mortem periods. J. Forensic Sci. 38: 686–690.

Lindahl, T., 1993. Instability and decay of the primary structure of DNA. Nature 362: 709–715.

Krawczak, M. & B. Bockel, 1991. DNA fingerprinting: a short note on mutation rates. Hum. Genet. 87: 632.

McNally, L., R.C. Shaler, M. Baird, I. Balazs, P. DeForest & L. Kobilinsky, 1989. Evaluation of deoxyribonucleic acid (DNA) isolated from human bloodstains exposed to ultravoilet light, heat, humidity and soil contamination. J. Forensic Sci. 34: 1059.

McNally, L., R.C. Shaler, M. Baird, I. Balazs, L. Kobilinsky & P. DeForest, 1989. The effects of environment and substrata on deoxyribonucleic acid (DNA): the use of casework samples from New York City. J. Forensic Sci. 34: 1070.

Presley, L.A., A.L. Baumstark & A. Dixon, 1993. The effects of specific latent fingerprint and questioned document examinations on the amplification and typing of the HLA DQ alpha gene region in forensic casework. J. Forensic Sci. 38: 1028–1035.

Roberts, H., J.A. Auer, K.A. Davey & S.J. Gutowski, 1992. Evaluation of experimental variation in DNA profiling and selection of appropriate match criteria. Proc. 11th. Int. Aust. NZ Forensic Sci. Symp.

Thompson, W.C., 1993. Evaluating the admissibility of new genetic identification tests: lessons from the 'DNA War'. J. Crim. Law & Crimin. 84: 22–104.

Thompson, W.C. & S. Ford, 1990. The meaning of a match: sources of ambiguity in the interpretation of DNA prints, in Forensic DNA Technology edited by M. Farley & J. Harrington.

Washio, K., S. Ueda & S. Misawa, 1990. Effects of cytosine methylation at restriction sites on deoxyribonucleic acid (DNA) typing. J. Forensic Sci. 35: 227–283.

PCR

Budowle, B., J.A. Lindsey, J.A. De Cou, B.W. Koons, A.M. Giusti & C.T. Comey, 1995. Validation and population studies of the loci LDLR, HYPA, HBGG, D7S8, and Gc (PM loci), and HLA-DQalpha. J. Forensic Sci. 40: 45–54.
Comey, C.T. & B. Budowle, 1991. Validation studies on the analysis of the HLA-DQα locus using the polymerase chain reaction. J. Forensic Sci. 36: 1633.
Comey, C.T., J.M. Jung & B. Budowle, 1991. Use of formamide to improve amplification of HLA-DQα sequences. BioTechniques 10: 60.
Jung, J.M., C.T. Comey, D.B. Baer & B. Budowle, 1991. Extraction strategy for obtaining DNA from bloodstains for PCR amplification and typing of the HLA-DQα Gene. Int. J. Legal Med. 104: 145.
Keohavong, P. & W.G. Thilly, 1989. Fidelity of DNA polymerases in DNA amplification. Proc. Natl. Acad. Sci. USA 86: 9253.
Krawczak, M., J. Reiss, J. Schmidtke & U. Rosler, 1989. Polymerase chain reaction: replication errors and reliability of gene diagnosis. Nucleic Acids Research 17: 2197.
Lygo, J.E., P.E. Johnson, D.J. Holdaway, S. Woodroffe, J.P. Whitaker, T.M. Clayton, C.P. Kimpton & P. Gill, 1994. The validation of short tandem repeat (STR) loci for the use in forensic casework. Int. J. Leg. Med. (in press).
Presley, L.A., A.L. Baumstark & A. Dixon, 1993. The effects of specific latent fingerprint and questioned document examinations on the amplification and typing of the HLA DQ alpha gene region in forensic casework. J. Forensic Sci. 38: 1028.

NRC report

Aldhous, P., 1992. Challenge to British Forensic databases. Nature 355: 191.
Aldhous, P., 1993. Geneticists attack NRC report as scientifically flawed. Science 259: 755. (See also: Weir, B.S. DNA fingerprinting report. Science 260: 473. Hartl, D.L. & R.C. Lewontin. DNA fingerprinting report. Science 260: 473).
Anderson, A., 1989. DNA fingerprinting on trial. Nature 342: 844.
Anderson, C., 1991. DNA fingerprinting discord. Nature 354: 500.
Anderson, C., 1992. Conflict concerns disrupt panels, cloud testimony. Nature 355: 753.
Anderson, C., 1992. Academy approves, critics still cry foul. Nature 356: 552.
Anderson, C., 1992. FBI attaches strings to its DNA database. Nature 357: 618.
Anderson, C., 1992. Courts reject DNA fingerprinting, citing controversy after NAS report. Nature 359: 349.
Budowle, B., 1992. Perspectives on the fixed bin method and the floor approach/ceiling principle. Proc. 1992 Int. Symp. Human Ident. Promega Corp. p. 391.
Budowle, B. & E.S. Lander, 1994. DNA fingerprinting dispute laid to rest. Nature 371: 735.
Chakraborty, R., 1992. NRC report on DNA typing. Science 260: 1059–1060.
Devlin, B., N. Risch & K. Roeder, 1993. Statistical evaluation of DNA fingerprints: A critique of the NRC's report. Science 259: 748–749.
Devlin, B., N. Risch & K. Roeder, 1992. NRC report on DNA typing. Science 260: 1057–1059.
Devlin, B., N. Risch & K. Roeder, 1994. Comments on the statistical aspects of the NRC's report on DNA typing. J. Forensic Sci. 39: 28–40.
Federal Bureau of Investigation, 1992. The FBI's responses to recommendations by the committee on DNA Technology in Forensic Science of the National Research Council, National Academy of Sciences. Crime Laboratory Digest 19: 49.
Hartl, D.L. & R.C. Lewontin. DNA fingerprinting report. Science 260: 473.
Hooper, C., 1992. Rancor precedes National Academy of Science's DNA fingerprinting report. J. NIH Res. 4: 76–80.
Kaye, D.H., 1995. The forensic debut of the NRC's DNA report: population structure, ceiling frequencies and the need for numbers. Genetica 96: 99–105.
Kobilinsky, L. & L. Levine, 1993. Commentary on the 'Ceiling principle'. J. Forensic Sci. 38: 1261–1262.
Lander, E.S., 1993. DNA Fingerprinting: The NRC Report. Science 260: 1221.
Lempert, R., 1993. DNA, science and the law: two cheers for the ceiling principle. Jurimetrics 34: 41–57.
Lewontin, R.C., 1992. The dream of the human genome. N.Y. Rev. of Books, May 28, 1992, pp. 31–40.

National Research Council, 1992. DNA Technology in Forensic Science. National Academy Press, Washington, D.C.
Roberts, L., 1992. DNA fingerprinting: Academy Reports. Science 256: 300.
Robertson, B., 1992. Why the NRC report on DNA is wrong, 1992. New Law J. 1619–1621.
Slimowitz, J.R. & J.E. Cohen, 1993. Violations of the ceiling principle: exact conditions and statistical evidence. Am. J. Human Genet. 53: 314–323.
Weir, B.S., 1992. Population genetics in the forensic DNA debate. Proc. Natl. Acad. Sci. USA, 89: 11654.
Weir, B.S., 1993. Forensic population genetics and the National Research Council (NRC). Am. J. Hum. Genet. 52: 437.
Weir, B.S., 1993. DNA fingerprinting report. Science 260: 473.

Admissibility

Altman, J.D., 1992. Admissibility of forensic DNA profiling evidence. Wash. Univ. J. Urban Contem. 44: 211.
Annas, G.J., 1992. Setting standards for the use of DNA-typing results in the courtroom – the state of the art. New Eng. J. Med. 36: 1641–1644.
Annot., Admissibility of DNA Identification Evidence, 84 A. Law R. 4th 312 (1991).
Beeler, L. & W.R. Wiebe, 1988. DNA identification tests and the courts. 63 Wash. Law Rev. 903–955.
Bezak, E.M., 1992. DNA profiling evidence: the need for a uniform and workable evidentiary standard of admissibility. 26 Val. U. Law Rev. 595.
Blair, C.T., 1990. Spencer v. Commonwealth and recent developments in the admissibility of DNA fingerprint evidence. 76 Va. Law Rev. 853–876.
Brashears-Macatee, S., 1992. A test both lawyers and scientists can live with. 71 Neb. Law Rev. 920–936.
Butler, L.G., 1992. State v. Davis: DNA evidence and the use of Frye in Missouri. 60 UMKC Law Rev. 577.
DiRusso, M.J., 1990. DNA 'Profiles' – The problems of technology transfer. 8 N.Y. Law Sch. J. Hum. Rts. 183.
Gass, D.A. & M.M. Shultz, 1992. An analysis of decisional law governing the use of DNA evidence, pp. 43–59 in DNA on Trial: Genetic Identification and Criminal Justice edited by P.R. Billings. Cold Spring Harbor Laboratory, Cold Spring Harbor, New York.
Giannelli, P.C., 1991. Criminal discovery, scientific evidence, and DNA. Vand. Law Rev. 44: 791.
Grover, •.•., 1989. A new twist in the double helix: admissibility of DNA 'fingerprinting' in California. Computer & High Tech. Law J. 5: 469.
Harmon, R.P., 1991. General admissibility considerations for DNA typing evidence: Let's learn from the past and let the scientists decide this time around, pp. 153–180 in M.A. Farley & J.J. Harrington, 1991, Forensic DNA Technology. Lewis, Chelsea, MI.
Harmon, R.P., 1993. Legal criticisms of DNA typing: where's the beef? J. Crim. Law & Crimin. 84: 175–880.
Hoeffel, J.C., 1990. The dark side of DNA profiling: unreliable scientific evidence meets the criminal defendant. Stan. Law Rev. 42: 465–538.
Hoke, 1990. DNA tests in criminal prosecutions: too much evidence or not enough? J. Legal Med. 11: 481.
Hymer, A.D., 1991. DNA testing in criminal cases: a defense perspective, pp. 181–199 in M.A. Farley & J.J. Harrington, 1991, Forensic DNA Technology. Lewis, Chelsea, MI.
Imwinkelried, E., 1991. The debate in the DNA cases over the foundation for the admission of scientific evidence: the importance of human error as a cause of forensic misanalysis. Wash. U. Law Q. 69: 19.
Jackson, D.B., 1989. DNA fingerprinting and proof of paternity. Fam. Law Rep. 15: 3007–3013.
Jakubaitis, J.L., 1991. 'Genetically' altered admissibility: legislative notice of DNA typing. Clev. St. Law Rev. 39: 415.
Kaye, D.H., 1991. The admissibility of DNA testing. Cardozo Law Rev. 13: 353–360.
Kramer, S.R., 1993. Admissibility of DNA statistical data: a proliferation of misconception. Cal. W. Law Rev. 30: 145–178.
Lempert, R., 1991. Some caveats concerning DNA as criminal identification evidence: with thanks to the Reverend Bayes. Cardozo Law Rev. 13: 303–341.
Liebeschuetz, J., 1991. Statutory control of DNA fingerprinting in Indiana. Ind. Law Rev. 25: 204.
McCormick on Evidence 205(B), 210–211 (J. Strong 4th ed, 1992) St. Paul MN: West Pub. Co.
McElfresh, K.C., D. Vining-Forde & I. Balazs, 1993. DNA-based identity testing in forensic science: court admissibility of DNA data has survived five years of strong challenges. Bioscience 43: 149.
Moenssens, A.M., 1990. DNA evidence and its critics – how valid are the challenges? Jurimetrics J. 31: 87–108.

Neufeld, P.J., 1993. Have you no sense of decency? J. Crim. Law & Crimin. 84: 189–202.

Norman, J.A., 1990. DNA fingerprinting: is it ready for trial? U. Miami Law Rev. 45: 243.

O'Brien, J.P. Jr., 1994. DNA fingerprinting: the Virginia approach. Wm. & Mary Law Rev. 35: 767–804.

Pearsall, A., 1989. DNA printing: the unexamined 'witness' in criminal trials. Calif. Law Rev. 77: 665–703.

Petrovich, 1990. DNA typing: a rush to judgment. Ga. Law Rev. 24: 669.

Petrosinelli, J.G., 1990. The admissibility of DNA typing: a new methodology. Geo. Law J. 79: 313–336.

Renskers, 1990. Trial by certainty: implications of genetic 'DNA fingerprints', Emory Law J. 39: 309.

Robins, S., 1993. The use and admissibility of DNA typing evidence: problems and prospects for its future place in New Jersey. Rutgers Law J. 24: 847–882.

Russell, G., 1994. A pathfinder on the admissibility of forensic DNA evidence in criminal cases. Legal Ref. Services Quart. 13: 19.

Schmitt, M.N. & L.H. Crocker, 1990. DNA typing: novel scientific evidence in the military courts. A.F. Law Rev. 32: 227.

Shank, C.G., 1992. DNA evidence in criminal trials: modifying the law's approach to protect the accused from prejudicial genetic evidence. Ariz. Law Rev. 34: 829–871.

Shultz, M.M., 1992. Reasons for doubt: legal isues in the use of DNA identification evidence, pp. 19–42 in DNA on Trial: Genetic Identification and Criminal Justice edited by P.R. Billings. Cold Spring Harbor Laboratory, Cold Spring Harbor, New York.

Stenson, •.•., 1989. Admit it! DNA fingerprinting is reliable. Hous. Law Rev. 26: 677.

Steventon, B., 1993. The ability to challenge DNA evidence, the Royal Commission on Criminal Justice Research Study No. 9.

Swafford, L.L., 1990. Admissibility of DNA genetic profiling evidence in criminal proceedings: the case for caution. Pepp. Law Rev. 18: 123.

Tande, 1989. DNA fingerprinting: a new investigatory tool. Duke Law J. 474.

Thaggard, L., 1991. DNA fingerprinting: overview of the impact of the genetic witness on the American system of criminal justice. Miss. Law J. 61: 423.

Thompson, W.C., 1993. Evaluating the admissibility of new genetic identification tests: lessons from the 'DNA War'. J. Crim. Law & Crimin. 84: 22–104.

Thompson, W.C. & S. Ford, 1989. DNA Typing: acceptance and weight of the new genetic identification tests. Va. Law Rev. 75: 45–108.

Thompson, W.C. & S. Ford, 1988. DNA Typing. Trial, Sept., 1988: 56–64.

Whitmore, L.D., 1993. The admissibility of DNA evidence in criminal proceedings. Wayne Law Rev. 39: 1411–1432.

Williams, C.L., 1987. DNA fingerprinting: A revolutionary technique in forensic science and its probable effects on criminal evidentiary law. Drake Law Rev. 37: 1–32.

White, •.•. & Greenwood, •.•., 1988. DNA fingerprinting and the law. Mod. Law Rev. 51: 145.

Statistical, genetic and general issues

Agresti, A. & D. Wackerly, 1977. Some exact conditional tests of independence for RxC crossclassification tables. Psychometrika 42: 111–125.

Ahrens, J.H. & U. Dieter, 1974. Computer methods for sampling from gamma, beta, Poisson and binomial distributions. Computing 12: 223–246.

Aickin, M., 1984. Some fallacies in the computation of paternity probabilities. Am. J. Hum. Genet. 36: 904–915.

Aickin, M. & D. Kaye, 1982. Some mathematical and legal considerations in using serological tests to prove paternity, pp. 155–168 in R.H. Walker (ed.). Inclusion probabilities in parentage testing. Am. Assoc. Blood Banks, Arlington, VA.

Aitken, C.G.G., 1987. The use of statistics in forensic science. J. Forensic Sci. Soc. 27: 113–115.

Akane, A., K. Matsubara, H. Shiono, I. Yuasa, S.-I. Yokota, M. Yamada & Y. Nakagome, 1990. Paternity testing: blood group systems and DNA analysis by variable number of tandem repeat markers. J. Forensic Sci. 35: 1217.

Albert, J., J. Wahlberg & M. Uhlén, 1993. Forensic evidence by DNA sequencing. Nature 361: 595–596.

Aldhous, P., 1993. Geneticists attack NRC report as scientifically flawed. Science 259: 755.

Allard, J.E., 1992. Murder in South London: A novel approach to DNA profiling. J. Forensic Sci. Soc. 32: 49–58.

Allen, M., T. Saldeen, U. Peterson & U. Gallensten, 1993. Genetic typing of HLA class II genes in Swedish populations: application to forensic analysis. J. Forensic Sci. 38: 554–570.

Allen, M.Z., 1992. Forensic uses of DNA. San Jose Studies 18: 3.

Allen, R.J., D.J. Balding & A. Stein, 1995. Probability and proof in state v. Skipper: an internet exchange. Jurimetrics 35: 277.

Anderson, A.J. & B.G. McGadgen, 1990. Prehistoric two-way voyaging between New Zealand and East Polynesia: Mayor Island obsidian on Raoul Island and possible Raoul Island obsidian in New Zealand. Acheology on Oceania 25: 24–37.

Anderson, A., 1989. DNA fingerprinting: Judge backs technique. Nature 340: 582.

Anderson, A., 1989. DNA fingerprinting on trial. Nature 342: 844.

Anderson, A., 1990. Forensic tests proved innocent. Nature 346: 499.

Anderson, C., 1991. DNA fingerprinting discord. Nature 354: 500.

Anderson, C., 1992. FBI gives in on genetics. Nature 355: 663.

Anderson, C., 1992. Conflict concerns disrupt panels, cloud testimony. Nature 355: 753–754.

Anderson, C., 1992. Coincidence or conspiracy? Nature 355: 753.

Anderson, C., 1992. FBI attaches strings to its DNA databases. Nature 355: 753.

Archer, F., 1810. Facts illustrating a disease peculiar to the female children of Negro slaves. Med. Reposit. 1: 319–323.

Armour, J.A.L. & A.J. Jeffreys, 1992. Biology and applications of human minisatellite loci. Curr. Opinin Genet. Devel. 2: 850– 856.

Atchison, B.A. & A.M. Georaglis, 1990. DNA profiling: a review of the techniques and interpretation of DNA testing. Leo Cusen Institute, Melbourne.

Austad, S.N., 1992. Forensic DNA typing. Science 255: 1050.

Ayala, F.J. & B. Black, 1993. Science and the courts. Am. Sci. 81: 230–239.

Baechtel, F.S., K.L. Monson, G.E. Forsen, B. Budowle & J.J. Kearney, 1991. Tracking violent criminal offender through DNA typing profiles – a national database system concept, pp. 356–360 in T. Burke, G. Dolf, A.J. Jeffreys & R. Wolff (eds.). DNA Fingerprinting: Approaches and Applications. Birkhauser, Basel.

Baer, A.S., 1993. New genetics in old boxes? Am. J. Hum. Genet. 53: 530. (see also: Chakraborty, R., R. Deka & R.E. Ferrell, 1993. Reply to Baer. Am. J. Hum. Genet. 53: 531).

Bär, W. & K. Hummel, 1991. DNA fingerprinting: its application in forensic casework, pp. 349–355 in T. Burke, G. Dolf, A.J. Jeffreys & R. Wolff (eds.). DNA Fingerprinting: Approaches and Applications. Birkhauser, Basel.

Baird, M., I. Balazs, A. Giusti, L. Miyazaki, L. Nicholas, K. Wexler, E. Kanter, J. Glassberg, F. Allen, P. Rubinstein & L. Sussman, 1986. Allele frequency distribution of two highly polymorphic DNA sequences in three ethnic groups and its application to the determination of paternity. Am. J. Hum. Genet. 39: 489–501.

Baird, M.L., 1991. Analysis of forensic DNA samples by single locus VNTR probes, pp. 39–49 in M.A. Farley & J.J. Harrington, 1991, Forensic DNA Technology. Lewis, Chelsea, MI.

Balasz, I., 1993. Population genetics of 14 ethnic groups using phenotypic data from VNTR loci, pp. 193–210 in DNA Fingerprinting: State of the Science, Edited by S.D.J. Pena, R. Chakraborty, J.T. Epplen & A.J. Jeffreys. Birkhauser Verlag Basel.

Balding, D.J. & P. Donnelly, 1994. The prosecutor's fallacy and DNA evidence. Crim. Law Rev. 1994: 711.

Balding, D.J., P. Donnelly & R.A. Nichols, 1994. Some causes for concern about DNA profiles. Stat. Sci. 9: 248.

Balszs, I., M. Baird, M. Clyne & E. Meade, 1989. Human population genetic studies of five hypervariable DNA loci. Am. J. Hum. Genet. 44: 182–190.

Ballantyne, J., G. Sensabaugh & J. Witkowski (Editors), 1989. DNA Technology and Forensic Science. Banbury Report 32. Cold Spring Harbor Press.

Balding, D.J. & P. Donnelly, 1994. Inference in forensic identification. To appear. J. Roy. Statist. Soc. 157.

Balding, D.J. & P. Donnelly, 1994. How convincing is DNA evidence? Nature 368: 285–286. (See also: Brookfield, J.F.Y., 1994. DNA profiling on trial. Nature 369: 351; Chakraborty, R., 1994. DNA profiling on trial. Nature 369: 351; Weir, B.S., 1994. DNA profiling on trial. Nature 369: 351).

Balding, D.J. & R.A. Nichols, 1994. DNA profile match probability calculation: how to allow for population stratification, relatedness, database selection and single bands. Forensic Sci. Int. 64: 125–140.

Balding, D.J. & R.A. Nicols, 1995. A method for characterizing differentiation between populations at multiallelic loci and its implications for establishing identity and paternity. Genetica 96: 3–12.

Barinaga, M., 1989. Pitfalls come to light. Nature 339: 89.

Bashinsky, J.S., 1991. Managing the implementation and use of DNA typing in the crime laboratory, pp. 201–235 in M.A. Farley & J.J. Harrington, 1991, Forensic DNA Technology. Lewis, Chelsea, MI.

Baur, M.P., C. Rittner & H.D. Wihner, 1981. The prior probability parameter in paternity testing. Its relevance and estimation by maximum likelihood, pp. 389–392 in Lectures of the Ninth International Congress of the Society for Forensic Hemogenetics, Bern.

Baur, M.P., R.C. Elston, H. Gurtler, K. Henningsen, K. Hummel, H. Matsumoto, W. Mayr, J.W. Morris, L. Niejenhuis, H. Polesky, D. Salmon, J. Valentin & R. Walker, 1986. No fallacies in the formulation of the paternity index. Am. J. Hum. Genet. 39: 528–536.

Berghaus, G., B. Brinkmann, C. Rittner & M. Staak (Editors), 1991. DNA-Technology and Its Forensic Application: Proceedings of an International Symposium. Springer-Verlag, Berlin.

Berkson, J., 1978. In dispraise of the exact test. J. Stat. Planning Inf. 2: 27–42.

Bernado, J.M., 1976. Algorithm AS 103 : psi (digamma) function. Appl. Stat. 25: 315–317.

Bernstam, V.A., 1992. pp. 54–57 in Handbook of Gene Level Diagnostics in Clinical Practice 54–57. CRC Press, Boca Raton, Florida.

Berry, D.A., 1990. DNA fingerprinting: What does it prove? Chance 3: 15.

Berry, D.A., 1991. Inferences using DNA profiling in forensic identification and paternity cases. Stat. Sci. 6: 175–205. (See also Comments following paper by: D.H. Kaye, K. Lange, I. Evett).

Berry, D.A., 1991. Probability of paternity, pp. 150–156 in C.G.G. Aitken & D.A. Stoney (eds.). The Use of Statistics in Forensic Science. Ellis Horwood, Chichester, UK.

Berry, D.A., 1994. DNA, statistics, and the Simpson case. Chance 7: 9–12.

Berry, D.A. & S. Geisser, 1986. Inferences in cases of disputed paternity, in M.H. DeGroot, S.E. Feinberg & J.B. Kadane, (Editors), Statistics and the Law. Wiley, New York.

Berry, D.A., I.W. Evett & R. Pinchin, 1992. Statistical inference in crime investigations using deoxyribonucleic acid profiling. Applied Stat. 41: 499–531.

Bever, R.A., M. DeGuglielmo, R.W. Staub, C.M. Kelly & R.S. Foster, 1992. Forensic DNA typing. Science 255: 1050.

Billings, P.R. (Editor), 1992. DNA on trial: Genetic Identification and Criminal Justice. Cold Spring Harbor Laboratory Press, New York.

Bockel, B., P. Nurnberg & M. Krawczak, 1992. Likelihoods of multilocus DNA fingerprints in extended families. Am. J. Human Genet. 51: 554–561.

Borowsky, R., 1988. HLA and the probability of paternity. Am. J. Hum. Genet. 42: 132–134.

Bowcock, A.M., A. Ruiz-Linares, J. Tomforhde, E. Minch, J.R. Kidd & L.L. Cavalli-Sforza, 1994. High resolution of human evolutionary trees with polymorphic microsatellites. Nature 369: 455.

Box, G.E.P. & G.C. Tiao, 1973. Bayesian inference in statistical analysis. Addison-Wesley, Reading, MA.

Briggs, T.J., 1978. The probative value of bloodstains on clothing. Medicine, Science and the Law 18: 79–83.

Brinkman, B., S. Rand & P. Wiegand, 1991. Population and family data of RFLPs using selected single- and multi-locus systems. Legal Medicine 104: 81–86.

Brookfield, J., 1989. Analysis of DNA fingerprinting data in cases of disputed paternity. IMA J. Math. Appl. Med. Biol. 6: 111–131.

Brookfield, J., 1992. Interpreting DNA Fingerprints. Nature 356: 483.

Brookfield, J., 1992. Law and probabilities. Nature 355: 207–208.

Brookfield, J., 1992. Reply to D'Eustachio. Nature 356: 483.

Brookfield, J.F.Y., 1992. The effect of population subdivision on estimates of the likelihood ratio in criminal cases using single-locus probes. Heredity 69: 97–100.

Brookfield, J.F.Y., 1994. DNA profiling on trial. Nature 369: 351.

Brookfield, J.F.Y., 1994. The effect of relatives on the likelihood ratio associated with DNA profile evidence in criminal cases. J. Forensic Sci. Soc. (In press).

Brookfield, J.F.Y., 1995. The effect of relatedness on likelihood ratios and the use of conservative estimates. Genetica 96: 13–19.

Brown, A.H.D., M.W. Feldman & E. Nevo, 1980. Multilocus structure of natural populations of *Hordeum spontaneum*. Genet. 96: 523–536.

Buckleton, J. & K. Walsh, 1991. A continuous model for interpreting the positions of bands in DNA locus-specific work. J. Forensic Sci. Soc. 31: 353.

Buckleton, J.S., K.A.J. Walsh & I.W. Evett, 1991. Who is 'random man?' J. Forensic Sci. Soc. 31: 463–468.

Buckleton, J.S., K.A.J. Walsh, G.A.F. Seber & D.G. Woodfield, 1987. A stratified approach to the compilation of blood group frequency surveys. J. For Sci. Soc. 27: 103–122.

Buckleton, J.S., K.A.J. Walsh & C.M. Triggs, 1991. A continuous model for interpreting the position of bands in DNA locus-specific work. J. For. Sci. Soc. 31: 353–363.

Buckleton, J.S., K.A.J. Walsh & C.M. Triggs, 1992. Discussion of the paper 'Statistical inference in crim investigations using deoxyribonucleic acid profiling'. Applied Stat. 41: 527–528.

Budowle, B., 1992. Reply to Green. Am. J. Hum. Genet. 50: 441– 443.

Budowle, B., 1995. The effects of inbreeding on DNA profile frequency estimates using PCR-based loci. Genetica 96: 21–25.

Budowle, B. & F.S. Baechtel, 1990. Modifications to improve effectiveness of restriction fragment length polymorphism typing. Appl. Theor. Electrophoresis 1: 181–187.

Budowle, B., F.S. Baechtel, A.M. Giusti & K.L. Monson, 1990. Applying highly polymorphic variable number of tandem repeats loci genetic markers to identify testing. Clinical Biochemistry 23: 287.

Budowle, B., H.A. Deadman, R.S. Murch & F.S. Baechtel, 1988. An introduction to the methods of DNA analysis under investigation. Crime Laboratory Digest 15: 498–505.

Budowle, B., A.M. Giusti, J.S. Waye, F.S. Baechtel, R.M. Fourney, D.E. Adams, L.A. Presley, H.A. Deadman & K.L. Monson, 1991. Fixed-bin analysis for statistical evaluation of continuous distributions of allelic data from VNTR loci, for use in forensic comparisons. Am. J. Hum. Genet. 48: 841–855.

Budowle, B. & E.S. Lander, 1994. DNA fingerprinting dispute laid to rest. Nature 371: 735.

Budowle, B. & K.L. Monson, 1989. A statistical approach for VNTR analysis. Proc. Int. Symp. Forensic Aspects of DNA.

Budowle, B. & K.L. Monson, 1993. The forensic significance of various reference population databases for estimating the variable number of tandem repeat (VNTR) loci profiles, pp. 177– 192, in DNA Fingerprinting: State of the Science, Edited by S.D.J. Pena, R. Chakraborty, J.T. Epplen & A.J. Jeffreys. Birkhauser Verlag Basel.

Budowle, B. & K.L. Monson, 1994. Greater differences in forensic DNA profile frequencies estimated from racial groups than from ethnic subgroups. Clinica Chimica Acta 228: 3.

Budowle, B., K.L. Monson, K.S. Anoe, F.S. Baechtel, D.L. Bergman, E. Buel, P.A. Campbell, M.E. Clement, H.W. Corey, L.A. Davis, A. Dixon, P. Fish, A.M. Guisti, T.L. Grant, T.M. Gronert, D.M. Hoover, L. Jankowski, A.J. Kilgore, W. Kimoto, W.H. Landrum, H. Leone, C.R. Longwell, D.C. MacLaren, L.E. Medlin, S.D. Narveson, M.L. Pierson, J.M. Pollock, R.J. Raquel, J.M. Reznicek, G.S. Rogers, J.E. Smerick & R.M. Thompson, 1991. A preliminary report on binned general population data on six VNTR loci in Caucasians, blacks and Hispanics from the United States. Crime Laboratory Digest 18: 9–26.

Budowle, B., K.L. Monson & A.M. Giusti, 1994. A reassessment of frequency estimates of Pvu II-generated VNTR profiles in a Finnish, and Italian, and a general United States Caucasian database: No evidence for ethnic subgroups afecting forensic estimates. Am. J. Hum. Genet. 55: 533–539.

Budowle, B., K.L. Monson, A.M. Giusti & B. Brown, 1994. The assessment of frequency estimates of Hae III-generated VNTR profiles in various reference databases. J. Forensic Sci. 39: 319–352.

Budowle, B., K.L. Monson, A.M. Giusti & B. Brown, 1994. The assessment of frequency estimates of Hing I-generated VNTR profiles in various reference databases. J. Forensic Sci. 39: 94–112.

Budowle, B. & J. Stafford, 1991. Response to expert Report by D.L. Hartl, submitted in the case of United States v. Yee. Crime Laboratory Digest 18: 101.

Budowle, B. & J. Stafford, 1991. Response to 'Population genetic problems in the forensic use of DNA profiles' by R.C. Lewontin, submitted in the case of United States v. Yee. Crime Laboratory Digest 18: 109.

Budowle, B., S. Sundaram & R.E. Wenk, 1985. Population data on the forensic genetic markers: phosphoglucomutase-I, esterase D, erythrocyte acid phosphatase and glycxylase I. Forensic Sci. Int. 28: 77–81.

Buffery, C., F. Burridge, M. Greenhalgh, S. Jones & G. Willo, 1991. Allele frequency distributions of four variable number tandem repeat (VNTR) loci in the London area. Forensic Sci. Int. 52: 53–64.

Burk, D.L., 1988. DNA fingerprinting: possibilities and pitfalls of a new technique, Jurimetrics J. 28: 455–471.

Burk, D.L., 1990. DNA identification: possibilities and pitfalls revisited. Jurimetrics J. 31: 53.

Burk, D.L., 1992. DNA identification: assessing the threat to privacy. U. Toledo Law Rev. 24: 87–102.

Burke, T., G. Dolf, A. Jeffreys & R. Wolf (Editors), 1991. DNA Fingerprinting: Approaches and Applications. Springer-Verlag, Basel.

Butler, D., 1994. UK to set up DNA database of criminals. Nature 370: 588–589.

Callen, D.F., A.D. Thompson, Y. Shen, H.A. Philips, R.I. Richards, J.C. Mulley & G.R. Sutherland, 1993. Incidence and origin of 'null' alleles in the (AC)n microsatellite markers. Am. J. Hum. Genet. 52: 922–927.

Camilli, G., 1990. The test of homogeneity for 2 × 2 contingency tables: a review of and some personal opinions on the controversy. Psychological Bulletin 108: 135–145.

Casella, G. & R.L. Berger, 1990. Statistical Inference. Pacific Grove, CA, Wadsworth and Brooks/Cole.

Cavalli-Sforza, L.L. & W.F. Bodmer, 1971. The Genetics of Human Populations. Freeman, San Francisco.

Cavalli-Sforza, L.L. & A. Piazza, 1993. Human genomic diversity in Europe: a summary of recent research and prospects for the future. Eur. J. Hum. Genet. 1: 3–18.

Chakraborty, R., 1981. The distribution of the number of heterozygous loci in an individual in natural populations. Genet. 98: 461–466.

Chakraborty, R., 1984. Detection of nonrandom association of alleles from the distribution of the number of heterozygous loci in a sample. Genetics. 108: 719–731.

Chakraborty, R., 1990. Genetic profile of cosmopolitan populations: effects of hidden subdivision. Anthrop. Anz. 48: 313–331.

Chakraborty, R., 1991. Inclusion of data on relatives for estimation of allele frequencies. Am. J. Hum. Genet. 49: 242–243.

Chakraborty, R., 1991. Statistical interpretation of DNA typing data. Am. J. Hum. Genet. 49: 895–897.

Chakraborty, R., 1992. Sample size requirements for addressing the population genetic issues of forensic use of DNA typing. Human Biology 64: 141–159.

Chakraborty, R., 1992. Effects of population subdivision and allele frequency differences on interpretation of DNA typing data for human identification, pp. 205–222 in Proc. Third Int. Symp. on Human Ident. Promega Corporation, Madison, Wisconsin.

Chakraborty, R., 1993. NRC report on DNA typing. Science 260: 1059–1060.

Chakraborty, R., 1993. A class of population genetic questions formulated as the generalized occupancy problem. Genetics 134: 953–958.

Chakraborty, R., 1994. DNA profiling on trial. Nature 369: 351.

Chakraborty, R. & S.P. Daiger, 1991. Polymorphisms at VNTR loci suggest homogeneity of the white population of Utah. Human Biology 63: 571–588.

Chakraborty, R., M. de Andrade, S.P. Daiger & B. Budowle, 1992. Apparent heterozygote deficiencies observed in DNA typing data and their implications in forensic applications. Annals of Human Genet. 56: 45–57.

Chakraborty, R., R. Deka, L. Jin & R.E. Ferrell, 1992. Allele sharing at six VNTR loci and genetic distances among three ethnically defined human populations. Human Biology 4: 387–397.

Chakraborty, R., R. Deka, L. Jin & R.E. Ferrell, 1993. Reply to Baer. Am. J. Hum. Genet. 53: 531–532.

Chakraborty, R., M. Fornage, R. Gueguen & E. Boerwinkle, 1991, pp. 127–143 in T. Burke, G. Dolf, A.J. Jeffreys & R. Wolff (eds.). DNA Fingerprinting: Approaches and Applications. Birkhauser, Basel.

Chakraborty, R. & P.W. Hedrick, 1983. Paternity exclusion and the paternity index for two linked loci. Hum. Hered. 33: 13–23.

Chakraborty, R. & L. Jin, 1992. Heterozygote deficiency, population substructure and their implications in DNA fingerprinting. Human Genet. 88: 267–272.

Chakraborty, R. & L. Jin, 1993. A unified approach to study hypervariable polymorphisms: statistical considerations of determining relatedness and population distances. In DNA Fingerprinting: State of the Science. pp. 153–175. ed. by S.D.J. Pena, R. Chakraborty, J.T. Epplen & A.J. Jeffreys. Birkhauser, Basel.

Chakraborty, R., L. Jin, Y. Zhong, M.R. Srinivasan & B. Budowle, 1993. On allele frequency computation from DNA typing data, 1993. Int. J. Legal Med. 106: 103.

Chakraborty, R., M.I. Kamboh, M. Nwankwo & R.E. Ferrell, 1992. Caucasian genes in American blacks: new data. Am. J. Hum. Genet. 50: 145–155.

Chakraborty, R. & K.K. Kidd, 1991. The utility of DNA typing in forensic work. Science 254: 1735–1739. (See also: Lewontin, R.C. & D.L. Hartl, 1991. Population genetics in forensic DNA typing. Science 254: 1745. Wills, C., 1992. Science 255: 1050, Austad, S.N., 1992. Science 255: 1050, Bever, R.A., R.W. Staub, C.M. Kelly & R.S. Foster, 1992. Science 255: 1051 (February, 1992), Yarborough, L.R., 1992. Science 255: 1052, Cleveland, D.W., 1992. Science 255: 1052, Koshland, D.E. Jr., 1992. Science 255: 1052, Chakraborty, R. & K.K. Kidd, 1992.

Science 255: 1053, Lewontin, R.C. & D.L. Hartl, 1992. Science 255: 1054).

Chakraborty, R. & Z. Li, 1995. Correlation of DNA fragment sizes within loci in the presence of non-detectable alleles. Genetica 96: 27–36.

Chakraborty, R. & A.K. Roychoudhury, 1975. Paternity exclusion by genetic markers in Indian populations. Indian J. Med. Res. 63: 162–169.

Chakraborty, R. & W.J. Schull, 1976. A note on the number of exclusions to be expected in paternity testing. Am. J. Hum. Genet. 28: 615–618.

Chakraborty, R., M. Shaw & W.J. Schull, 1974. Exclusion of paternity: the current state of the art. Am. J. Hum. Genet. 26: 477–488.

Chakraborty, R., M.R. Srinivasan & S.P. Diager, 1993. Evaluation of standard error and confidence interval of estimated multilocus genotype probabilities, and their implications in DNA forensics. Am. J. Hum. Genet. 52: 60–70.

Chakraborty, R., M.R. Srinivasan & M. de Andrade, 1993. Intraclass and interclass correlations of allelic sizes within and between loci in DNA typing data. Genetics 133: 411–419.

Chakraborty, R., Y. Zhong, L. Jin & B. Budowle, 1994. Nondetectability of restriction fragments and independence of DNA-fragment sizes within and between loci in RFLP typing of DNA. Am. J. Hum. Genet. (In press).

Chakraborty, R. & Y. Zhong, 1994. Human Heredity 44: 1–9.

Cheng, R.C.H. & G.M. Feast, 1979. Some simple gamma variate generators. Appl. Stat. 28: 290–295.

Chernoff, H., 1992. Stat. Sci. 6: 192–196.

Chimera, J.A., C.R. Harris & M. Litt, 1989. Population genetics of the highly polymorphic locus D16S7 and its use in paternity evaluation. Am. J. Human Genet. 45: 926–931. de Gorgey, 1988. The advent of DNA databanks: implications for individual privacy. Am. J. Law & Med. 16: 109.

Chow, S.T., W.F. Tan, K.H. Yap & T.L. Ng, 1993. The development of DNA profiling database in an Hae III based RFLP system for Chinese, Malays, and Indians in Singapore. J. Forensic Sci. 38: 874.

Church, G.M. & W. Gilbert, 1984. Genomic Sequencing. Proc. Natl. Acad. Sci. USA 81: 1991–1995.

Clark, A.G., J.F. Hamilton & G.K. Chambers, 1995. Inference of population subdivision from the VNTR distributions of New Zealanders. Genetica 96: 37–49.

Cleveland, D.W., 1992. Forensic DNA typing. Science 255: 1050.

Cochran, W.G., 1954. Some methods for strengthening the common χ^2 tests. Biometrics 10: 417–451.

Cockerham, C.C., 1969. Variance of gene frequencies. Evolution 23: 72–84.

Cockerham, C.C. & B.S. Weir, 1973. Descent measures for two loci with some applications. Theor. Pop. Biol. 4: 300–330.

Cockerham, C.C. & B.S. Weir, 1993. Estimation of gene flow from F-statistics. Evolution 47: 855–863.

Cohen, J.E., 1990. DNA fingerprinting: what (really) are the odds? 1990. Chance 3: 26.

Cohen, J.E., 1990. DNA fingerprinting for forensic identification: Potential effects on data interpretation of subpopulation heterogeneity and band number variability. Am. J. Hum. Genet. 46: 358–368. (See also: Chakraborty, R., 1991. Statistical interpretation of DNA typing data. Am. J. Hum. Genet. 49: 895–897).

Cohen, J.E., 1992. The ceiling principle is not always conservative, 1992. Assigning genotype frequencies for forensic DNA testing. Am. J. Hum. Genet. 51: 1165–1168.

Cohen, J.E., M. Lynch & C.E. Taylor, 1991. Forensic DNA tests and Hardy-Weinberg equilibrium. Science 25: 1037–1039.

Collins, A. & N.E. Morton, 1994. Likelihood ratios for DNA identification. Proc. Natl. Acad. Sci. USA. 91: 6007–6011.

Cotterman, C.W., 1940. A calculus for statico-genetics. Ph.D. Thesis, Ohio State University. Reprinted in P. Ballonof (ed.), Genetics and Social Structure (1974). Dowden, Hutchinson & Ross, Inc., Stroudsburg, Pennsylvania.

Cressie, N. & T.R.C. Read, 1984. Multinomial goodness of fit tests. J. Roy Stat. Soc. B46: 440–464.

Cressie, N. & T.R.C. Read, 1989. Pearson's X^2 and the log likelihood ratio statistic G^2: a comparative review. International Statistical Review 57: 19–43.

Crow, J.F. & K. Aoki, 1984. Group selection for a polygenic behavioral trait: estimating the degree of subdivision. Proc. Natl. Acad. Sci. USA. 81: 6073–6077.

Crow, J.F. & C. Denniston, 1993. Population genetics as it relates to human identification. In: The Fourth International Symposium on Human Identification, Promega Corporation, Madison, WI.

Crow, J.F. & M. Kimura, 1970. An Introduction to Population Genetic Theory. New York, Harper and Row.

Curnow, R.N., 1991. DNA fingerprinting, pp. 146–150 in The Use of Statistics in Forensic Science, edited by C.G.G. Aitken and D.A. Storey, Ellis Horwood, New York.

Curnow, R.N., 1995. Conditioning on the number of bands in interpreting matches of multilocus DNA profiles. Genetica 96: 51–53.

Curnow, R.N. & Wheeler, 1993. Probabilities of incorrect decisions in paternity cases using multilocus DNA probes. J.R. Stat. Soc. (A) 156: 207–223.

D'Agostino, R.B., W. Chase & A. Belanger, 1988. The appropriateness of some common procedures for testing the equality of two independent binomial populations. Am. Statist. 42: 198–202.

Dallapiccola, B., G. Novelli & A. Spinella, 1991. PCR DNA typing for forensics. Nature 354: 179.

Davidson, J.M., 1987. Maori origins, pp. 13–29 in The Prehistory of New Zealand, Longman Paul Limited, Auckland.

Dawson, T.C., 1993. DNA profiling: evidence for the presecution. J. Forensic Sci. Soc. 33: 238–242.

Debenham, P.G., 1991. DNA fingerprinting; a biotechnology in business. pp. 342–348 in T. Burke, G. Dolf, A.J. Jeffreys & R. Wolff (eds.). DNA Fingerprinting: Approaches and Applications. Birkhauser, Basel.

Debenham, P.G., 1994. Genetics leaves no bones unturned. Nature Genet. 6: 113–114.

Decorte, R. & J.-J. Cassiman, 1991. Detection of amplified VNTR alleles by direct chemiluminescence: application to the genetic identification of biological samples in forensic cases, pp. 371–390 in T. Burke, G. Dolf, A.J. Jeffreys & R. Wolff (eds.). DNA Fingerprinting: Approaches and Applications. Birkhauser, Basel.

Deka, R., R. Chakraborty & R.E. Ferrell, 1991. A population genetic study of six VNTR loci in three ethnically defined populations. Genomics 11: 83–92.

Deka, R., R. Chakraborty, S. DeCroo, F. Rothhammer, S.A. Barton & R.E. Ferrell, 1992. Characteristics of polymorphism at a VNTR locus 3′ to the apolipoprotein B gene in five human populations. Am. J. Hum. Genet. 51: 1325–1333.

D'Eustachio, P., 1992. Interpreting DNA fingerprints. Nature 356: 483. (See also; Brookfield, J.J.Y., 1992. Reply. Nature 356: 483).

Devlin, B., 1993. Forensic inference from genetic markers. Stat. Methods in Med. Res. 2: 241–262.

Devlin, B., T. Krontiris & N. Risch, 1993. Population genetics of the HRAS1 minisatellite locus. Am. J. Hum. Genet. 53: 1298–1305.

Devlin, B. & N. Risch, 1992. Ethnic differentiation at VNTR loci, with special reference to forensic applications. Am. J. Hum. Genet. 51: 534–548.

Devlin, B. & N. Risch, 1992. A note on Hardy-Weinberg equilibrium of VNTR data by using the Federal Bureau of Investigation's fixed-bin method. Am. J. Hum. Genet. 51: 549–553.

Devlin, B. & N. Risch, 1993. Physical properties of VNTR data, and their impact on a test of allelic independence. Am. J. Hum. Genet. 53: 324–329.

Devlin, B., N. Risch & K. Roeder, 1990. No excess of homozygosity at loci used for DNA fingerprinting. Science 249: 1416–1420. (See also: Cohen, J.E., M. Lynch & C.E. Taylor; P. Green & E.S. Lander, 1991. Forensic DNA tests and Hardy-Weinberg equilibrium. Science 253: 1037–1039; Devlin, B., N. Risch & K. Roeder, 1991. Response to Cohen et al. Science 253: 1039–1041).

Devlin, B., N. Risch & K. Roeder, 1991. Forensic DNA tests and Hardy-Weinberg equilibrium. Science 253: 1039–1041.

Devlin, B., N. Risch & K. Roeder, 1991. Estimation of allele frequencies for VNTR loci. Am. J. Hum. Genet. 48: 662–676.

Devlin, B., N. Risch & K. Roeder, 1992. Forensic inference from DNA fingerprints. J. Am. Stat. Assoc. 87: 337–350.

Devlin, B., N. Risch & K. Roeder, 1993. Statistical evaluation of DNA fingerprinting: a critique of the CRC's report. Science 259: 748–750. (See also: Hartl, D.L. & R.C. Lewontin, 1993. Response to Devlin et al. Science 260: 473–474; Devlin, B., N. Risch & K. Roeder, 1993. NRC report on DNA typing. Science 260: 1057–1059).

Devlin, B., N. Risch & K. Roeder, 1993. NRC report on DNA typing. Science 260: 1057–1059.

Devlin, B., N. Risch & K. Roeder, 1993. Comments on the statistical aspects of the NRC's report on DNA typing. J. Forensic Sci. 39: 28–40.

Dickson, D., 1993. Science faces new treatment in British courts. UK. Nature 364: 178.

Dickson, D., 1993. DNA database proposal gets cautious welcome in UK. Nature 364: 179.

DiLonardo, A.M., P. Darlu, M. Baur, C. Orrego & M.C. King, 1984. Human genetics and human rights: identifying the families of kidnapped children. Am. J. Forensic Med. Pathol. 5: 339–347.

Donnelly, P., 1992. The non-independence of matches at different loci in single locus DNA profiles, cited in Discussion of the Paper by Bery, Evett & Pinchin, Applied Statistics 41: 521–525.

Donnelly, P., 1995. Nonindependence of matches at different loci in DNA profiles: quantifying the effect of close relatives on the match probability. Heredity 75: 26–34.

Donnelly, P., 1995. Match probability calculations for multi- locus DNA profiles. Genetica 96: 55–67.

Dykes, D.D., 1987. Parentage testing using restriction fragment length polymorphisms (RFLPs), pp. 59–85 in Clinical Application of Genetic Engineering, edited by L.C. Lasky & J.M. Edwards- Moulds. American Association of Blood Banks, Arlington, VA.

Edwards, A., H.A. Hammond, L. Jin, C.T. Caskey & R. Chakraborty, 1992. Genetic variation at five trimeric and tetrameric tandem repeat loci in four human populations. Genomics 12: 241–253.

Ellman, I.M. & D. Kaye, 1979. Probabilities and proof; Can HLA and blood group testing prove paternity? New York Univ. Law School J. 54: 1131–1162.

Elston, R., 1986. Probability and paternity testing. Am. J. Hum. Genet. 39: 112–122.

Elston, R.C. & R. Forthofer, 1977. Testing for Hardy Weinberg equilibrium in small samples. Biometrics 33: 536–542.

Emigh, T.H., 1978. The power of tests for random mating in genetics. Biometrics 34: 730.

Emigh, T.H., 1980. A comparison of tests for Hardy-Weinberg equilibrium. Biometrics 36: 627–642.

Emigh, T.H. & O. Kempthorne, 1975. A note on goodness-of-fit of a population to Hardy-Weinberg structure. Am. J. Hum. Genet. 27: 778–783.

Eriksen, B., A. Bertelsen & O. Svensmark, 1992. Statistical Analysis of the Measurement of errors in the determination of fragment length in DNA-RFLP analysis. Forensic Sci. Int. 52: 181–191.

Eriksen, B. & O. Svensmark, 1991. Analysis of a Danish Caucasian population sample of single locus DNA-profiles, allele frequencies, frequencies of DNA-profiles and heterozygosity. Forensic Sci. Int. 59: 119.

Eriksen, B. & O. Svensmark, 1993. DNA – Profiling of stains in criminal cases: analysis of measurement errors and band shift. Forensic Sci. Int. 61: 21–34.

Eriksen, B. & O. Svensmark, 1994. The effect of sample size on the estimation of the frequency of DNA profiles in RFP analysis. Forensic Sci. Int. 65: 195.

Essen-Möller, E., 1938. Die Beweiskraft der Aehnlichkeit im Vaterschaftsnachweis; theoretische Grundlagen. Mitt. Anthrop. Ges. (Wien) 68: 9–53.

Essen-Möller, E. & C.E. Quensel, 1939. Zur Theorie des Vaterschaftsnachweises auf grund von Aehnlichkeitsbefunden. Dtsch. Z. Gerichtl. Med. 31: 70–96.

Evett, I.W., 1977. The interpretation of refractive index measurements. Forensic Science 9: 209–217.

Evett, I.W., 1983. What is the probability that this blood came from that person? A meaningful question? J. Forensic Sci. Soc. 23: 35.

Evett, I.W., 1984. A quantitative theory for interpreting transfer evidence in criminal cases. Appl. Stat. 33: 25–32.

Evett, I.W., 1986. A quantitative theory for interpreting transfer evidence in criminal cases. Applied Statistics 33: 25–32.

Evett, I.W., 1986. A Bayesian Approach to the problem of interpreting glass evidence in forensic science casework. J. Forensic Sci. Soc. 26: 3–18.

Evett, I.W. 1987. Bayesian inference and forensic science: problems and perspectives. The Statistician 36: 99–105.

Evett, I.W., 1987. On meaningful questions: a two-trace transfer problem. J. Forensic Sci. Soc. 27: 375–381.

Evett, I.W., 1990. Analysis of DNA multilocus profiles in a paternity case in which the child's profile may be partial. J. Forensic Sci. Soc. 30: 293.

Evett, I.W., 1990. The theory of interpreting scientific transfer evidence, pp. 141–180 in A. Maehley & R.L. Williams (eds.). Forensic Science Progress 4. Springer-Verlag, Berlin.

Evett, I.W., 1991. Interpretation: a personal odyssey, pp. 9–22 in C.G.G. Aitken & D.A. Stoney (eds.). The Use of Statistics in Forensic Science. Ellis Horwood, Chichester, UK.

Evett, I.W., 1991. Evaluation of DNA profiles: sense and nonsense. J. Forensic Sci. Soc. 31: 205.

Evett, I.W., 1991. Trivial Error. Nature 354: 114.

Evett, I.W., 1992. DNA statistics: putting the problems into perspective. Justice of the Peace 156: 583–586.

Evett, I.W., 1992. DNA statistics: putting the problems into perspective. Jurimetrics J. 33: 139–145.

Evett, I.W., 1992. Evaluating DNA profiles in a case where the defence is 'It was my brother'. J. Forensic Sci. Soc. 32: 5–14.

Evett, I.W., 1993. Establishing the evidential value of a small quantity of material found at a crime scene. J. Forensic Sci. Soc. 33: 83–86.

Evett, I.W., 1995. Avoiding the transposed conditional. Science & Justice 35: 127–131.

Evett, I.W. & J.S. Buckleton, 1989. Some aspects of the Bayesian approach to evidence evaluation. J. Forensic Sci. Soc. 29: 317–324.

Evett, I.W. & J.S. Buckleton, 1990. The interpretation of glass evidence. A practical approach. J. Forensic Sci. Soc. 30: 215–223.

Evett, I.W., J.S. Buckleton, A. Raymond & H. Roberts, 1993. The evidential value of DNA profiles. J. For. Sci. Soc. 33: 243–244.

Evett, I.W., C. Buffery, G. Willot & D. Stoney, 1991. A guide to interpreting single locus profiles of DNA mixtures in forensic cases. J. Forensic Sci. Soc. 31: 41–47.

Evett, I.W., P.E. Cage & C.G.G. Aitken, 1987. Evaluation of the Likelihood Ratio for Fibre Transfer Evidence in Criminal Evidence. Applied Statistics 36: 174–180.

Evett, I.W. & P. Gill, 1991. A discusion of the robustness of methods for assessing the evidential value of DNA single locus profiles in crime investigations. Electrophoresis 12: 226–230.

Evett, I.W., P.D. Gill, J.K. Scranage & B.S. Weir, 1995. Establishing the robustness of STR statistics for forensic applications. (submitted).

Evett, I.W. & R. Pinchin, 1992. DNA single locus profiles: tests for the robustness of statistical procedures within the context of forensic science. Int. J. Leg. Med. 104: 267–272.

Evett, I.W., R. Pinchin & C. Buffery, 1992. An investigation of the feasibility of infering ethnic origin from DNA profiles. J. For. Sci. Soc. 32: 301–306.

Evett. I.W., J.K. Scranage & R. Pinchin, 1992. An efficient statistical procedure for interpreting DNA single locus profiling data in crime cases. J. For. Sci. Soc. 32: 307–326.

Evett, I.W., J. Scranage & R. Pinchin, 1993. An illustration of the advantages of efficient statistical methods for RFLP analysis in forensic science. Am. J. Hum. Genet. 52: 498–505.

Evett, I.W. & B.S. Weir, 1991. Flawed reasoning in court. Chance 4: 19–21.

Evett, I.W. & D.J. Werrett, 1990. Bayesian analysis of single locus DNA profiles. Int. Symp. Human Ident. Promega Corporation.

Evett, I.W., D.J. Werrett & J.S. Buckleton, 1989. Paternity calculations from DNA multilocus profiles. J. Forensic Sci. Soc. 29: 249–254.

Evett, I.W., D.J. Werrett, P. Gill & J.S. Buckleton, 1989. DNA Fingerprinting on Trial. Nature 340: 435.

Evett, I.W., D.J. Werrett & A.F.M. Smith, 1989. Probabilistic analysis of DNA profiles. J. Forensic Sci. Soc. 29: 191–196.

Ewens, W.J., 1979. Mathematical Population Genet. Springer- Verlag, Berlin.

Fairley, W.B., 1973. Probabilistic analysis of identification evidence. J. Legal Studies 2: 493–513.

Fairley, W.B., 1983. Statistics in law. In Kotz and Johnson, Encyclopaedia of Statistical Sciences. Wiley, New York.

Fairley, W.B. & F. Mosteller, 1974. A conversation about Collins. Univ. Chicago Law Review 41: 242–253.

Farr, C.J. & P.N. Goodfellow, 1991. DNA fingerprinting: New variations on the theme. Nature 354: 184.

Farley, M.A. & J.J. Harrington (Editors), 1991. Forensic DNA Technology. Lewis, chelsea, Michigan.

Federal Bureau of Investigation, 1993. VNTR Population Data: A Worldwide Study. Volumes I–IV. Forensic Science Research and Training Center, Quantico, VA.

Feinberg, S.E., 1979. The use of chi-square statistics for categorical data problems. J. Roy. Stat. Soc. B41: 54–64.

Feinberg, S.E. & M.J. Schervish, 1986. The relevance of Bayesian inference for the presentation of evidence and for legal decision making. Boston Univ. Lae Rev. 66: 771–789.

Felsenstein, J., 1989. PHYLiP – Phylogeny Inference Package (Version 3.2). Cladistics 5: 164–166.

Ferguson, T.S., 1967. Mathematical Statistics: A Decision Theoretic Approach. Academic Press, New York.

Findlay, D.R., 1993. DNA profiling and criminal law – a merger or a takeover. J. Forensic Sci. Soc. 33: 234–237.

Finkelstein, M.O. & W.B. Fairley, 1970. A Bayesian approach to identification evidence. Harvard Law Review 83: 489–517.

Finkelstein, M.O. & W.B. Fairley, 1971. A comment on 'Trial by Mathematics'. Harvard Law Review 84: 1801–1809.

Finney, D.J., 1977. Probabilities based on circumstantial evidence. J. Am. Stat. Assoc. 72: 316–319.

Fisher, R.A., 1924. The conditions under which χ^2 measures the discrepancy between observation and hypothesis. J. Roy. Stat. Soc. 87: 442–450.

Fisher, R.A., 1935. The logic of inductive inference. J. Roy. Stat. Soc. 98: 39–54.

Fisher, R.A., 1951. Standard calculation for evaluating a blood-group system. Heredity 5: 95–102.

Flint, J., A.J. Boyce, J.J. Martinson & J.B. Clegg, 1989. Population bottlenecks in Polynesia revealed by minisatellites. Hum. Genet. 83: 257–263.

Fowler, C., 1991. DNA Profiling and Forensic Science. Today's Life Science. 56–60.

Fowler, S.J., P. Gill, D.J. Werrett & D.R. Higgs, 1988. Individual specific DNA fingerprints from a hypervariable region probe: Alpha-globin 3′ HVR. Hum. Genet. 79: 142.

Franklin-Barbajosa, C., 1992. DNA typing: The new science of identity. National Geographic 181: 112–124.

Fu, Y.X. & J. Arnold, 1992. A table of exact sample sizes for use with Fisher's exact test for 2×2 tables. Biometrics 48: 1103–1112.

Freckleton, I., 1989. DNA Profiling: optimism and realism. Law Institute J. 63: 360–363.

Gaensslen, R.E., S.C. Bell & H.C. Lee, 1987. Distributions of genetic markers in United States populations: I. Blood group and secretor systems. J. Forensic Sci. 32: 1016–1058.

Gaensslen, R.E., S.C. Bell & H.C. Lee, 1987. Distributions of genetic markers in United States populations: II. Isoenzyme systems. J. Forensic Sci. 32: 1348–1381.

Gaensslen, R.E., S.C. Bell & H.C. Lee, 1987. Distributions of genetic markers in United States populations. III. Serum group systems and hemoglobin variants. J. Forensic Sci. 32.

Gart, J.J. & J. Nam, 1984. A score test for the possible presence of recessive alleles in generalized ABO-like genetic systems. Biometrics 40: 887–894.

Gart, J.J. & J. Nam, 1988. The equivalence of two tests and models for HLA data with no observed double blanks. Biometrics 44: 869–873.

Gasparini, P., P. Mandich, G. Novelli, E. Bellone, F. Sangiulo, F. DeStefano, L. Potenza, E. Trabetti, M. Marigo, P.F. Pignatti, B. Dallapiccola & F. Ajmar, 1991. Forensic applications of molecular genetic analysis: an Italian collaborative study on paternity testing by the determination of variable number of tandem repeat DNA polymorphisms. Hum. Hered. 41: 174.

Gettinby, G., 1984. An empirical approach to estimating the probability of innocently acquiring bloodstains of different ABO groups on clothing. J. Forensic Sci. Soc. 24: 221–227.

Gaudette, B.S., 1986. Evaluation of associative physical evidence. J. Forensic Sci. Soc. 26: 117–118.

Gaudette, B.S., 1986. Evaluation of associative physical evidence. J. Forensic Sci. Soc. 26: 117–118.

Geisser, S., 1990. Some remarks on DNA fingerprinting. Chance 3: 8.

Geisser, S., 1992. Some statistical issues in medicine and forensics. J. Am. Stat. Assoc. 87: 607–614.

Geisser, S. & W. Johnson, 1992. Testing Hardy-Weinberg equilibrium on allellic data from CNTR loci. Am. J. Hum. Genet. 51: 1084–1088.

Geisser, S. & W. Johnson, 1993. Testing independence of fragment lengths within VNTR loci. Am. J. Hum. Genet. 53: 1103–1106 (see also: Weir, B.S., 1993. Independence tests for VNTR alleles defined as quantile bins. Am. J. Hum. Genet. 53: 1107–1113).

Gettinby, G., M. Peterson & N. Watson, 1993. Statistical interpretation of DNA evidence. J. Forensic Sci. Soc. 33: 212–217.

Gibbons, J.D. & J.W. Pratt, 1975. P-values: interpretation and methodology. American Statistician 29: 20–25.

Gill, P. & I.W. Evett, 1995. Population genetics of short tandem repeat (STR) loci. Genetica 96: 69–87.

Gill, P., I.W. Evett & S. Woodroffe, 1991. Databases, quality control and interpretation of DNA profiling in the Home Office Forensic Science Service. Electrophoresis 12: 204.

Gill, P., P.L. Ivanov, C. Kimpton, R. Piercy, N. Benson, G. Tulley, I. Evett, E. Hagelberg & K. Sullivan, 1994. Identification of the remains of the Romanov family by DNA analysis. Nature Genet. 6: 130–135.

Gill, P., J.E. Lygo, S.J. Fowler & D.J. Werrett, 1987. An evaluation of DNA fingerprinting for forensic purposes. Electrophoresis 8: 38–44.

Gill, P., K. Sullivan & D.J. Werrett, 1990. The analysis of hypervariable DNA profiles: problems associated with the objective determination of the probability of a match. Human Genet. 85: 75–79.

Gill, P. & D.J. Werrett, 1990. Interpretation of DNA profiles using a computerized database. Electrophoresis 11: 444–448.

Gill, P., S. Woodroffe, J.E. Lygo & E.S. Millican, 1991. Population genetics of four hypervariable loci. Int. J. Legal Med. 104: 221–227.

Gjertson, D.W., J. Hopfield, P.A. Lachembruch, M.R. Mickey, T. Sublet, C. Yuge & P.I. Terasaki, 1990. Measurement error in determination of band size for highly polymorphic single-locus DNA markers, pp. 3–11 in Advances in Forensic Haemogenetics 3, edited by H.F. Polesky & W.R. Mayr. Springer-Verlag, Heidelberg.

Gjertson, D.W., M.R. Mickey, J. Hopfield, T. Takenouchi & P.I. Terasaki, 1988. Calculation of probability of paternity using DNA sequences. Am. J. Human Genet. 43: 860–869.

Gjertson, D.W. & J.W. Morris, 1995. Assessing probability of paternity and the product rule in DNA systems. Genetica 96: 89–98.

Goldgar, D. & E.A. Thompson, 1988. Bayesian interval estimation of genetic relationships: application to paternity testing. Am. J. Hum. Genet. 42: 135–142.

Good, I.J., 1965. The Estimation of Probabilities: An Essay on Modern Bayesian Methods. MIT Press, Cambridge, Massachusetts.

Good, I.J., 1991. Weight of evidence and the Bayesian likelihood ratio, pp. 85–106 in C.G.G. Aitken & D.A. Stoney (eds.). The Use of Statistics in Forensic Science. Ellis Horwood, Chichester, UK.

Good, P., 1994. Permutation Tests: A Practical Guide to Resampling for Testing Hypotheses: Springer-Verlag, New York.

Green, P., 1992. Population genetic issues in DNA fingerprinting. Am. J. Hum. Genet. 50: 440–441. (See also: Budowle, B., 1992. Reply to Green. Am. J. Hum. Genet. 50: 441).

Green, P. & E.S. Lander, 1991. Forensic DNA tests and Hardy- Weinberg equilibrium. Science 253: 1038–1039.

Greenland, S., 1991. On the logical justification of conditional tests for two-by-two contingency tables. American Statistician 45: 248–251.

Gregg, P. & E.S. Lander, 1991. Forensic DNA tests and Hardy- Weinberg equilibrium. Science 253: 1038–1039.

Grove, D.M., 1980. The interpretation of forensic evidence using a likelihood ratio. Biometrika 67: 243–246.

Grubb, A., 1993. Legal aspects of DNA profiling. J. Forensic Sci. Soc. 33: 228–233.

Grubb, A. & D.S. Pearl, 1990. Blood Testing, AIDS and DNA Profiling; Law and Policy. Jordan & Sons Ltd., Bristol.

Grunbaum, B.W., S. Selvin, B.A. Myrhe & N. Pace, 1980. Distribution of gene frequencies and discrimination probabilities for 22 human blood genetic systems in four racial groups. J. Forensic Sci. 25: 428–444.

Gunn, P.R., 1989. Identity testing by DNA profiling. Aust. J. For. Sci. 22: 27–36.

Guo, S.-W. & E.A. Thompson, 1992. Performing the exact test of Hardy-Weinberg proportion for multiple alleles. Biometrics 48: 361–372.

Gurtler, H., 1956. Principles of blood group statistical evaluation of paternity cases at the University Institute of Forensic Medicine Copenhagen. Acta Med. Leg. Soc. (Liege) 9: 83– 94.

Haber, M., 1981. Exact significance levels of goodness-of-fit tests for the Hardy-Weinberg equilibrium. Human Heredity 31: 161–166.

Hagelberg, E., I.C. Gray & A.J. Jeffreys, 1991. Identification of the skeletal remains of a murder victim by DNA analysis. Nature 352: 427–429.

Hagerman, P.J., 1990. DNA typing in the forensic arena. Am. J. Hum. Genet. 47: 876–877.

Haldane, J.B.S., 1964. A defense of beanbag genetics. Perspect.

Hamilton, J.F., 1994. Multi-locus and single-locus DNA profiling in New Zealand. Ph.D. Thesis, Victoria University of Wellington, New Zealand.

Hammond, V.A. & D.D. Garner, 1991. Scientific and legal aspects of DNA typing. Disputed Paternity Proceedings (1991).

Harding, R.M., 1992. VNTRs in review. Evol. Anthrop. 1: 62–71.

Harding, R.M., A.J. Boyce & J.B. Clegg, 1992. The evolution of tandemly repetitive DNA: recombination rules. Genetics 132: 847– 859.

Harding, R.M., A.J. Boyce, J.J. Martinson, J. Flint & J.B. Cleg, 1993. A computer simulation model of VNTR population genetics: constrained recombination rules out the infinite alleles model. Genetics 135: 911–922.

Hardy, G.H., 1908. Mendelian proportions in a mixed population. J. Genet. 58: 237–242.

Harmon, R.P., 1993. DNA evidence. Science 261: 13.

Hart, S.D., C. Webster & R. Menzies, 1993. A note on portraying the accuracy of violence predictions. Law & Human Behav. 17: 695–700.

Hartl, D.L. & R.C. Lewontin, 1993. Response to Devlin et al. Science 260: 473–474.

Hartl, D.E. & R.C. Lewontin, 1994. DNA fingerprinting. Science 226: 201. (See Also: Koshland, D.E., 1994. Response. Science 266: 202–203.)

Hartman, J., R. Keister, B. Houlihan, L. Thompson, R. Baldwin, E. Buse, B. Driver & M. Kuo. 1995. Diversity of ethnic and racial VNTR RFLP fixed-bin frequency distributions. Am. J. Hum. Genet. 55: 1268–1278.

Haskey, J., 1991. The ethnic minority populations resident in private households – estimates by county and metropolitan district of England and Wales. Population Trends 63, HMSO: 22– 35.

Helminen, P., C. Ehnholm, M.-L. Lokki & L. Peltonen, 1988. Application of DNA 'fingerprints' to paternity determinations. The Lancet, Mar. 12, 1988: 574–576.

Helminen, P., V. Johnsson, C. Ehnholm & L. Peltonen, 1991. Proving paternity of children with deceased fathers. Human Genet. 87: 657–660.

Helminen, P., A. Sajantila, V. Johnsson, M. Lukka, C. Ehnholm & L. Peltonen, 1992. Amplification of three hypervariable regions by polymerase chain reaction in paternity determinations: comparison with conventional methods and DNA fingerprinting. Mol. Cell Probes 6: 21–26.

Helmuth, R., N. Fildes, E. Blake, M.C. Luce, J. Chimera, R. Madej, C. Gorodezky, M. Stoneking, N. Schmill, W. Klitz & H.A. Erlich, 1990. L HLA DQα allele and genotype frequencies in various human populations, determined by using enzymatic amplification and oligonucleotide probes. Am. J. Hum. Genet. 47: 515–523.

Henke, L., S. Cleef, M. Zakrzewska & J. Henke, 1990. BamHI polymorphism of locus D2S44 in a West German population as revealed by VNTR probe YNH24. Int. J. Legal Med. 104: 33.

Hernández, J.L. & B.S. Weir, 1989. A disequilibrium coefficient approach to Hardy-Weinberg testing. Biometrics 45: 53–70.

Herrera & Tracey, 1994. DNA Fingerprinting. Jones & Bartlett, New York.

Herrin, G. Jr., 1992. A comparison of models used for calculation of RFLP pattern frequencies. J. Forensic Sci. 37: 1640–1651.

Herrin, G. Jr., 1993. Probability of matching RFLP patterns from unrelated individuals. Am. J. Hum. Genet. 52: 491–497.

Hicks, J., 1992. FBI's case for genetics. Nature 357: 355.

Higuchi, R., 1991. Human error in forensic DNA typing. Am. J. Human Genet. 48: 1215–1216.

Hill, A.V.S., D.F. O'Shaughnessy & J.B. Cleg, 1988. The colonization of the Pacific: some current hypotheses, pp. 246– 285, in The Colonization of the Pacific: A Genetic Trail, edited by A.V.S. Hill & S.W. Serjeantson. Clarendon Press, Oxford.

Hill, W.G., 1986. DNA fingerprint analysis in immigration test- cases. Nature 322: 290. (See also: Jeffreys, A.J., J.F.Y. Brookfield & R. Semeonoff, 1986. Reply to Hill. Nature 322: 290–291).

Hille, E., 1959. Analytic Function Theory Vol. 1. Blaisdell Ginn, New York.

Hochmeister, M.N., B. Budowle, U.V. Borer & R. Dirnhofer, 1994. Swiss population data on the loci HLA-DQα, LDLR, GYPA, HBGG, D7S8, Gc and D1S80. Forensic Sci. Int. (In press).

Honma, M. & I. Ishiyama, 1990. Application of DNA fingerprinting to parentage and extended family relationship testing. Hum. Heredity 40: 356–362.

Horn, G.T., B. Richards, J.J. Merrill & K.W. Klinger, 1990. Characterization and rapid diagnostic analysis of DNA polymorphisms closely linked to the cystic fibrosis locus. Clin. Chem. 36: 1614–1619.

Howlett, R., 1989. DNA forensics and the FBI. Nature 341: 182– 183.

Huang, N.E. & B. Budowle, 1994. Chinese population data on the PCR-based loci HLA-DQα, LDLR, HBGG, D7S8, and Gc. Human Heredity (In press).

Hudson, R.R., 1990. Gene genealogies and the coalescent process. In D. Futuyma & J. Antonovics (Eds.) Oxford Surveys in Evolutionary Biology, pp. 1–44.

Hummel, L., O. Kundinger & A. Karl, 1981. The realistic prior probability from blood group findings for cases involving one or more men. Part II. Determining the realistic prior probability in one-man cases (forensic cases) in Frieberg, Muncih, East Berlin, Austria, Switzerland, Denmark, and Sweden, pp. 81–87 in Biomathematical Evidence of Paternity, edited by K. Hummel & J. Gerchow. Springer-Verlag, Berlin.

Ihm, P., 1961. Die mathematischen Grundlagen, vor allem für die statistiche Auswertung des serologischen und antropologischen Gutachtens, pp. 128–145 in Die Medizinische Vaterschaftsbegutachtung mit biostatistichem Beweis, edited by K. Hummel. Fischer, Stuttgart.

Ihm, P., 1981. The problem of paternity in the light of decision theory, pp. 53–68 in Biomathematical Evidence of Paternity, edited by K. Hummel & J. Gerchow. Springer-Verlag, Berlin.

Immanishi, T., T. Akaza, A. Kimura, K. Tokunaga & T. Gojobori, 1992. Allele and haplotype frequencies for HLA and complement loci in various ethnic groups. In: HLA 1991, Vol. I. pp. 1065– 1220. ed. by K. Tsuji, M. Aizawa & T. Sasazuki. Oxford Univ. Press, Oxford.

Ivanov, P.L., L.V. Verbovaya & S.V. Gurtovaya, 1991. Use of DNA printing for diagnosis of monozygotic twins. Sud.-Med. Ekspert. 34: 32.

Jackson, D.B., 1989. DNA fingerprinting and proof of paternity. Fam. Law Rep. 15: 3007–3013.

Jarjoura, D., J. Jamison & S. Androulakakis, 1994. Likelihood ratios for deoxyribonucleic acid (DNA) typing in criminal cases. J. Forensic Sci. 39: 64–73.

Jarjoura, D., J. Jamison & S. Anroulakakis, 1994. Likelihood ratios for DNA typing in criminal cases. J. Forensic Sci. 39: 64–73.

Jeffreys, A.J., 1993. DNA typing: approaches and applications. J. Forensic Sci. Soc. 33: 204–211.

Jeffreys, A.J., J.F.Y. Brookfield & R. Semeonoff, 1985. Positive identification of an immigrant test-case using human DNA fingerprints. Nature 317: 818–819. (See also: Hill, W.G., 1986. DNA fingerprint analysis in immigration test-cases. Nature 322: 290; Jeffreys, A.J., J.F.Y. Brookfield & R. Semeonoff, 1986. Reply to Hill. Nature 322: 290–291).

Jeffreys, A.J., N.J. Royle, V. Wilson & Z. Wong, 1988. Spontaneous mutation rates to new length alleles at tandem- repetitive hypervariable loci in human DNA. Nature 332: 278–281.

Jeffreys, A.J., M. Turner & P.G. Debenham, 1991. The efficiency of multilocus DNA fingerprint probes for individualization and establishment of family relationships, determined from extensive casework. Am. J. Hum. Genet. 48: 824– 840.

Jeffreys, A.J., V. Wilson, S.L. Thein, D.J. Weatherall & B.A.J. Ponder, 1986. DNA 'fingerprints' and segregation analysis of multiple markers in human pedigrees. Am. J. Hum. Genet. 39: 11–24.

Jin, L. & R. Chakraborty, 1995. Population structure, stepwise mutations, heterozygote dificiency and their implications in DNA forensics. Heredity 74: 274–285.

Jones, D.A., 1972. Blood samples: probability of discrimination. J. Forensic Sci. Soc. 12: 355–359.

Joyce, C., 1990. High profile: DNA in court again. New Scientist 127: 10–11.

Kahn, R., 1991. An introduction to DNA structure and genome organization, pp. 25–38 in M.A. Farley & J.J. Harrington, 1991, Forensic DNA Technology. Lewis, Chelsea, MI.

Karlin, S., E.C. Cameron & P.T. Williams, 1981. Sibling and parent-offspring correlation estimation with variable family size. Proc. Natl. Acad. Sci. USA 78: 2664–2668.

Kaye, D., Statistical evidence of discrimination (and comments). J. Am. Stat. Assoc. 77: 773–783.

Kaye, D.H., 1989. The probability of an ultimate issue; the strange cases of paternity testing. Iowa Law Review 1: 75–109.

Kaye, D.H., 1990. DNA paternity probabilities. Fam. Law Q. 24: 279–304.

Kaye, D.H., 1993. DNA evidence: probability, population genetics, and the courts. Harvard J. Law Tech. 7: 101–172.

Kaye, D.H., 1995. The forensic debut of the NRCs DNA report: population structure, ceiling frequencies and the need for numbers. Genetica 96: 99–105.

Kaye, D.H. & R. Kanwischer, 1988. Admissibility of genetic testing in paternity litigation: a survey of state statutes. Fam. Law Q. 22: 109–116.

Kidd, R.J., F.L. Black, K.M. Weiss, I. Balazs & K.K. Kidd, 1991. Studies of three Amerindian populations using nuclear DNA polymorphisms. Human Biology 63: 775.

King, M.-C., 1989. Genetic testing of identity and relationship. Am. J. Hum. Genet. 44: 179–181.

Kingman, J.F.C., 1993. Poisson Processes. Oxford University Press, Oxford.

Kingston, C.R., 1965. Applications of probability theory in criminalistics. J. Am. Stat. Assoc. 60: 70–80.

Kingston, C.R., 1965. Applications of probability theory in criminalistics – II. J. Am. Stat. Assoc. 60: 1028–1034.

Kingston, C., 1989. A perspective on probability and physical evidence. J. Forensic Sci. 34: 1336.

Kirby, L.T., 1990. DNA Fingerprinting: An Introduction. Stockton Press, New York.

Kelly, K.F., J.J. Rankin & R.C. Wink, 1987. Method and applications of DNA fingerprinting: A guide for the non- scientist. Crim. Law Rev. 105–110.

Kobilinsky, L. & L. Levine, 1988. Recent application of DNA analysis to issues of paternity. J. Forensic Sci. 33: 1107–1108.

Kobilinsky, L. & L. Levine, 1989. Discussion of 'Recent application of DNA analysis to issues of paternity'. J. Forensic Sci. 35: 5–6.

Koehler, J.J., 1992. Probabilities in the courtroom: an evaluation of objections and policies, pp. 167–183 in D.K. Kagehior & W.S. Laufer (eds.), Handbook of Psychology and Law. Springer-Verlag, New York.

Koehler, J.J., 1993. DNA matches and statistics: Important questions, surprising answers. Judicature 76: 222–229.

Koehler, J.J., 1993. Error and exaggeration in the presentation of DNA evidence at trial. Jrimetrics 34: 21–35.

Koehler, K. & K. Larntz, 1980. An empirical investigation of goodness-of-fit statistics for sparse multinomials. J. Am. Stat. Assoc. 75: 336–344.

Koehler, K.J., 1986. Goodness of fit tests for log linear models in sparse contingency tables. J. Am. Stat. Assoc. 81: 483–493.

Korey, D.J., G.A. Bishop & G.W. McCaughan, 1993. Allele non- amplification: a source of confusion in linkage studies employing microsatellite polymorphisms. Hum. Mol. Genet. 2: 289–291.

Koshland, D.E., 1992. DNA fingerprinting and eyewitness testimony. Science 256: 593.

Koshland, D.E., 1994. The DNA fingerprint story (continued). Science 265: 1015.

Koshland, D.E., 1994. Response. Science 266: 202–203.

Krane, D.E., R.W. Allen, S.A. Sawyer, D.A. Petrov & D.L. Hartl, 1992. Genetic differences at four DNA typing loci in Finnish, Italian and mixed Caucasian populations. Proc. Natl. Acad. Sci. USA 89: 10583–10587. (See also: Budowle, B., K.L. Monson & A.M. Giusti, 1994. A reassessment of frequency estimates of Pvu II-generated VNTR profiles in a Finnish, and Italian, and a general United States Caucasian database: No evidence for ethnic subgroups affecting forensic estimates. Am. J. Hum. Genet. 55: 533–539).

Krawczak, M. & B. Bockel, 1992. A genetic factor model for the statistical analysis of multilocus DNA fingerprints. Electrophoresis 13: 10.

Krawczak, M. & J. Schmidtke, 1992. The decision theory of paternity disputes: optimization considerations applied to multilocus DNA fingerprinting. J. Forensic Sci. 37: 1525–1533.

Kullback, S., 1959. Information Theory and Statistics. Wiley, New York.

Kullback, S., 1985. Minimum discrimination information (MDI) estimation. In Encyclopedia of Statistical Sciences 5: 527–529, Ed. S. Kotz & N.L. Johnson. New York, Wiley.

Kuo, M.C., 1982. Linking a bloodstain to a missing person by genetic inheritance. J. Forensic Sci. 27: 438–444.

Kurosaki, K., T. Matsushita & S. Ueda, 1993. Individual DNA identification from ancient DNA remains. Am. J. Hum. Genet. 53: 638–643.

Lander, E.S., 1989. DNA fingerprinting on trial. Nature 339: 501–505. (See also: Evett, I.W., D.J. Werrett, P. Gill & J.S. Buckleton, 1989. DNA fingerprinting on trial. Nature 340: 435; Taylor, G., 1989. DNA fingerprinting. Nature 240: 672).

Lander, E.S., 1989. Population genetic considerations in the forensic use of DNA typing, pp. 143–156 in Banbury Report 32: DNA Technology and Forensic Science. Cold Spring Harbor Laboratory Press, Cold Spring Harbor, NY.

Lander, E.S., 1991. Research on DNA typing catching up with courtroom application. Am. J. Hum. Genet. 48: 819–823. (See also: Harmon, R.P., 1991. Please leave law to the lawyers. Am. J. Hum. Genet. 49: 891; Wooley, J.R., 1991. A response to Lander: The Courtroom Perspective. Am. J. Hum. Genet. 49: 892. Caskey, C.T., 1991; Comments on DNA-Based forensic analysis. Am. J. Hum. Genet. 49: 893. Chakraborty, R., 1991. Statistical interpretation of DNA typing data. Am. J. Hum. Genet. 49: 895. Daiger, S.P., 1991. DNA fingerprinting. Am. J. Hum. Genet. 49: 897; Cotton, R.W., L. Forman & C.J. Word, 1991. Research on DNA typing validated in the literature. Am. J. Hum. Genet. 49: 898. Lander, E.S., 1991. Lander Reply. Am. J. Hum. Genet. 49: 899).

Lander, E.S., 1993. DNA fingerprinting: the NRC report. Science 260: 1221.

Lange, K., 1987. Am. J. Hum. Genet. 39: 148–150.

Lange, K., 1991. Comment on 'Inferences using DNA profiling in forensic identification and paternity cases' by D.A. Berry. Stat. Sci. 6: 190–192.

Lange, K., 1993. Match probabilities in racially admixed populations. Am. J. Hum. Genet. 52: 305–311.

Lange, K., 1995. Applications of the Dirichlet distribution to forensic match probabilities. Genetica 96: 107–117.

Larntz, K., 1978. Small sample comparisons of exact levels for chi-squared goodness-of-fit statistics. J. Am. Stat. Assoc. 73: 253–263.

Lee, H.C. & R.E. Gaensslen, 1990. DNA and other polymorphisms in forensic science. Year Book Medical Publishers Inc., Chicago.

Lee, H.C., C. Ladd & F. Timady, 1994. DNA typing in forensic science. Am. J. Forensic Med. 15: 269.

Lee, P.M., 1989. Bayesian Statistics: An Introduction. Edward Arnold, London.

Lehrman, S., 1994. US may regulate DNA testing laboratories. Science 370: 588.

Leimar, O. & N. Ryman, 1984. The distribution of the paternity index as a basis for evaluation of sequential testing in paternity analysis. Hum. Hered. 34: 46–58.

Lempert, R., 1991. Some caveats concerning DNAs as criminal identification evidence: with thanks to the Reverend Bayes. Cardozo Law Rev. 13: 303–341.

Lempert, R., 1993. DNA, science and the law: two cheers for the ceiling principle. Jurimetrics 34: 41–57.

Lempert, R., 1993. The suspect population and DNA identification. Jurimetrics 34: 1–7.

Lempert, R., 1995. The honest scientist's guide to DNA evidence. Genetica 96: 119–124.

Lenth, R.V., 1986. On identification by probability. J. Forensic Sci. Soc. 26: 197–213.

Letuth, R.V., 1986. On identification by probability. J. Forensic Sci. Soc. 26: 197–213.

Levene, H., 1949. On a matching problem arising in genetics. Ann. Math. Stat. 20: 91–94.

Lewin, R., DNA typing on the witness stand. Science 244: 1033– 1035.

Lewontin, R.C., 1972. The apportionment of human diversity. Evol. Biol. 6: 381–398.

Lewontin, R.C., 1982. Human variety, pp. 1–13 in Human Diversity. Scientific American Books, W.H. Freeman and Company, San Francisco, CA.

Lewontin, R.C., 1993. Which Population? Am. J. Human Genet. 52: 205 (See also: Weir, B.S. & I.W. Evett, 1992. Whose DNA? Am. J. Human Genet. 50: 869; Weir, B.S. & I.W. Evett, 1993. Reply to Lewontin. Am. J. Human Genet. 52: 206).

Lewontin, R.C., 1994. The use of DNA profiles in forensic contexts. Stat. Sci. 9: 59.

Lewontin, R.C. & C.C. Cockerham, 1959. The goodness-of-fit test for detecting natural selection in random mating populations. Evolution 13: 561–564.

Lewontin, R.C. & D.L. Hartl, 1991. Population genetics in forensic DNA typing. Science 254: 1745–1750. (See also: Chakraborty, R. & K.K. Kidd, 1991. The utility of DNA typing in forensic work. Science 254: 1050, Bever, R.A., R.W. Staub, C.M. Kelly & R.S. Foster, 1992. Science 255: 1051 (February, 1992); Yarborough, L.R., 1992. Science 255: 1052; Cleveland, D.W., 1992. Science 255: 1052; Koshland, D.E. Jr., 1992. Science 255: 1052; Chakraborty, R. & Kidd, 1992. Science 255: 1053; Lewontin, R.C. & D.L. Hartl, 1992. Science 255: 1054; Also see: Budowle, B. & J. Staford, 1991. Response to expert Report by D.L. Hartl, submitted in the case of United States v. Yee. Crime Laboratory Digest 18: 101. Budowle, B. & J. Stafford, J., 1991. Response to 'Population genetic problems in the forensic use of DNA profiles' by R.C. Lewontin, submitted in the case of United States v. Yee. Crime Laboratory Digest 18: 109).

Lewontin, R.C. & D.L. Hartl, 1992. Response. Science 25: 1054–1055.

Li, C.C. & A. Chakravarti, 1985. Basic fallacies in the formulation of the paternity index. Am. J. Hum. Genet. 37: 809– 818. (See also: Valentin, J., 1985. Am. J. Hum. Genet. 38: 582– 585; Li, C.C. & A. Chakravarti, 1985. Am. J. Hum. Genet. 38: 586–589).

Li, C.C. & A. Chakravarti, 1988. An expository review of two methods of calculating the paternity probability. Am. J. Hum. Genet. 43: 197–205.

Li, C.C. & A. Chakravarti, 1994. DNA profile similarity in a subdivided population. Human Heredity 44: 100–109.

Lindley, D.V., 1977. A problem in forensic science. Biometrika 64: 207–213.

Lindley, D.V., 1990. The present position in Bayesian statistics. Statist. Sci. 5: 44–89.

Little, R.J.A., 1989. Testing the equality of two independent binomial proportions. American Statistician 43: 283–288.

Louis, E.J. & E.R. Dempster, 1987. An exact test for Hardy- Weinberg equilibrium and multiple alleles. Biometrics 43: 805–811.

Lovell, W.S., 1994. DNA fingerprinting. Science 266: 201–202. (See also: Koshland, D.E., 1994. Response. Science 266: 202–203).

Ludes, B.P., P.D. Mangin, D.J. Malicer, A.N. Chalumeau, A.J. Chaumont, 1991. Parentage determination on aborted fetal material through deoxyribonucleic acid (DNA) profiling. J. Forensic Sci. 36: 1219.

Lynch, E., 1993. Ruling in case State of Minnesota v. Robert Joseph Guevara. State of Minnesota District Court, First Judicial District, Court File K9-92-1873.

Lynch, M., 1988. Estimation of relatedness by DNA fingerprinting. Mol. Biol. Evol. 5: 584–599.

Lynch, M., 1991. Analysis of population genetic structure by DNA fingerprinting, pp. 113–126 in T. Burke, G. Dolf, A.J. Jeffreys & R. Wolff (eds.). DNA Fingerprinting: Approaches and Applications. Birkhauser, Basel.

Magnusson, E., 1993. Incomprehension and miscomprehension of statistical evidence: an experimental study. Australian Institute of Criminology Conference on Law, Medicine and Criminal Justice. Queensland, July 1993.

Maha, G.C., J.M. Mason, G.M. Stuhmiller & U. Heine, 1995. Reply to Pena. Am. J. Hum. Genet. 56: 1506–1507.

Maiste, P.J., 1993. Comparison of Statistical Tests for Independence at Genetic Loci with Many Alleles. Ph.D. Thesis, North Carolina State University, Raleigh, NC.

Maiste, P.J. & B.S. Weir, 1995. A comparison of Tests for Independence in the FBI RFLP Databases. Genetica 96: 125–138.

Majumder, P.P., 1984. On the positive identification of paternity. Hum. Hered. 34: 261–262.

Majumder, P.P. & M. Nei, 1983. A note on the positive identification of paternity by using genetic markers. Hum. Hered. 33: 29–35.

Malécot, G., 1948. Les Mathématiques de l'hérédité. Masson, Paris.

Manly, B.F.J., 1991. Randomization and Monte Carlo Methods in Biology. Chapman and Hall, New York.

Markowicz, K.R., L.A. Tonelli, M.B. Anderson, D.J. Green, G.L. Herrin, R.W. Cotton, J.L. Gottschall & D.D. Garner, 1990. Use of deoxyribonucleic acid (DNA) fingerprints for identity determination: comparison with traditional paternity testing methods – Part II. J. For. Sci. 35: 1270. (See also: Tonelli, L.A., K.R. Markowicz, M.B. Anderson, D.J. Green, G.L. Herrin, R.W. Cotton, D.D. Dykes & D.D. Garner, 1990. Use of deoxyribonucleic acid (DNA) fingerprints for identity determination: comparison with traditional paternity testing methods – Part I. J. Forensic Sci. 35: 1265).

Martin, W., 1983. Consideration of 'silent genes' in the statistical evaluation of blood group findings in paternity testing, pp. 245–257 in Inclusion Probabilities in Parentage Testing, edited by R.H. Walker. American Association of Blood Banks, Arlinton, Virginia.

Martinson, J.J., R.M. Harding, G. Philippon, F. Flye Saint Marie, J. Roux, A.J. Boyce & J.B. Clegg, 1993. Demographic reductions and genetic bottlenecks in humans; minisatellite allele distribution in Oceania. Hum. Genet. 91: 445–450.

Marx, J.L., 1988. DNA fingerprinting takes the witness stand. Science 240: 1616–1618.

Mayr, W.R., 1991. Biostatistics in paternity testing using genetic systems with linkage disequilibrium. Am. J. Hum. Genet. 48: 636–637.

McCormick, C., 1992. McCormick on Evidence, Volume 1, 4th edition, edited by J. Srong. West Publishing Company, Minneapolis, Minnesota.

McEwen, J.E., 1995. Forensic DNA data banking by state crime laboratories. Am. J. Hum. Genet. 56: 1487–1492.

McEwen, J.E. & P.R. Reilly, 1995. A survey of DNA diagnostic laboratories regarding DNA banking. Am. J. Hum. Genet. 56: 1477–1486.

McQuillan, J. & K. Edgar, 1992. A survey of the distribution of glass on clothing. J. Forensic Sci. Soc. 32: 333–348.

Menoti-Raymond, M. & S.J. O'Brien, 1993. Dating the genetic bottleneck of the African cheetah. Proc. Natl. Acad. Sci. USA 90: 3172–3176.

Mickey, M.R., J. Tiwari, J. Bond, D. Gjertson & P.I. Terasaki, 1983. Paternity probability calculations for mixed races, pp. 325–347 in Inclusion Probabilities in Parentage Testing, edited by R.H. Walker. American Association of Blood Banks, Arlington, Virginia.

Mickey, M.R., D.W. Gjertson & P.I. Terasaki, 1986. Empirical validation of the Essen-Möller probability of paternity. Am. J. Hum. Genet. 39: 123–132.

Miller, K.S., 1987. Some Eclectic Matrix Theory. Robert E. Krieger Publishing, Malabar, Florida.

Mills, H., 1991. The Birmingham Six case – Vital scientific evidence kept from defence. The (London) Independent, 28 March 1991, p. 2.

Moenssens, A.A., 1990. DNA evidence and its critics — How valid are the challenges? Jurimetrics 31: 87–108.

Monson, K.L. & B. Budowle, 1993. A comparison of the fixed bin method with the floating bin and direct count methods: effect of VNTR profile frequency estimation and reference population. J. Forensic Sci. 38: 1037.

Moore, D.S., 1986. Tests of chi-squared type. In Goodness-of- fit techniques, (eds) R.B. D'Agostino & M.A. Stephens, pp. 63–95, New York, Marcel Dekker.

Morris, J.W., 1989. Experimental validation of paternity probability. Transfusion 29: 281.

Morris, J.W. & C.H. Brenner, 1988. Bloodstain classification errors revisited. J. Forensic Sci. Soc. 28: 49–53.

Morris, J.W., R.A. Garber, J. d'Autremont & C.H. Brenner, 1988. The avuncular index and the incest index, pp. 607–611 in Advances in Forensic Haemogenetics 1. Springer-Verlag, Berlin.

Morris, J.W. & D.W. Gjertson, 1993. The paternity index, population heterogeneity, and the product rule. Adv. Forensic Haemogenet. 5. Springer-Verlag, Berlin.

Morris, J.W. & D.W. Gjertson, 1994. Population genetics issues in disputed parentage, pp. 63–66 in Proceedings from the Fourth International Symposium on Human Identification. Promega Corporation, Madison, Wisconsin.

Morris, J.W., I. Sanda & J. Glassberg, 1989. Biostatistical evaluation of evidence from continuous allele frequency distribution deoxyribonucleic acid

(DNA) probes in reference to disputed paternity and identity. J. Forensic Sci. 34: 1311–1317.

Morton, N.E., (Editor), 1973. Genetic Structure of Populations. University Press, Honolulu, Hawaii.

Morton, N.E., 1992. Genetic structure of forensic populations. Proc. Natl. Acad. Sci. USA. 89: 2556–2560.

Morton, N.E., 1993. DNA in court. Eur. J. Hum. Genet. 1: 172–178.

Morton, N.E., 1993. Kinship bioassay on hypervariable loci in blacks and caucasians. Proc. Natl. Acad. Sci. USA 90: 1892–1896.

Morton, N.E., 1995. Alternative approaches to population structure. Genetica 96: 139–144.

Morton, N.E., 1995. Genetic structure of forensic populations. Am. J. Hum. Genet. 55: 587–588.

Morton, N.E., A. Collins & I. Balazs, 1993. Kinship bioassay on hypervariable loci in blacks and caucasians. Proc. Natl. Acad. Sci. USA 90: 1892–1896.

Morton, N.E. & B. Keats, 1976. Human microdifferentiation in the Western Pacific. In Kirk, R.L. & A.G. Thorne (eds.), The Origin of the Australians. Human Biology Series 6, Australian Institute of Aboriginal Studies, Canberra. Humanities Press, New Jersey.

Mosimann, J.E., 1962. On the compound multinomial distribution, the multivariate β-distribution, and correlations among proportions. Biometrika 49: 65–82.

Mott, R.F., T.B.L. Kirkwood & R.N. Curnow, 1989. A test for the statistical significance of DNA sequence similarities for application in data bank searches. Comp. App. Bosci. 5: 123–131.

Mott, R.F., T.B.L. Kirkwood & R.N. Curnow, 1990. Test for the statistical significance of protein sequence similarities in data bank searches. Protein Engineer 4: 149–154.

Mourant, A.E., 1954. The Distribution of the Human Blood Groups. Blackwell, Oxford.

Mourant, A.E., A.C. Kopec & K. Domaniewska-Sobczak, 1976. The Distribution of the Human Blood Groups and Other Polymorphisms. Oxford University Press. (See also: Tills, D., A. Kopec & R.E. Tills, 1983. The Distribution of the Human Blood Groups and Other Polymorphisms. Suppl. 1. Oxford University Press).

Mueller, L.D., 1991. Population genetics of hypervariable human DNA, pp. 51–62 in M.A. Farley & J.J. Harrington, 1991, Forensic DNA Technology. Lewis, Chelsea, MI.

Mueller, L.D., 1993. The use of DNA typing in forensic science. Accountability in Research 2: 1–13.

Nata, M., T. Yokoi, Y. Aoki, K. Sagisaka, K. Hiraiwa & T. Takatori, 1991. Paternity test with single locus DNA probes. Jpn. J. Leg. Med. 45: 138.

Nei, M., 1972. Genetic distance between populations. Am. Nat. 106: 283–292.

Nei, M., 1973. Analysis of gene diversity in subdivided populations. Proc. Natl. Acad. Sci. USA 70: 3321–3323.

Nei, M., 1977. F-statistics and analysis of gene diversity in subdivided populations. Ann. Hum. Genet. 41: 225–233.

Nei, M., 1984. On the positive identification of paternity. Hum. Hered. 34: 258–260.

Nelson, S.L., 1978. Nomograph for samples having zero defectives. J. Quality Technology 10: 42–43.

Neufield, P. & N. Colman, 1990. When science takes the witness stand. Sci. Am., May 1990: 46–53.

Neyman, J., 1949. Contribution to the theory of the χ^2 test. Proc. 1st Berkeley Symposium Math. Statist. Prob., pp. 239–273.

Neyman, J. & E.S. Pearson, 1928. On the use and interpretation of certain test criteria for purposes of statistical inference. Biometrika 20A: 175–263.

Nichols, R.A. & D.J. Balding, 1991. Effects of population structure on DNA fingerprint analysis in forensic science. Heredity 66: 297–302.

Nisbett, R.E. & L. Ross, 1980. Human inference: Strategies and shortcomings of social judgment. Prentice-Hall, Englewood Cliffs, New Jersey.

Norman, C., 1989. Maine case deals blow to DNA fingerprinting. Science 246: 1556–1558.

Nowak, R., 1994. Forensic DNA goes to court with O.J. Science 265: 1352–1354.

Odelberg, S.J., D.B. Demers, E.H. Westin & A.A. Hossaini, 1988. Establishing paternity using minisatellite DNA probes when the putative father is unavailable for testing. J. Forensic Sci. 33: 921.

Odelberg, S.J., R. Plaetke, J.R. Eldridge, L. Ballard, P. O'Connel, Y. Nakamura, M. Leppert, J.-M. Lalouel & R. White, 1989. Characterization of eight VNTR loci by agarose gel electrophoresis. Genomics 5: 915–924.

Oden, N.L., 1991. Allocation of effort in Monte Carlo simulation for power of permutation tests. J. Am. Stat. Assoc. 86: 1074–1076.

Owen, G.W. & K.W. Smalldon, 1975. Blood and semen stains on outer clothing and shoes not related to crime: Report of a survey using presumptive tests. J. Forensic Sci. Soc. 20: 391–403.

Parker, J.B., 1970. A statistical treatment of identification problems. J. Forensic Sci. Soc. 6: 33–39.

Pascali, V.L., E. d'Aloja & M. Dobosz, 1991. Estimating allele frequencies of hypervariable DNA systems. Forensic Sci. Int. 51: 271.

Pascali, V.L., E. d'Aloja, M. Dubosz & M. Pescarmona, 1991. Estimating allele frequencies of hypervariable DNA systems. Forensic Sci. Int. 51: 273.

Pascali, V.L., E. d'Aloja & M. Dobosz, 1992. Letter to the Editor. Int. J. Leg. Med. 105: 57.

Pearson, E.F., R.W. May & M.G.D. Dabbs, 1971. Glass and paint fragments found in men's outer clothing – a report of a survey. J. Forensic Sci. 13: 283–302.

Pearson, K., 1900. On the criterion that a given system of deviations from the probable in the case of a correlated system of variables is such that it can be reasonable supposed to have arisen from random sampling. Philosophy Magazine Series (5) 50: 157–172.

Pena, S.D.J., 1995. Pitfalls of paternity testing based solely on PCR typing of minisatellites and microsatellites. Am. J. Hum. Genet. 56: 1503–1504. (See also: Maha, G.C., J.M. Mason, G.M. Stuhmiller & U. Heine, 1995. Reply to Pena. Am. J. Hum. Genet. 56: 1506–1507).

Pena, S.D.J., R. Chakraborty, J.T. Epplen & A.J. Jeffreys. (Editors), 1993. DNA Fingerprinting: State of the Science. Birkhäuser, Basel.

People v. Castro, 545 N.Y.S.2d 985 (N.Y. Sup. Ct., 1989).

People v. Collins, 1968. California Reporter 66: 497.

People v. Keene, 591 N.Y.S.2d 733 (N.Y. Sup. Ct., 1992).

Polesky, H. & D. Dykes, 1989. The results of parentage testing with a single locus DNA probe, 4 pp. in American Association of Blood Banks' National Conference on DNA for Parentage Testing: Current State of the Art, Apr. 17–18, 1989, Leesburg, Virginia.

Potthof, R.F. & M. Whittinghill, 1965. Maximum-likelihood estimation of the proportion of non-paternity. Am. J. Hum. Genet. 17: 480–494.

Press, W.H., S.A. Teukolsky, W.T. Vetteling & B.P. Flannery, 1988. Numerical Recipes in C. The Art of Scientific Computing. 2nd Edition. Cambridge Univ. Press, New York.

Promega Corporation, 1989. Data Acquisition and Statistical Analysis for DNA Laboratories. Proceedings of the International Symposium on Human Identification.

Promega Corporation, 1991. New Technologies, Standardization of Methods and Data Sharing for DNA Typing Laboratories. Proceedings of the Second International Symposium on Human Identification.

Promega Corporation, 1992. Application of new technologies, standardization and validation of methods, data sharing, and legal issues affecting the DNA typing community. Proceedings of the Third International Symposium on Human Identification.

Rao, C.R., 1973. Linear Statistical Inference and Its Applications. Wiley, New York.

Radlow, R. & E.F. Alf, 1975. An alternate multinomial assessment of the accuracy of the χ^2 test of goodness-of-fit. J. Am. Stat. Assoc. 70: 811–813.

Read, T.R.C., 1982. On choosing a goodness-of-fit test. Unpublished PhD Thesis, Flinders University, South Australia.

Reynolds, J., B.S. Weir & C.C. Cockerham, 1983. Estimation of the coancestry coefficient: Basis for a short term genetic distance. Genetics 105: 767–779.

Richards, •.•., 1989. DNA fingerprinting and paternity testing. U. Cal. Davis Law Rev. 22: 609.

Risch, N. & B. Devlin, 1992. On the probability of matching DNA Fingerprints. Science 255: 717–720.

Risch, N.J. & B. Devlin, 1992. DNA Fingerprint matches. Science 256: 1744–1745.

Roberts, H., 1993. DNA profiling: towards the identification of individuals. Aust. J. Forensic Sci. 25: 43.

Roberts, L., 1991. Fight Erupts over DNA Fingerprinting. Science 254: 1721–1723.

Robertson, A. & W.G. Hill, 1984. Deviation from Hardy-Weinberg proportions: sampling variances and use in estimation of inbreeding coefficients. Genetics 107: 703–718.

Robertson, B. & G.A. Vignaux, 1992. Expert evidence: law, practice and probability. Oxford J. Legal Studies 12: 392–403.

Robertson, B. & G.A. Vignaux, 1993. Probability — the logic of the law. Oxford J. Legal Studies 13: 457–478.

Robertson, B. & G.A. Vignaux, 1995. DNA evidence: wrong answers or wrong questions? Genetica 96: 145–152.

Robertson, J., A.M. Ross & L.A. Burgoyne, 1990. DNA in Forensic Science: Theory, Techniques, and Application. Ellis Horwood, Chichester.

Robertson, J., J. Ziegle, M. Kronick, D. Madden & B. Budowle, 1991. Genetic typing using automated electrophoresis and fluorescence detection, pp. 391–398 in T. Burke, G. Dolf, A.J. Jeffreys & R. Wolff (eds.). DNA Fingerprinting: Approaches and Applications. Birkhauser, Basel.

Roe, A., 1993. Correlations and interactions in random walks and population genetics. Ph.D. Thesis, University of London.

Roeder, K., 1994. DNA Fingerprinting: a review of the Controversy. Statist. Sci. 9: 222–278.

Rothwell, T.J., 1984. The frequency of occurrence of various human blood groups in the United Kingdom, with observations on their regional variation. J. Forensic Sci. Soc. 25: 135–144.

Rothwell, T.J., 1993. DNA profiling and crime investigation – the European context. J. Forensic Sci. Soc. 33: 226–227.

Roychoudhury, A.K. & M. Nei, 1988. Human Polymorphic Genes. World Distribution. Oxford University Press.

Sajantila, A., K. Makkonen, C. Ehnholm & L. Peltonen, 1992. DNA profiling in a genetically isolated population using three hypervariable DNA markers. Hum. Hered. 42: 372.

Saks, M.J. & J.J. Koehler, 191. What DNA 'Fingerprinting' can teach the law about the rest of forensic science. Cardozo Law Rev. 13: 361–372.

Salmon, D. & C. Salmon, 1980. Blood groups and genetic markers polymorphisms and probability of paternity. Transfusion 20: 684–694.

Sampson, A.R. & R.L. Smith, 1985. An information theory model for the evaluation of circumstantial evidence. IEEE Transactions on Systems. Man and Cybernetics. SMC-15: 9–16.

Schatkin, S.B., 1984. Disputed Paternity Proceedings, Vols I and II, 4th ed., rev., Matthew Bender, New York.

Schneider, B.E., 1978. Algorithm AS 121: trigamma function. Appl. Stat. 27: 97–99.

Schneider, P.M., R. Fimmers & J. Bertrams, 1992. Biostatistical basis of individualization and segregation analysis using the multilocus DNA probe MZ 1.3: results of a collaborative study. Forensic Sci. Int. 55: 45.

Seber, G.A.F., 1985. Use of blood samples in paternity testing and forensic science. J. Roy. Soc. New Zealand 15: 157–168.

Seber, G.A.F., 1991. Evidence from blood, pp. 134–146 in C.G.G. Aitken & D.A. Stoney (eds.). The Use of Statistics in Forensic Science. Ellis Horwood, Chichester, UK.

Seheult, A., 1978. On a problem in forensic science. Biometrika 65: 646–648.

Selvin, S., 1983. Some statistical properties of the paternity ratio, pp. 77–88 in Inclusion Probabilities in Parentage Testing, edited by R.H. Walker. American Association of Blood Banks, Arlington, Virginia.

Selvin, S., B.W. Grunbaum & B.A. Myhre, 1983. The probability of exclusion or likelihood of guilt of an accused: paternity. J. Forensic Sci. Soc. 23: 19.

Selvin, S., B.W. Grunbaum & B.A. Myhre, 1983. The probability of non-discrimination or likelihood of guilt of an accused: criminal identification. J. Forensic Sci. Soc. 23: 27–33.

Sensabaugh, G.F., 1981. Uses of polymorphic red cell enzymes in forensic sciences. Clinics in Haematology 10: 185–205.

Sensabaugh, G.F. & C. von Beroldingen, 1991. The polymerase chain reaction: application to the analysis of biological evidence, pp. 63–82 in M.A. Farley & J.J. Harrington, 1991, Forensic DNA Technology. Lewis, Chelsea, MI.

Serjeantson, S.W., 1989. HLA genes and antigens, pp. 120–173 in The Colonization of the Pacific: A Genetic Trail, edited by A.V.S. Hill & S.W. Serjeantson. Clarendon Press, Oxford.

Serjeantston, S.W. & A.V.S. Hill, 1989. The colonization of the Pacific: a genetic trail, pp. 286–294 in The Colonization of the Pacific: A Genetic Trail, edited by A.V.S. Hill & S.W. Serjeantson. Clarendon Press, Oxford.

Sessions, W.S., 1989. Federal Bureau of Investigation. J. Forensic Sci. 34: 1051–1053.

Shafer, G., 1982. Lindley's Paradox (and comments). J. Am. Stat. Assoc. 77: 325–334.

Shapiro, M.M., 1991. Imprints on DNA fingerprints. Nature 353: 121–122. (See also: Evett, I.W., 1991. Trivial Error. Nature 354: 114).

Shields, W.M., 1992. Forensic DNA typing as evidence in criminal proceedings: some problems and potential solutions, pp. 1–50 in Proc. Third Int. Symp. Human Ident. Promega Corporation, Madison, Wisconsin.

Shriver, M.D., L. Jin, R. Chakraborty & E. Boerwinkle, 1993. VNTR allele frequency distributions under the stepwise mutation model: a computer simulation approach. Genetics 134: 983–993.

Sinnock, P. & C.F. Sing, 1972a. Analysis of multilocus genetic systems in Tecumseh, Michigan. I. Definition of the data set and tests for goodness-of-fit to expectations based on gene, gamete and single-locus phenotype frequencies. Am. J. Hum. Genet. 24: 381–392.

Sinnock, P. & C.F. Sing, 1972a. Analysis of multilocus genetic systems in Tecumseh, Michigan. II.

Consideration of the correlation between nonalleles in gametes. Am. J. Hum. Genet. 24: 393–415.

Slimowitz, J.R. & J.E. Cohen, 1993. Violations of the ceiling principle: exact conditions and statistical evidence. Am. J. Human Genet. 53: 314–323.

Smith, J.C., R. Anwar, J. Riley, D. Jenner, A.F. Markham & A.J. Jeffreys, 1990. Highly polymorphic minisatellite sequences: allele frequencies and mutation rates for five locus-specific probes in a Caucasian population. J. Forensic Sci. Soc. 30: 19–32.

Smith, J.C., C.R. Newton, A. Alves, R. Anwar, D. Jenner & A.F. Markham, 1990. Highly polymorphic minisatellite DNA probes: further evaluation for individual identification and paternity testing. J. Forensic Sci. Soc. 30: 3.

Smouse, P. & R. Chakraborty, 1986. The use of restriction fragment length polymorphisms in paternity analysis. Am. J. Human Genet. 38: 918.

Society for Forensic Haemogeneics, 1989. Statement concerning DNA polymorphisms. Forensic Sci. Int. 43: 109.

Sokal, R.R., N.L. Oden & C. Wilson, 1992. Genetic evidence for the spread of agriculture in Europe by demic diffusion. Nature 351: 143.

State v. Futch, 860 p2d 264 (Ore, 1993).

State v. Jobe, 486 N.W.2d 407 (Minn, 1992).

Stedman, R., 1972. Human population frequencies of immigrant and indigenous populations from South East England. J. Forensic Sci. Soc. 25: 95–134.

Stedman, R., 1984. Statistics applicable to the inference of a victim's blood type from familial testing. J. Forensic Sci. Soc. 24: 9–22.

Steinberger, E.M., L.D. Thompson & J.M. Hartmann, 1993. On the use of excess homozygosity for subpopulation detection. Am. J. Hum. Genet. 52: 1275–1277.

Stone, R., 1994. NAS takes fresh look at DNA fingerprinting. Science 265: 1163.

Stoney, D.A., 1984. Statistics applicable to the inference of a victim's blood type. J. Forensic Sci. Soc. 24: 9–22.

Stoney, D.A., 1991. Transfer evidence, pp. 107–138 in C.G. G. Aitken & D.A. Stoney (eds.). The Use of Statistics in Forensic Science. Ellis Horwood, Chichester, UK.

Stoney, D.A., 1992. Reporting of highly individual genetic typing results: a practical approach. J. Forensic Sci. 37: 373.

Storer, B.E. & C. Kim, 1990. Exact properties of some exact test statistics for comparing two binomial proportions. J. Am. Stat. Assoc. 85: 146–155.

Sudbury, A.W. & J. Marinopoulos, 1993. Assessing the evidential value of DNA profiles matching without using the assumption of independent loci. J. Forensic Sci. Soc. 33: 73–82.

Suissa, S. & J.J. Shuster, 1985. Exact unconditional sample sizes for the 2×2 binomial trial. J. Roy. Stat. Soc. A 148: 317–327.

Sullivan, P.J., 1992. DNA fingerprint matches. Science 256: 1743. (See also: Risch, R. & B. Devlin, 1992. Science 256: 1744, Budowle, B., 1992. Science 256: 1746).

Sussman, L.N. & B.K. Gilja, 1981. Blood group tests for paternity and nonpaternity. New York State J. Med. 81: 34–346.

Syndercomb Court, D., T. Fedor & M. Gouldstone, 1992. Investigation of the between-gel and within-gel variation in fragment size determinations found when using single locus DNA probes. Forensic Sci. Int. 53: 173.

Taylor, G., 1989. DNA fingerprinting. Nature 340: 672.

Thompson, W.C., 1989. Are juries competent to evaluate statistical evidence? Law Contemp. Problems 52: 9–41.

Thompson, W.C., 1993. Evaluating the admisibility of new genetic identification tests: lessons from the 'DNA war'. J. Criminal Law and Criminology 84: 701–781.

Thompson, W.C., 1995. Subjective interpretation, laboratory error and the value of forensic evidence: three case studies. Genetica 96: 153–168.

Thompson, W.C. & S. Ford, 1989. DNA typing: Acceptance and weight of the new genetic identification tests. Virginia Law Review 75: 45–108.

Thompson, W.C. & S. Ford, 1990. Is DNA fingerprinting ready for the courts? New Scientist 125: 20–25.

Thompson, W.C. & S. Ford, 1991. The meaning of a match: sources of ambiguity in the interpretation of DNA prints, pp. 93–152 in Forensic DNA Technology, edited by M. Farley & J. Harrington.

Thompson, W.C. & E.L. Schumann, 1987. Interpretation of statistical evidence in criminal trials. The prosecutor's fallacy and the defense attorney's fallacy. Law & Human Behavior 11: 167.

Thompson, Y. & R. Williams, 1991. Blood group frequencies of the population of Trinidad and Tabago, West Indies. J. Forensic Sci. Soc. 31: 441–447.

Tills, D., A. Kopec & R.E. Tills, 1983. The Distribution of the Human Blood Groups and Other Polymorphisms. Suppl. 1. Oxford University Press.

Tonelli, L.A., K.R. Markowicz, M.B. Anderson, D.J. Green, G.L. Herrin, R.W. Cotton, D.D. Dykes & D.D. Garner, 1990. Use of deoxyribonucleic acid (DNA) fingerprints for identity determination: comparison with traditional paternity testing methods — Part I. J. Forensic Sci. 35: 1265. (See also: Markowicz, K.R., L.A. Tonelli, M.B. Anderson, D.J. Green, G.L. Herrin, R.W. Cotton, J.L. Gottschall & D.D. Garner, 1990. Use of deoxyribonucleic acid (DNA) fingerprints for identity determination: comparison with traditional paternity testing methods — Part II. J. For. Sci. 35: 1270).

Tribe, L.H., 1971. A further critique of mathematical proof. Harvard Law Review 84: 1810–1820.

United States Department of Health, Education and Welfare, 1980. Selected genetic markers of blood and secretions for youths, 12–17 years of age. DHEW Publication number (PHS) 80- 1664. National Center for Health Statistics, Hyattsville, MD.

United States v. Yee, 1991. 134 F.R.D. 161 (N.D. Ohio).

Upton, G.J.G., 1982. A comparison of alternative tests for the 2×2 comparative trial. J. Roy. Stat. Soc. A 145: 86–105.

Urquhart, A. & P. Gill, 1993. Tandem-repeat internal mapping (TRIM) of the Involucrin Gene: repeat number and repeat-pattern polymorphism within a coding region in human populations. Am. J. Hum. Genet. 53: 279–286.

Valdes, A.M., M. Slatkin & N.B. Freimer, 1993. Allele frequencies at microsatellite loci: the stepwise mutation model revisited. Genetics 133: 737–749.

Valentin, J., 1984. Paternity index and attribution of paternity. Hum. Hered. 34: 255–257.

Wahlund, S., 1928. Zuzammensetzung von Populationen und Korrelationserscheinungen vom Standpunkt der Vererbungslehre aus betrechtet. Hereditas 11: 65–106.

Wainscoat, J.S., S. Pilkington, T.E.A. Peto, J.I. Bell & D.R. Higgs, 1987. Allele specific DNA identity patterns. Human Genet. 75: 384–387.

Wald, A., 1947. Sequential Analysis. Wiley, New York.

Walker, R.H., (Editor), 1983. Inclusion Probabilities in Parentage Testing. American Association of Blood Banks, Arlington, Virginia.

Walker, R., 1992. Molecular biology in paternity testing. Laboratory Med. 23: 752–757.

Wall, W.J., R. Williamson, M. Petrou, D. Papaioannou & B.H. Parkin, 1993. Variation of short tandem repeats within and between populations. Hum. Molec. Genet. 2: 1023–1029.

Walsh, J.J., 1992. The population genetics of forensic DNA typing: "Could it have been someone else?" Crim. Law Quart. 34: 468.

Walsh, K.A.J. & J.S. Buckleton, 1986. On the problem of assessing the evidential value of glass fragments embedded in footwear. J. Forensic Sci. Soc. 26: 55–60.

Walsh, K.A.J. & J.S. Buckleton, 1988. A discussion of the law of mutual independence and its application to blood group frequency data. J. For. Sci. Soc. 28: 95–98.

Walsh, K. & J.S. Buckleton, 1991. Calculating the frequency of occurrence of a blood type for a 'random man'. J. Forensic Sci. Soc. 31: 49–58.

Walsh, K.A.J., J.S. Buckleton & C.M. Triggs, 1994. Assessing prior probabilities considering geography. J. Forensic Sci. Soc. 34: 47–51.

Walsh, K.A.J., M.E. Lawton, J.S. Buckleton, G.A.F. Seber & D.G. Woodfield, 1986. Comments on the use of blood marker frequency data. For. Sci. Int. 32: 131–133.

Ward, R.H. & C.F. Sing, 1970. A consideration of the power of the chi-square test to detect inbreeding effects in natural populations. Am. Naturalist 104: 355–366.

Waye, J.S., 1994. Allelic stability of a VNTR locus 3′ αHVR: linkage disequilibrium with the common α- thalassaemia-1 deletion of South-East Asia (-SEA). Hum. Heredity 44: 61–67.

Weedn, V.W. & R.K. Roby, 1993. Forensic DNA testing. Arch. Path. Lab. Med. 117: 486.

Weinberg, W., 1908. Uber den Nackweis der Vererburg beim Menschen. Jahresh. Verein f. Vaterl Naturk. Wurtemb. 64: 368–382 (English translation in S.H. Boyer (ed.) Papers on Human Genetics, 1963. Prentice-Hall, Englewood Cliffs, New Jersey.

Weir, B.S., 1990. Genetic Data Analysis. Sinauer, Sunderland, MA.

Weir, B.S., 1990. Intraspecific variation, pp. 373–410 in Molecular Systematics and Evolution, edited by D. Hillis & C. Moritz. Sinauer Associates, Sunderland, Massachusetts. (See also: Second edition, 1995).

Weir, B.S., 1992. Independence of VNTR alleles defined as fixed bins. Genetics 130: 873–887.

Weir, B.S., 1992. Independence of VNTR alleles defined as floating bins. Am. J. Hum. Genet. 51: 992–997.

Weir, B.S., 1992. Population genetics in the forensic DNA debate. Proc. Natl. Acad. Sci. USA 89: 11654–11659.

Weir, B.S. 1992. Discussion of 'Statistical inference in crime investigations using DNA profiling' by D.A. Berry et al. Applied Statistics 41: 528–529.

Weir, B.S., 1993. Forensic population genetics and the National Research Council (NRC). Am. J. Hum. Genet. 52: 437.

Weir, B.S., 1993. Independence tests for VNTR alleles defined as quantile bins. Am. J. Hum. Genet. 53: 1107–1113.

Weir, B.S., 1993. The status of DNA fingerprinting. The World & I. November, 1993, 214–219.

Weir, B.S., 1994. The effects of inbreeding on forensic calculations. Ann. Rev. Genetics 28: 597–621.

Weir, B.S., 1994. DNA profiling on trial. Nature 369: 351.

Weir, B.S., 1994. Discussion of 'DNA fingerprinting: a review of the controversy' by K. Roeder, Statistical Science 9: 222– 278.

Weir, B.S., 1994. Discusion of 'Inference in forensic identification' by D.J. Balding & P. Donnelly, J. Roy. Stat. Soc. (In press).

Weir, B.S. (Editor), 1995. Human Identification. Kluwer Academic Publishers, Dordrecht.

Weir, B.S., 1995. A bibliography for the use of DNA in human identification. pp. 179–213 in Human Identification: The use of DNA markers, edited by B.S. Weir. Kluwer Academic Publishers, Dordrecht.

Weir, B.S. & C.C. Cockerham, 1978. Testing hypotheses about linkage disequilibrium with multiple alleles. Heredity 42: 105– 111.

Weir, B.S. & C.C. Cockerham, 1984. Estimating *F*-statistics for the analysis of population structure. Evolution 38: 1358–1370.

Weir, B.S. & C.C. Cockerham, 1989. Complete characterization of disequilibrium at two loci. In M.W. Feldman (Ed.) Mathematical Evolutionary Theory, Princeton Univ. Press, Princeton, pp. 86– 110.

Weir, B.S. & I.W. Evett, 1992. Whose DNA? Am. J. Hum. Genet. 50: 869 (See also: Lewontin, R.C., 1993. Which Population? Am. J. Hum. Genet. 52: 205; Weir, B.S. & I.W. Evett, 1993. Reply to Lewontin. Am. J. Hum. Genet. 52: 206).

Weir, B.S. & I.W. Evett, 1993. Reply to Lewontin. Am. J. Human Genet. 52: 206. (See also: Lewontin, R.C., 1993. Which Population? Am. J. Human Genet. 52: 205).

Weir, B.S. & B.S. Gaut, 1993. Matching and binning of VNTRs in forensic science. Jurimetrics J. 34: 9–19.

Weir, B.S. & W.G. Hill, 1993. Population genetics of DNA profiles. J. Forensic Sci. Soc. 33: 219–226.

Wiener, A.S., 1976. Likelihood of parentage, pp. 124–131 in L.M. Seideman (ed.). Paternity Testing by Blood Grouping. 2nd ed., Charles C. Thomas, Springfield, IL.

Wills, C., 1992. Forensic DNA typing. Science 255: 1050.

Witkowski, J., Fingerprinting of the future? The New England Biolabs J. 2: 1–7.

Witkowski, J., 1991. Milestones in the development of DNA technology, pp. 1–23 in M.A. Farley & J.J. Harrington, 1991, Forensic DNA Technology. Lewis, Chelsea, MI.

Wooley, J.R., 1991. A response to Lander: The Courtroom Perspective. Am. J. Hum. Genet. 49: 892–893.

Wooley, J. & R.P. Harmon, 1992. The forensic DNA brouhaha: science or debate. Am. J. Hum. Genet. 51: 1164–1165.

Wright, S., 1921. Systems of mating. Genetics 6: 111–178.

Wright, S., 1922. Coefficients of inbreeding and relationship. Amer. Nat. 56: 30–338.

Wright, S., 1943. Isolation by distance. Genetics 28: 114–138.

Wright, S., 1965. The interpretation of population structure by F-statistics with special regard to systems of mating. Evolution 19: 395–420.

Wright, S., 1969. Evolution and the Genetics of Populations, Vol. 2, University of Chicago Press, Chicago.

Yarborough, L.R., 1992. Forensic DNA typing. Science 255: 1050.

Yassourdis, A. & J.T. Epplen, 1991. On paternity determination from multilocus DNA profiles. Electrophoresis 12: 221–226.

Yasuda, N., 1968. An extension of Wahlund's principle to evaluate mating type frequency. Am. J. Hum. Genet. 20: 1–23.

Yates, F., 1984. Tests of significance for 2×2 contingency tables (with discussion). J. Roy. Stat. Soc. A 147: 426–463.

Zaykin, D., L.A. Zhivotovsky & B.S. Weir, 1995. Exact tests for association between alleles at arbitrary numbers of loci. Genetica 96: 169–178.

Zhang, X.-W., L. Lan, Z.-Y. Huo, B.Z. Duan & L. Koblinsky, 1991. Restriction fragment length polymorphism analysis of Forensic science casework in the People's Republic of China. J. Forensic Sci. 36: 531.

Special applications

Blackett, R.S. & P. Keim, 1992. Big game species identification by deoxyribonucleic acid (DNA) probes. J. Forensic Sci. 37: 590–596.

Burke, T. & M.W. Bruford, 1987. DNA fingerprinting in birds. Nature 327: 149–150.

Burke, T., G. Dolf, A.J. Jeffreys & R. Wolff (eds.). DNA Fingerprinting: Approaches and Applications. Birkhauser, Basel.

Cano, R.J., H.N. Poinar, J. Norman, J. Pieniazek, A. Acra & G.O. Poinar Jr., 1993. Amplification and sequencing of DNA from a 120–135-million-year-old weevil. Nature 363: 536–538.

Cherfas, J., 1991. Ancient DNA: still busy after death. Science 253: 1354.

DeSalle, R., J. Gatesy, W. Wheeler & D. Grimaldi, 1992. DNA sequences from a fossil termite in oligmiocene amber and phylogenetic implications. Science 257: 1933–1936.

Dix, J.D., S.D. Stout & J. Mosley, 1991. Bones, blood, pellets, glass, and no body. J. Forensic Sci. 36: 949–952.

Fisher, D., M.M. Holland, L. Mitchell, P.S. Sledzik, A.W. Wilcox, M. Wadhams & V.W. Weedn, 1993. Extraction, evaluation, and amplification of DNA from decalcified and undecalcified United States Civil War Bone, 38. J. Forensic Sci. 60–68.

Guglich, E.A., P.J. Wilson & B.N. White, 1993. Application of DNA fingerprinting to enforcement of hunting regulations in Ontario. J. Forensic Sci. 38: 48–59.

Guglich, E.A., P.J. Wilson & B.N. White, 1994. Forensic application of repetitive DNA markers to the species identification of animal tissues. J. Forensic Sci. 39: 353–361.

Hagelberg, E. & J.B. Clegg, 1993. Genetic polymorphisms in prehistoric Pacific islanders determined by amplification of ancient bone DNA. Proc. Roy. Soc. B. 252: 163–170.

Hagelberg, E. & B. Sykes, 1989. Ancient bone DNA amplified. Nature 342: 485.

Holmes, E.C., A.J. Leigh Brown & P. Simmonds, 1993. Sequence data as evidence. Nature 364: 766.

Jeffreys, A.J., V. Wilson & D.B. Morton, 1987. DNA fingerprints of dogs and cats. Animal Genet. 18: 1–15.

Kurosaki, K., T. Masushita & S. Ueda, 1993. Individual DNA identification from ancient human remains. Am. J. Hum. Genet. 53: 638–643.

Lynch, M., 1988. Estimation of relatedness by DNA fingerprinting. Mol. Biol. Evol. 5: 584–599.

Marx, 1989. Science gives ivory a sense of identity. Science 245: 1120.

Morell, V., 1992. 30-million-year-old DNA boosts an emerging field. Science 257: 1860–1862.

Paabo, S., 1985. Molecular cloning on ancient Egyptian mummy DNA. Nature 314: 644–645.

Paabo, S. & J.A. Gifford, 1988. Mitochondrial DNA sequence from a 7000-year old brain. Nucleic Acids Res. 16: 9775–9787.

Piercy, R. & K.M. Sullivan, 1993. The application of mitochondrial DNA typing to study of white Caucasian identification. Int. J. Legal Med. 106: 85–90.

Ross, P.E., 1992. Eloquent remains. Sci. Am., May 1992: 144–125.

Schwartz, T.R., E.A. Schwartz, L. Mieszerski, L. McNally & L. Kobilinsky, 1991. Characterization of deoxyribonucleic Acid (DNA) obtained from teeth subjected to various environmental conditions. J. Forensic Sci. 36: 979–990.

Stacey, G.N., B.J. Bolton & A. Doyle, 1992. DNA fingerprinting transforms the art of cell authentication. Nature 357: 261–262.

Whiting, B., 1993. Tree's DNA 'fingerprint' splinters killer's defense. Ariz. Republic, May 28, 1993: p. A1.

Contemporary Issues in Genetics and Evolution

1. J.F. McDonald (ed.): *Transposable Elements and Evolution.* 1993 ISBN 0-7923-2338-6
2. Th.A. Markow (ed.): *Developmental Instability: Its Origins and Evolutionary Implications.* Proceedings of the International Conference (Tempe, Arizona, June 1993). 1994 ISBN 0-7923-2678-4
3. M.R. Rose and C.E. Finch (eds.): *Genetics and Evolution of Aging.* 1994 ISBN 0-7923-2902-3
4. B.S. Weir (ed.): *Human Identification: The Use of DNA Markers.* 1995 ISBN 0-7923-3520-1

KLUWER ACADEMIC PUBLISHERS – DORDRECHT / BOSTON / LONDON

FSC
www.fsc.org
MIX
Papier aus verantwortungsvollen Quellen
Paper from responsible sources
FSC® C105338

If you have any concerns about our products,
you can contact us on
ProductSafety@springernature.com

In case Publisher is established outside the EU,
the EU authorized representative is:
Springer Nature Customer Service Center GmbH
Europaplatz 3, 69115 Heidelberg, Germany

Printed by Libri Plureos GmbH
in Hamburg, Germany